MY STORY

A Memoir

By

Gene Cunneely Sr.

Special thanks to my son, Gene Jr.,

my granddaughter, Kanani Duncan,

and my granddaughter-in-law, Mallory Cunneely.

With their help and commitment, my words—

my story—became this published book.

DEDICATION

Alice Marie Curry, New York, 1951

This memoir, composed by me, is dedicated to my wife Alice Marie. Reviewing my work, which includes years of falling in love, starting a great family, raising five wondrous children, seeing to their health and growth, supporting their education, and providing to them parents who loved and cared for each other.

At a prior period in my life, after I had retired from work, the idea came to me that, I had been given overall, a fabulous, wonderful life. So many good things graced my life even from the earliest days.

I clearly realized that I had received so many amazing happy events during both my early years and continued through my later years. As I looked back, I was astonished at how blessed I had been throughout my years, and how much more grateful I should have been. And so, I began writing an autobiography. And, as I wrote, it became more unbelievable at how many wondrous things came to me, without any of my own intentions. Many lovable humans entered my life, and how closely the Good Lord watched over me.

I decided to always put my trust in the Lord. And I have never been disappointed.

My family provided me the happiest years, but also challenged me the most. As I reviewed my life up to this point, I realized the happiest years began after I proposed marriage to Alice Marie, and she consented. My wife, Alice Marie, provided me with 61 years of love and contentment. My wife was a special human being; a devoted wife and mother, and a woman with superior common sense and an amazing capability to give her husband and children beautiful love and devotion. I became dependent on her to manage the many problems that seemed never to end. Even during the many months and years that I was away from home and Alice Marie had to handle managing the home and five kids, including their various activities and schooling.

Alice Marie brought so much love and caring into my life. I cannot recall any domestic problems, as my wife always had a positive and agreeable disposition. My love for her grew and grew, as I realized how fortunate I had been when I decided it was the correct moment to ask for her hand, and she said yes. It was the making of a good marriage as I could not have found a better spouse and beloved one. I thank God for His watching over me and providing me a wonderful wife, who was always there to support me. She now rests in Peace. I have no doubts that Alice Marie is close to the Lord in heaven. May God bless her. And I dedicate this life story to Alice Marie, who I will always love and pray for her happiness in heaven. I dedicate this memoir to her for always helping others.

SPECIAL ACKNOWLEDGMENT

Mary Cunneely: The Backbone of the Cunneely Family

The most influential individual from the Cunneely Family was Mary (Fitzmaurice) Cunneely, without dispute. Mary Cunneely was born in County Mayo, Ireland, in the 1860's. Mary, with her husband, James, immigrated to the United States and settled in the New York City borough of Queens. James soon found employment in the Calvary Cemetery where he worked for many years. James died December 23, 1928, thus leaving Mary to become the head of the family for many more years. The Cunneely family grew to reach eight, including Mary's sons and their families. Thankfully, my father, Mary's son, was well employed with the Long Island Railroad. After my parents split, my father insisted on taking custody of my brother and me, however, my mother also declined any other to do so herself. From that point on, Mary Cunneely became a true mother figure for my brother and me. Although Mary was in her 70's, she continued to raise her daughters, two sons and grandsons as her own. As her adopted boys, we always looked upon our grandmother as our mom. Mary passed away April 1953, when I was in Korea.

"THE FIGHTER PILOT"

By: Friar Tuck

Published in The Interceptor, February 1968

The Fighter Pilot

Say what you will about him: arrogant, cocky, boisterous, and a fun-loving fool to boot - he has earned his place in the sun. Across the span of fifty years he has given this country some of its proudest moments and most cherished military traditions. But fame is short-lived and little the world remembers. Almost forgotten are the 1400 fighter pilots who stood alone against the might of Hitler's Germany during the dark summer of 1940 - and in the words of Sir Winston Churchill gave England "Its Finest Hour." Gone from the hardstands of Duxford are the 51's with their checkerboard noses that terrorized the finest fighter squadrons the Luftwaffe had. Dimly remembered - the 4th Fighter Group that gave Americans some of their few proud moments in the skies over Korea. How fresh in recall are the Air Commandos who valiantly struck the VC with their aging "Skyraiders" in the rain- and blood-soaked valley called A Shau? And how long will be remembered the "Thuds" over "Route Pack Six" and the flak-filled skies above Hanoi? So here's a "nickle on the grass" to you, my friend, for your spirit, enthusiasm, sacrifice, and courage - but most of all to your friendship. Yours is a dying breed and when you are gone - the world will be a lesser place.

Friar Tuck

Table of Contents

INTRODUCTION

On the following pages I have tried to write a narrative of my life from my earliest days to this date. I term it as a "memoir" because it is a report of happenings in my life's journey only as I remembered them and interpreted them. Surely, in many instances, my memory was not perfect, and my interpretation of the events may not gain a consensus from those who were there. But, in every case I did my research to the best of my ability and tried to be fair to all sides. And in all honesty, time had played havoc with my memory cells, and in some case names and dates seemed to have faded somewhat.

A logical question to be asked is: "Why did you bother to write this story?"

It's a fair question because writing this manuscript took close to 18 months of my time, and who is going to read it anyway. Possibly very few if anyone will bother to do so. However, in writing this piece, I found it worthwhile in several ways. Each of us, as we grow into our later years looks back on our lives with wonderment and awe. We begin to doubt and question whether we, to any extent, came close to fulfilling a purpose or a goal that our Creator might have laid out for us. Did it make any difference at all that we entered the world? Did we bring any joy or comfort to our fellow humans? Did we leave something behind that was lasting or worthwhile to someone or anyone who will benefit? These are philosophical questions, or perhaps, spiritual questions, which haunt us as we come closer to leaving this mortal life for whatever lies beyond. Writing about my life, which included my successes and failures, wins and losses, moments of joy and depression, the good and bad times, my good deeds and my terrible faults, my wise decisions, and my awful ones, gave me a perspective of how difficult life can be as all of us struggle to find happiness and at the same time do what is right.

Fortunately, looking back I was guided by wonderful people who entered my life at just the right time, when I needed them the most. When my mother left us when I was only five, my grandmother took us in and replaced her with a love and dedication unmatched. Of course, my father was a wonderful man, full of common sense and devotion to his sons and later to his grandchildren. I was blessed with marvelous teachers: the sisters at Blessed Sacrament School and later, the Marianist Brothers and Priests at Chaminade High School and the Jesuit Fathers at Fordham University. I found in Alice Marie Curry a beautiful Catholic woman who became my wife and the mother of our five children. I can never describe what a difference she made in my life, and I hate to think where I would have been if she had not said yes when I asked for her hand. There have been many others who gave me wise counsel during my life's journey for which I will be forever grateful. I am also grateful for my three sons and two daughters who have given me such joy and pride during my life. To be a father and now a grandfather is a special honor and privilege and a role that I can only thank God for every day. I can only express my heartfelt thanks to the Almighty for preserving me as He did so that I am able to enjoy these wonderful days with my extended family and watch them as they pursue their own lives. Perhaps I can help them in some small ways.

My Story is far from a perfect read. No doubt I have forgotten many episodes I should have included and probably included some that were not worth being in it. Also, I probably left out many embarrassing episodes that I conveniently forgot. I did not intentionally do so. The truth is, though I certainly had more than my share of screw-ups, all in all, my successes were more numerous than my foul ups. During my close to 5,000 hours of flying, I probably made 5,000 pilot errors; fortunately, none killed me. It pays to be a little lucky at those speeds. There was not enough room in this book to include all the mistakes I made in my life. There is not enough room either to include all my sins, but the Good Lord knows about all of them and that's bad enough.

For those of you who dare to read further, good luck. This is just one person's story. It's a story about a boy who started out in the depression years, lost his mother early but never went without a meal or a loving home life. Through hard work and luck, he got a great education and ended up in the greatest air force in the world flying the world's best fighter aircraft. It's a success story against the odds, and a happy story because I was always the recipient of love. And I always tried to return that love through my actions.

Gene Cunneely, Sr.
Spokane, Washington, 2023

PART I:
CHILDHOOD AND EARLY YEARS

CHAPTER ONE: IN THE BEGINNING

When I was a child, I spoke as a child,
I felt as a child, I thought as a child.

-St. Paul

Early Years

Although being born is surely one of the most momentous events in a person's life, no one remembers it. We all recall with vividness other significant happenings, such as our first day at school, graduations, successful athletic achievements, wedding day, and so on, still no one I have ever met can remember being born. Yet, we all experience it, and our birth is commemorated yearly until the very last one. And sometimes for special people, it is remembered and celebrated after death. But this most vital happening in our life (outside of conception, I suppose) is a blank. I guess we can be thankful that the pain and trauma associated with leaving the warmth and comfort of the womb and entering the cruel, cold world are not recalled.

So, I can only relate that I came out of my mother's womb into the harsh world at two- thirty AM, May 24, 1929, in St. John's Hospital, Long Island City, New York. This I know from the records and from what I was told by my parents. However, I do not remember it at all. Just as well perhaps as most babes that arrive appear not to be pleased by it at all. I was the second son of my parents Joseph Laurence and June Margaret (Bernhardt) Cunneely. My older brother, Joseph Junior, had arrived October 26, 1925. My birth date was four months and 13 days before the crash on Wall Street that was the defining moment of the Great Depression that persisted until World War II I have lived my life referred to interminably as a "depression baby". Which, for sure, I was.

It was my father's decision to name me Eugene, a name not applied prior to anyone in either family. It means "of noble birth," which could be one reason the name was not used before in either family. My father just liked the name. My middle name, Paul, was in honor of my father's good friend Father Joseph Paul Weist, the "singing priest." I was baptized on June 17, 1929, at St. Ambrose's Church, Brooklyn, New York by Father Weist. My godparents were my uncle Frank Cunneely and Anna Gilbert, a friend of the family.

My earliest memories were centered on our home in Kew Gardens. My father was a station agent for the Long Island RR. He had joined the railroad at age 15, having completed one year at Bryant High School. He was able to get a railroad job thanks to the intercession of his older brother, my Uncle Tom, who was a switchman for that line. My father and his brothers went to work early in their lives in order to support the family. College education was out of the question for them; a good steady job was by far more important. I'm not sure that any of them completed high school.

We lived in a small apartment at 8266 Austin Street near the Kew Gardens railroad station where my father was the agent. The apartment was a walkup over a store restaurant. It was not much of a place, two bedrooms and one bath, but certainly convenient for my father who left the home about five o'clock each morning. He did this seven days a week. Despite the small size of our home, we always had a dog. A big dog, a German shepherd to be exact, named Jerry. We fed him from the scraps provided by the restaurant below. He was a beautiful animal and gentle with us, except when eating. You gave him plenty of space during feeding time. My earliest memories include many hours with Jerry romping around the railroad station where there was plenty of open grass area.

Immigrant Grandparents

My father was the fourth son and sixth child of James and Mary (Fitzmaurice) Cunneely. There were seven total: Thomas, James, Anna, Francis, Mary, Joseph, and Lillian. My grandparents lost a child in infancy, Stephen I believe. The first two were born in Ireland, but in the late 1880's, they immigrated to the United States. My father was born in New York, October

26, 1892. My grandfather, whom I never knew, was a grave digger at Calvary Cemetery in Long Island City, one the largest cemeteries in the world. He eventually was a section foreman there, and as you might guess, is buried there. He died in 1928, so neither my brother nor I ever knew him or he, us. But his picture was widely displayed in our homes. He was a fine-looking man with a handlebar moustache. His photos show a stern man who clearly would be head of the house and a disciplinarian. My father's tales of his boyhood confirm that his father was not one to be trifled with. There were rules about mealtimes, proper dress, language, and such, which one was foolhardy to neglect. My picture of him from family stories depicted a strong family man, very Catholic, and devoted to his wife and children.

My grandmother was another story altogether. I came to know her well as she eventually raised my brother and me as her own. She was a short lady, and when I first saw her, she had long white hair, which sometimes was up in a bun as they called it, or other times curled in a tight way after having had a perm. She married my grandfather when she was only sixteen in Ballyhannis, County Mayo. I know that she came from a farming family, and she had at least one brother who became a distinguished veterinary in Michigan. I recall how strong her hands were from the many years of housework, cooking, and other chores that women had to perform daily in those years before the many conveniences we know today. She was a gentlewoman, but firm and devoted to her family. Till she died she continued to live with her two daughters Anna and Lillian, neither of whom ever married. Despite her hard life, she had a wondrous sense of humor, enjoyed a good story, and gatherings of the clan. For these get-togethers, she always was the head cook, and she was unequaled in the kitchen. She was a master baker of cakes, pies, and cookies that in those times had to be brought to life from scratch, that is, with raw materials and manual labor.

Dad Was a Railroad Man and a Democrat

Some of my memories of the Kew Gardens' years were about the railroad station where my father worked. I spent many hours in his office. The place intrigued me. First there was the huge clock with the long pendulum that clicked each second loudly and then chimed the hour. Time was of extreme importance in railroads, and that clock was always accurate to the second. Next to the window that opened into the waiting room was a case containing the tickets that my father sold to the travelers. They were small yellow cards on which was printed the destination and a serial number. There were also round-trip tickets and weekly and monthly commutation tickets that made things more interesting. Next to the window was a date stamp machine, shaped as a large stapler that had a split tennis ball on the top. When a ticket was sold, the agent inserted the ticket between the claws of this machine and came down on the ball with his fist. (Two times for round trip tickets.) The machine stamped the date on the back of the ticket. I was enthralled with this device and in slow hours Dad would let me validate tickets with it as he sold them. I always gave that tennis ball my best shot.

My father was well known in Kew Gardens. Many people, who enjoyed living in the posh suburb of Kew Gardens, commuted to Manhattan by LIRR. Kew Gardens was the home of many well to do people and celebrities. I can remember Marjorie Gates and Jerome Cowan from the movies, and Gene Fowler, the journalist and great friend of Jimmy Walker all of whom my father knew well. There were also many politicians and city officials whom he got to know well and who sometimes did him favors. There were often many visitors in the station office, especially in the afternoons when things slowed down. They would come in and sit with Dad and talk about everything under the sun.

My father was a great conversationalist and had a fine sense of humor. He also usually had on hand a good assortment of Havana cigars that were always in use. The aroma of cigars was one of the earliest recollections I had of my father. Even today, when I smell cigar smoke, I remember father.

My Mother

My recollections of my mother in these early years are few. My mother went by the name of June, although I am fairly certain she was baptized Julia. She was only about 19 when she married my father who was nearly 13 years older. She came from a German family who lived in the Greenpoint section of Brooklyn. I have no idea of how they met, but the Cunneely family did have members living in the Greenpoint area. My mother was a lovely brunette with an ample figure. I am sure she loved her children but, early on, it became clear that she was not enthralled with housework and the interminable chores of mothering two young boys. She did marry a man who had a steady, if not a well-remunerated job, and who also owned a car. There were not too many of those around in those days and it was a great luxury. But Joseph was a man who was ready to settle down and very much tuned into the responsibilities of raising a family in challenging times. Therefore, my memories include problems between them having to do with the home and the children. She had a trying time with my brother Joseph, who was an active and often frustrating boy. The drudgery of housework in a small and not too well-appointed apartment and chasing around after two rascally boys weighed on her. The fun days of courtship and partying were clearly over. My mother began to feel cheated out of the years in her early womanhood that should have brought more good times, parties, dancing, etc. The leap from courtship to married life and then motherhood was beyond what she was prepared for. And it began to show. And the arguments between them became more frequent and heated.

As a small boy, these things were worrisome and not understandable. It is a frightening thing for a small child to see and hear his parents fight. My world was centered and totally dependent on my parents and to feel them tearing apart with an unknown result was something I still remember. My mother's reaction was to become more strident and impatient with my brother and me. On occasion she would sneak out with her friends and head for the nightclubs of Manhattan. When returning in the early morning hours, another eruption would take place. My father's temper would flare as he berated his young wife for leaving her home and kids for the good times. And she would return his criticisms in full kind. I faintly recall one night when I believe he slapped her fairly hard. Things at home became progressively worse.

While I was in Kindergarten at P.S 99, my brother Joe, who was always called Junior, was sent to school at Our Lady Queen of Martyrs in Forest Hills. My father rather hoped that the Sisters there could put some manners on him. To go to school, Junior, would take the LIRR one stop, and then walk about a block to the school. He seemed to have accomplished this well for a young boy, but the Sisters did have their time with him. He always seemed to get into fights with other boys. Apparently, he more than held his own in fisticuffs, but the wear and tear on his school uniform was something else. And costly. My father kept a close eye on him.

During these years, my father became active in politics. He was a member of the Thomas Jefferson Democratic Club. Most of the Cunneely's were Democrats mostly due to the links that Tammany Hall established with Irish immigrants as they got off the boat in New York. I'm sure they set my grandfather up with his job at Calvary Cemetery. My father loved Jimmy Walker, New York's mayor until he was forced to resign in 1932, and often told me stories of meeting him and listening to his eloquent speeches (delivered even when the Mayor was heavily under the influence of old man alcohol). Being three sheets to the wind did not seem to dim his capability to sway the masses and make them love him even more.

My father was a staunch Democrat all his life. He later became president of the Thomas Jefferson Democratic Club, and I recall helping out in some of the elections placing literature on doorsteps and under car windshield wipers. The club served as a source of social interaction that he lost especially after his marriage came apart. He enjoyed his hours there, the camaraderie, and his efforts on behalf of candidates he believed in. It was in the club's headquarters on Metropolitan Avenue where he honed his game of pool at which he was very adept until his very late years. These were times when unions were beginning to gain strength, and surely one was needed for railroad workers. But the depression was still deep and wide, jobs were scarce, and workers feared pushing too hard lest they would be out on the street. My father used his association with the Democratic Party to advance this cause.

The End of the Family

My reflections on my earliest years are pitifully vague and sparse. Despite my parents' difficulties, I do not look upon them as unhappy times. I had a warm bed, a good father who spent his spare time with his sons, and three meals a day, although the depression years were in full swing. I owned a tricycle and a great companion, my dog Jerry. But those years ended soon and badly.

In 1935, when I was five, my father and mother separated permanently. There had been a series of verbal clashes and it was clear that my mom was unable to dedicate herself to homemaking and mothering. And that was exactly what my father wanted from her. My father was about 13 years older than my mother. She was only 20 years old when she gave birth to my brother. She clearly was not prepared for the burden of two children in a small apartment and the end of a life of parties and dance for which she had hoped. My dad made it clear that he would take custody of his sons and she told him to take them and "go to hell with them". Even at my early age, I knew what that meant. My mother left the home. My brother and I were bewildered by it all, and my dad had a serious problem on his hands. As he did so often in his life, he turned to his own mother for help.

CHAPTER TWO:
OUR FAMILY COMES APART

All happy families resemble each other, each unhappy family is unhappy in its own way.

-Tolstoy

My New Family

In today's world, much is spoken and written about broken families. The term "single parent" and "dysfunctional family" are familiar in today's society. They refer to a widespread, serious problem, far greater than it was when I was a child, and one that was not foreseen. But I never heard those terms in my early years. Nor, probably, did my father or mother. Divorce in the thirties was rare, and then, only for the wealthy. To get a divorce, one party had to give evidence of adultery on the part of the other spouse. These were staged events using a private investigator, a hotel room, a photographer, and a no contest on the part of the supposed offender. These episodes made the front pages of the new papers. They had to be terribly embarrassing, but the parties involved would do these things to obtain freedom. Little was said about the effect these divorces had on children. In later years, we found out with awful results.

My father never considered divorce. He was raised in the Catholic faith that did not permit divorce. He had no conflict that the words, "What God has joined together, let no man put asunder" meant exactly that. Married parents, who were experiencing problems, usually worked through them for the sake of the children. Later, popular psychologists rejected that solution, but the devastation that ensued on the children of broken homes, has caused a major review of that approach. My father knew which way to go from the first. Though only 43 years old at the time of the separation, divorce and possibly remarriage never occurred to him. His sons came first.

And so, my mother moved back to her parents' home in Greenpoint, Brooklyn and my father brought us to his mother's house in Elmhurst. We were apart.

My grandmother, with her unmarried daughters, Anna and Lillian, and unmarried son, my Uncle Frank, lived in a two-bedroom apartment on the second floor of a house at 40-18 Warren Street. Downstairs lived the Reed's, consisting of my father's sister Mae, her husband Jim, and my two cousins, Lou and James. Uncle Jim Reed was a round man, quite bald but a man with great humor. He worked most of his life for the American Rope Company in Brooklyn. My cousin Lou was about six years older than me. He was a precocious youth, extremely creative and imaginative. Very smart, he was also industrious and inventive. He was also a great tease, not just to me but to his aunts, uncles, and grandmother as well. He called her Ganger at his earliest talking age and the name stuck. Cousin James was closer to my brother's age, a year older, I believe. He was blonde haired and a good athlete. He was exceedingly popular with his classmates and always fun to be around.

It was an awful squeeze for the three of us to live with my grandmother. There were only two bedrooms and one bathroom. My brother and I shared one room, and my father and Uncle Frank had the other. My grandmother and Aunt Anna slept together on a bed in what would have been a dining room. My poor Aunt Lil slept on a cot in the alcove; a room at the front of the house that was more like an enclosed porch. There was little privacy for anyone, and much difficulty in meeting the demands on the one bathroom. Fortunately, my father left the house very early and my Aunt Anna shortly after for her work in Manhattan. All her life, she worked for the Otis Elevator Company and her earnings were what supported her mother and her sister. Aunt Lil was a frail woman, having suffered from polio in her early years. She helped her mother with housework and her major joy in life came from her dog Dolly. When we left our home in Kew Gardens, my father had to give away our dog Jerry. This was a heartbreaking experience for me, and by no means was Dolly anywhere a suitable substitute, she being pretty much of a housedog and elderly to boot.

My Uncle Frank was only a sometime help. He was severely addicted to bouts of alcoholism, only worked intermittently, and was unreliable for family support activities. At the same time, my Uncle Frank was one of the greatest characters I have known. He was in his forties when we came to live in Elmhurst, but wounds suffered in WW I, (some from chemical warfare), and the distress of the depression, caused him to turn to booze. He fought in the trenches in France, and after the war served with the CCC (Civilian Conservation Corps) on several projects across the company. His stories of those years were fascinating and a part of our national history. He was a fantastic teller of funny stories, of which he seemed to have an unlimited number. And he could tell these stories in any dialect called for. At parties and other gatherings of the family and friends, my Uncle Frank was in demand for entertainment. But his life was an unhappy one overall. He came and went from the house, usually on binges. He went from job to job. Although he was a handyman and quite talented in many areas, he usually lost his position due to alcohol related absences.

Our move to Elmhurst and living with my grandmother and my aunts and uncle, was a huge transition. I was only six at the time, but I sensed that something terrible had taken place and that things would never be the same again. I was right.

But fortunately, for my father and his two sons, his mother came to the rescue. Only in later years have I fully recognized the incredible sacrifice this was for her. In her middle seventies, widowed, and still caring for two daughters and an alcoholic son, she said yes to taking us in and being a mother to two energetic boys of five and nine. And she accomplished this enormous undertaking with skill, grace, and love and understanding beyond my comprehension. She became my mother in every aspect of that term. And this was at a time when my world was very insecure indeed, and my need for stability and family sustenance was very great.

My grandmother now rose early to get her two new charges dressed, fed, and off to Public School 89 on time and with their lunches and books in hand. Her housework was never ending. Beds had to be made, furniture dusted, breakfast dishes washed and put away, grocery shopping done, and then preparations for feeding dinner for seven people. Of course, my brother and I returned from school in the midafternoon hungry for snacks that we knew Grandma always had handy. There were also dirty clothes to do, especially for us boys. There were no washing machines in homes in those days. Grandma would do many of our clothes in the sink with a bar of soap scrubbing the dirt out of them on the washboard. These were hung out to dry on the clothesline, which stretched from the rear bedroom window to the electric pole in the back of the yard. Though strenuous to use, these lines were very efficient for drying clothes except when the rains came. And of course, many of these clothes had to be ironed as well.

I remind myself of how hard my grandmother had to work to take care of her extended family at the time of her life when her children should have been taking care of her. She performed these daily tasks with never a word of complaint. She performed them skillfully because she was experienced and intelligent. But mostly she performed them through the force of love she had for her children and her children's children. She accepted that this was what God wanted her to do and she was His maidservant like her namesake Mary, the Mother of Jesus. My grandmother was the first saint I ever knew.

P.S. 89

My brother and I attended P.S.89 together with our cousins, Lou and James. We walked to school each day, a distance of about three quarters of a mile. I was in first grade and my brother in fourth grade. My brother quickly established his reputation as someone not to be trifled with. He engaged in frequent affairs of fisticuffs and wrestling with schoolmates for which he was often detailed to the office of the principal. The cause of these disagreements I never really knew. P.S. 89 was a good school and I completed up to fifth grade there. My brother and my cousins graduated from there and went on to Newtown High School. I especially recall my third-grade teacher Mrs. Blanche Stephens who taught me the enjoyments one could derive from reading. Mrs. Stephens was a round woman of good humor and vigor. I still possess the book, *Sonny Elephant*, that she read to us in class and that she gave to me for my perfect attendance. Through her enthusiasm, I began a lifelong love for books that continues to the present.

A second benefit I received from my years at P.S.89 was the assemblies. Once a month all the students assembled in the auditorium for a program. This program always began with the Pledge of Allegiance and a reading from the Bible. Yes, that is right, from the Bible, something not permissible today. These readings were usually from the Book of Psalms. To this day I can recite Psalm 23 by heart. (*The Lord is my shepherd*) P.S. 89 by today's standards would be an outstanding public school. Few of today's current public schools would match it for discipline and academic excellence. It gave to me a love of learning that carried me through my later years in Catholic schools and college. Oh, that we could revive that kind of public school for today's children.

Stability

My years at Warren Street with my grandmother brought some stability into my life and that of my father and brother. We now had a peaceful home that, was well kept and orderly. My father's income helped out immeasurably to pay the rent ($40/month) and put groceries on the table. We still had his car and on weekends would drive his mother and sisters on trips to see various points of interest on Long Island. We were certainly not well to do, but in those depression years, both my aunt and father, were working full time and bringing in a paycheck. We were close to our cousins, who were great companions, and we enjoyed wonderful times together on the ball fields. Also, our holidays included scrumptious meals prepared by my grandmother and Aunt Mae attended by both families. Despite the hardness of the times, I have to say, my childhood was a happy one, full of attentive love from my family. I began to understand what family should be and how it was meant to be the cornerstone of our lives. In those years, there was little money for vacations. But with my father's car, we did manage to visit the Long Island beaches. The Cunneely's loved the beach. Years earlier when the economy was healthier, they vacationed almost annually in Atlantic City, New Jersey. The beach and boardwalk there were world famous. This, of course, was well before the casinos came into being. There are numerous photos of my family having a splendid time there, enjoying the sun and the waves. Our new beach of preference was Rockaway Beach on the south shore of Queens County. In later years, with my father's income added to the household, we began to rent cottages there, first for a few weeks in the summer, and then later for the whole summer. It was in Rockaway Beach, years later, where I first met Alice Marie Curry.

Were We Poor?

The neighborhood where we lived was centered in the Borough of Queens. It would be described today as lower middle class, that is, the families who lived there had some income (someone in the house had a job). Few owned cars. Most did not take vacations, nor possess any items that would be classified luxury. The homes were two story duplexes or fourplexes. Mostly they were wood sided although some were brick. There was an alley between the structures, but only a handful had garages. The front of the house usually had a small green area surrounded by hedges. There were still many lots scattered around on which there was some grass, but mostly clay. It was on these that we played our ballgames. They were called sandlots. Later, those houses and that neighborhood was the locale for Archie Bunker's family in the TV series "All in the Family".

At the end of our street was the elevator train. Mounted on steel girders, smack over the street (Roosevelt Avenue), were the tracks of the Interborough Rapid Transit system (the IRT). This was a convenience for all the working folk of the neighborhood who took the El to their jobs daily. The IRT (Flushing line) connected our community to Long Island City, Brooklyn, Manhattan, the LIRR at the Woodside Station, and even the Bronx. The fare was a nickel. The trains regularly ran all day and you rarely had to wait more than 15 minutes. They were comfortable and fast, except at rush hours when the commuters packed themselves in, holding on to the straps, swaying back and forth at each curve and lurching over at stalls and stops. They were also noisy. From our house, one short block away from the El, the noise from the trains was deafening. Conversations came to a pause when a train came through, even in the back of the house. We were all used to this and unconsciously stopped and restarted our conversations at the appropriate intervals.

When playing stickball, a smash off the side of the elevator structure was a triple, a shot over the tracks and on to the other side was a home run. The IRT played a major role in our lives because it brought us closer to our work and schools, shopping at Macy's, and the Yankee Stadium; however, it also brought us noise, congestion, and a terrible eyesore that was always there.

One Sunday morning, after Sunday school at St. Bartholomew's, I waited for my father to pick me up. But after an hour or so, I trudged home.

There I discovered that my grandmother had fallen on the church steps and had broken her hip. She was in a cast and unable to walk. This was a grave setback for our family because Grandma was the center figure in every family activity. She healed quicker than anyone predicted including the doctors. For the first time, though, we all had to pitch in and help. It was an eye-opening experience.

Catholic School

We lived in this house about three years. Then Aunt Mae, Uncle Jim and my two cousins moved to 94th Street. This was only three blocks away but on the other side of Roosevelt Avenue and the El. We also moved to a home, still on Warren Street, but also on the other side of the El. This home was similar, but slightly roomier and of brick. We only lived in this house a few years before we moved again to 94th Street, a few houses up from the Reed Family. This neighborhood consisted mainly of Irish Catholic families.

Previously we had been in St. Bartholomew Parish, and it was there that my brother and I attended Sunday school. It was at St. Bart's that I received my first confession and communion. But my brother had graduated from P.S. 89 and would be attending Newtown High School. My father then made a decision that would affect my life immeasurably. He informed me that beginning in the fall, I would be attending Blessed Sacrament School. He was turning my education over to the nuns, Sisters of the Sacred Heart, and I had no say in the matter. Little did I know what a difference this would be. I soon found out.

CHAPTER THREE: MY EDUCATION BEGINS ANEW

A little education is a dangerous thing.

-Pope

Education has for its object the formation of character.

-Herbert Spencer

The Sisters

In the fall of 1940, I transferred to Blessed Sacrament School. The school was a two-story, yellow brick building adjacent to the convent where the Sisters lived. The school was less than ten minutes from our home on 94th Street in Jackson Heights. It became a whole new experience for me, one that required a severe adjustment in my attitude about authority and discipline. The Sisters of Blessed Sacrament School were well up to the task. When my father informed me of his decision, I did not consider it any big deal. I liked the idea of attending a parochial school because that meant no more Sunday school, which was a pain. Also, we would have no school on Holy Days. But there were some things about which I did not know. One, the Sisters ran their school differently from the public school. They enforced strict discipline and handed out much homework. The latter was to be done neatly and handed in on time. We all wore school uniforms and were expected to be well groomed and punctual. Transgressions were dealt with firmly.

I walked to school, and since the school was close, I left myself no time to spare when leaving the house. In the first week, I arrived at the school well after the bell had sounded. Sister Superior, the Principal, spotted me from her office window sauntering down the street toward the rear entrance. She knocked on the window of her office to get my attention and to urge me to get going. I paid her little notice. By the time I arrived at the students' entrance, she was already there. To say that I was startled to see her there is to understate it.

"Mister Cunneely, you're late for school," she said.

"Yeh, I know," I replied, neither addressing her properly nor removing my baseball cap. With that, Sister backhanded the cap cleanly off my head, contacting part of my forehead in the one motion. Her hand speed in this gesture was not lost on me.

"You will remove your hat inside the school," she stated and then, while striking my upper ribs forcefully with her knuckles, she continued: "And from now on you will arrive on time, and address me as Sister Superior, won't you?"

"Yes, Sister Superior," I answered in a most shaky voice.

And I meant it sincerely. I had never been so roughly treated before, in school or anywhere else for that matter. My ribs were aching from her well-placed jabs. But I understood that things were different now than before and there was little I could do to change it. I was not about to take on this person who came across as completely in control of things. And I was not about to tell my father about this confrontation, as I was quite sure there would be no sympathy coming from him. Wasn't this the reason he was sending me to parochial school in the first place? So, my life at school was forever changed and it was up to me to live with it the best way I could.

And I did. My sixth-grade teacher was Sister Mary Matthew. She could not have been more than 25 years of age. (It was hard to guess the age of nuns because they were clothed from top to bottom in flowing robes. Their foreheads and neck areas were covered with white starched linens so that only their faces showed. Sisters had to turn their heads to look to either side.) But Sister Mary Matthew had a young face, quite pleasant to look at. She was a genuinely happy person and she brought enthusiasm and humor to her class. Despite my initial disgruntlement with the nuns, I found myself beginning to truly like Sister Mary Matthew. One reason was that she made it clear that she very much liked all of her students. I also discovered that parochial school was a far different proposition than public school.

In sixth grade, with Sister Mary Matthew, I began to truly enjoy learning. For the first time I wanted to know as much as I could about everything. Sister, in addition to the standard curriculum, gave us music... classical music. Behold, I liked it. She would crank the old victrola and play 78 RPM records of Strauss, Mendelssohn, and McDowell. The current music of that era was Swing and the Big Bands. While I was not yet a teenager, my brother and cousins were much into this modern music. And I became so too. Yet my early captivation with classical music stuck with me all of my life. Sister also shared her love of poetry with the class and even had us attempt writing some. My eventual choice to major in English in college had its roots in the sixth grade with Sister Mary Matthew.

In the sixth grade I applied to become an altar boy. This was a well coveted position and several boys competed for it. To earn selection, you had to become proficient in reciting the Missal prayers in Latin. As none of us had ever studied Latin, we were given a study book which provided the prayers with phonetic assistance. Thus, we learned to recite the prayers in Latin without being quite sure of what it all meant. We were given a competitive test by Father Dlag. To pass, you had to recite the responses in Latin with proper pronunciation and with alacrity. I was one of those selected. So, I began my career as an altar boy at Blessed Sacrament Parish.

Over the next two years I served hundreds of daily and Sunday Masses, funeral Masses, weddings, and even a Confirmation with the Bishop of the Brooklyn Diocese presiding. I took great pride in my service. Especially delightful was when I was called out of class to serve at a funeral or wedding. Frequently I received tips from the family for my serving. Another thing I loved and do so to this day is the incense. I was especially good in using the incense dispenser and the distinct pungent odor of incense awakens in me a special feeling of reverence for the sacraments and the church. To this day I am able to recite the Latin response from the Tridentine mass, which, unfortunately is rarely used since Vatican II.

I also enjoyed serving at the weekday 6:30 AM masses. Grandma would give me some money and when returning from Church I would buy some doughnuts at the bakery. At seven in the morning, you could smell the aroma of the freshly baked goods emanating from the bakery. The fresh, still warm doughnuts highlighted my breakfast, which in those days, and even until today, began with Wheaties, the Breakfast of Champions.

Serving on the altar instilled in me a fierce love for the Church and its sacraments. It also made me consider the priesthood as a future vocation (alongside becoming a professional baseball player.) The priests I served as a boy were all good men and servants of God and His people. I especially admired Fr. Spengler, who would often join us in the playground and play stickball with us. He was a good ballplayer too and an exemplary role model for us.

My three years at Blessed Sacrament were crucial to my upbringing. I made many good friends, played on many teams, and honed my athletic skills in several sports. I also began to appreciate learning. The nuns, who brought to the classroom an energy and enthusiasm that was infectious, inspired me. I became an avid reader and user of the library. I wrote poetry and short stories and sought to increase my vocabulary through use of the dictionary. I was the best speller in class and usually carried the day for the boys in our spelling competitions versus the girls. My life was full and happy. I had a stable and peaceful home. My Grandmother and aunts provided the feminine attention that I needed in the absence of my mother. My father was working hard and long, but he had a steady job that was beginning to pay better. We had a car, which most families did not, and we were able to take some trips around Long Island and occasionally further. In 1939, we took a car trip to Michigan where we visited my grandmother's brother and cousins. It was a great trip and we stayed on a farm that was a whole new experience for my brother and me. We fed the pigs and worked with the cows. We pitched hay and rode some horses. We tried to milk the cows with little success and found that fresh milk, direct from the cow, was no bargain. Ugh! We swam in Lake Huron and toured Detroit. We stayed in boarding houses going and coming, and from it all, began to realize that the world was a lot bigger than what I had thought. It was a wonderful experience for a kid from the streets of New York.

History Intervenes

In my sixth-grade year, another event happened that would profoundly change my life and the lives of everyone around me. It happened on that infamous date, December 7, 1941. It was a Sunday afternoon that word came on the radio that the Japanese had attacked America. I was not certain where Pearl Harbor was, but I quickly realized that our country was about to embark on a truly horrendous piece of history.

CHAPTER FOUR: GROWING UP IN A WORLD AT WAR

War is Hell

-General Tecumseh Sherman

World War II

I was 12 years old when America entered its second great war of the 20th century. My brother Joseph was sixteen. Neither of us clearly understood what it all meant. We were to learn soon enough. However, my father and the rest of our family foresaw more clearly that our world was going to change significantly and not for the good. My father and my two uncles all served in the First World War. Frank saw combat and was wounded by mustard gas from the Germans. It was a great disruption in their lives and of course, my grandmother and the rest of the family suffered years of anguish and worry. Casualties were high and their relief was great when the war ended in November 1918. That was to be the "war to end all wars". But less than 25 years after, the United States was again heavily involved in a foreign war, this time on two fronts, against enemies armed with the latest weapons for aggressive war on the sea, on the ground, and in the air. Our elders knew that our family would once again be personally active and this time, the results would be more deadly and widespread.

Soon Enough, the War Came into Our Lives

First, it was clear that the new generation of young men would be called to serve. What is now referred to as the "greatest generation" was then declared the unluckiest generation. All males, eighteen and older, were registered to be drafted by lottery. Those rejected for service for whatever reason were often singled out for derision. To be listed as 4-F, physically unable to serve, was especially degrading. As it was a matter of time before you were called, many men chose to enlist immediately. My cousin Lou who was just finishing college went into the Navy as an officer candidate. He survived the war despite seeing much action. His brother James was drafted into the Army. He too came through unscathed seeing mostly stateside duty. My brother Joe, who was sixteen at the time of Pearl Harbor, had one year left in high school. On graduation, with the permission of our parents, he enlisted in the Navy. In August 1943, my father and I went with him to Manhattan and saw him off. It was a sad day indeed. The early war reports were not good, and casualties were building up. Every street had multiple homes with a son or father in the service. A blue star was hung in the window to show that that home had a man in service. Later, many of those stars were turned to gold indicating a life lost in the war. It was a most depressing time that lasted many years. Thankfully, my brother survived the war although he saw action in the Battle of Leyte Gulf. His ship, a baby carrier, the *USS Kitkun Bay*, was struck by a Kamikaze fighter and severely damaged. It listed so badly that the Captain ordered the ship to be abandoned. But it did not sink. They reboarded and limped back to Hawaii for repairs. Many men died that day.

Those of us at home saw many changes in our lives as well. While not life threatening, they brought the war into our very homes on a daily basis. Our Armed Forces needed metal, rubber, chemicals, and food. Those at home had to provide them as best we could by giving up such items as radios, bicycles, typewriters, toasters, phonographs, rubber boots, balls, and tires. No outdoor lighting at Christmas, or tinsel on the tree. There was a shortage of turkeys, cranberry sauce, sliced bread, and whipped cream.

To prevent hoarding, the Government instituted a system of rationing. Every man, woman, and child received a ration book which provided points. These points were used to purchase such items as meat, cheese, oil, canned goods, and butter. When I went to the grocers or the butcher for my grandma, she would give me the coupons from the ration books that l gave to the clerk for these items. No coupon, no item. Gasoline was especially curtailed. Essential workers such as doctors (who made house calls in those days), defense workers, and ministers received an A card which was good for three gallons a week. Others received a B or C card good for two gallons. As you might guess there was lots of trading going on and even some counterfeiting of cards. Carpooling was strongly encouraged, and I recall signs saying, “When you drive alone, you drive with Hitler.” Folks on the home front were mostly supportive though because this war was a terrible thing that affected all of us. No family was left unscathed. It was a time of increased spirituality, letter writing, listening to the radio, and volunteer work. Women entered the factories manning the assembly lines building tanks, planes, and guns. A famous poster of the time, by Norman Rockwell, I believe, depicted an attractive "Rosie the Riveter". Fathers and grandfathers were air raid wardens going out each night to keeps our cities dark and conducting air raid exercises. Young women felt the shortage of young men in their lives leading to loneliness and depression. It was a bad time, a tragic time. We looked to our leaders for strength. Thank God we had them. From FDR down, the nation brought forth its absolute best to guide us through this crisis. Some names stand out: Generals George Marshall, David Eisenhower, Omar Bradley, George Patton, Hap Arnold, Curtis LeMay, Douglas MacArthur and Admirals Nimitz, Sprague, and Halsey. Places we never heard of became known for the various battles fought there: Midway, Okinawa, Guadalcanal, Iwo Jima, Leyte Gulf, Aleutian Islands, Normandy, Dunkirk, the Bulge, North Africa, and many more. And always that question, was my son there? God forbid.

The war brought a whole new unexplored world to Jackson Heights, New York as it did to every hamlet and home in the free world. Nothing would be the same again. The safety and security of the free nations as well as the enslaved countries were at stake. I began to take an interest in the world beyond my home and school and pals. Major events were taking place and I realized for the first time that history was expanding at a pace never before experienced.

As the war progressed, certain terrible things that were happening or had happened came to light. They were awful, evil things that many of us never thought could be perpetrated by some humans on others. Largely populated cities were devastated, huge battles were fought on land, sea and air, and the extent of casualties of these events began to leak out. Treaties between nations, thought to be sacrosanct were abrogated. Betrayals were commonplace. Racial crimes, the systematic extinction of members of a race, religion, or nationality, were rumored, with more evidence coming out daily. The effect on those of us at home was numbing and frightening. And always, we lived in fear that our loved ones would become victims. We never prayed so much or so ardently.

In short, the war shortened my childhood and that of all my pals. We continued our school, our play, and all the other things kids do, but always there was that cloud hovering over and our thoughts rarely strayed too long from what was happening to our world.

My years at Blessed Sacrament flew by. Perhaps it was the seriousness of the times that affected me, but I became an avid reader, and I began to take my studies very seriously. I wanted to learn as much as I could about everything. I began weekly visits to the Elmhurst Library, which was a mile walk, and I would bring home the maximum number of books you were allowed. There was no television in those times, so after dinner and homework was done, I read. I read Mark Twain, Jack London, Charles Dickens, Robert Louis Stevenson, and the adventures in deep sea diving of Commander Ellsberg. I read fiction constantly and often dreamed that someday I would become a world-famous author. I also authored essays, short stories, and poetry. In this, I was greatly encouraged by my teachers, the Sisters of Blessed Sacrament. Each year I had to recite a poem at an assembly performance attended by students, teachers, and parents. These performances were nerve racking, but I usually came out well. I can say that beginning in those years I came to enjoy reading and writing that I still have today.

Then of course, I also considered myself a great athlete. I began by playing baseball, a game my father introduced me to from the earliest years of my life. He bought my brother and me gloves and he would play catch with us all the time. Later he would hit fungoes to us by the hour and we learned to judge fly balls and handle even the meanest grounders on sandlots splattered with stones and other debris. I played on our school team and in addition to being a consistent hitter, I eventually became the pitching ace. I had great control and rarely walked a batter. My fastball had good bounce and I learned to throw a curve ball that I could locate on either side of the plate. I won many games, and I was soon asked to pitch for a team of high school age kids who played on Sundays at the local ball field. Though I was by far the youngest player, I won most of my games. However, my biggest disappointment came in the seventh grade when I pitched against the eighth graders in the annual school picnic game. A huge crowd was there, and I desperately wanted to lead the first seventh grade team to beat the about to be grads. But we lost, 1- 0, when our centerfielder failed to make a catch on an easy fly ball. I tasted defeat, and I did not like it. It took a long time for that loss to fade in my memory.

The New York Yankees

My father was an avid Yankee fan. He always spoke with reverence about the great players of that time. Names like Christy Mathewson, Babe Ruth, Lou Gehrig, Three-Fingered-Brown, were just a few of many great heroes he had seen and loved to talk about. He told me about the New York Yankees and their Murderer's Row, a lineup that included Tony Lazarri, Bill Dickey, Lou Gehrig, Babe Ruth, and Red Rolfe. He took me to my first game to see these powerful and unbeatable players. The Boston Red Sox were the opposition, and their pitcher was one Lefty Grove. Grove shutout the Yankees with one hit (a dribbler by Gehrig). I was totally confused as I was promised that I would see the most deadly hitting team in the world in action. My first lesson was learned: good pitching will defeat good hitting.

I also began to play basketball. I was built small, but I learned how to pass, dribble and shoot. (Two handed, of course.) We played in the schoolyard where the backboards were metal and there were no nets. You formed a two- or three-manned team and waited for your turn to play. Winners stayed on, losers went off and went to the back of the line. The play was very competitive, with many arguments over fouls, and whether the ball actually went through the rim or not. (No nets.) Also, basketballs in those days were leather. It did not take long before a leather ball became smooth from hitting the pavement.

Basketballs were hard to come by and any boy who owned a ball was everybody's friend. The owner also got to play more than anyone else. And, of course, when the ball owner had to leave, the day's play was over for everyone. When I was twelve, I began to play on my brother's team, and we traveled into the city (Manhattan) and around Queens playing school or parish teams. I was just learning, but I learned fast that even though I was smaller than the rest, by smart positioning, anticipation, and hustle, I could help my team. These games gave me needed experience for my high school years.

I See the Country for the First Time

While I was in grade eight, my father took me on a trip to Los Angeles, California. My Uncle Tom was living out there. He was a bachelor and retired from the railroad. But he was an eccentric man, very solemn and prone to fits of depression. My father was concerned, and some friends of my Uncle Tom thought my father should come out and visit him. I really did not want to go. It was in the spring, and so much was going on that I did not want to miss. My father loved railroading and he planned the trip. My father was an expert in railroad travel and knew how to plan and ticket interline travel across the country connecting to many different rail lines. We left New York's Penn Station and trained to St. Louis. It was my first introduction to Pullman berths and eating in dining cars. I loved it. From St. Louis we traveled through the heartlands to Denver, then through the Rockies to Salt Lake City, across the salt flats to the Sierras and into California. We took the ferry across Oakland Bay into San Francisco and the train down the California coast to Los Angeles. It was my first glimpse of our great country. I was very moved by it. I also had a chance to visit with people from all areas, listen to their accents, and I found out that people are people regardless of where they come from, what they do, and what they look like.

After a not very happy visit with my uncle, we returned to New York via Phoenix, San Antonio, New Orleans, Charleston, Washington, D.C., and Philadelphia. It was a trip I will always recall with fond memories. The aroma of the Pullmans, the nighttime hoot of the steam whistles, the gracious attentiveness of the Pullman porters, the wonderful meals in the dining car as we made our way across the country, and subtle and then stark contrasts in the countryside as we crossed from one geographic area to the next produced vivid scenes that I still recall. From sea to shining sea, I saw what a spectacular country we live in. The things I read in my geography classes came to life. This trip planted the seeds in me that later new places and exploring whatever the area I came to offered. Years later when I flew across our country, as I did many, many times, I was able to again view our country from the air seeing the cities, farms, forests, mountains, great rivers, lakes, ranch lands, and glaciers. I traveled to every state and ended up living in many of them. It made me proud and happy to have been born an American and to have had the opportunity to defend it and the freedoms it provides.

My years at Blessed Sacrament School were golden. I had numerous friends with whom I palled with, played sports with, and shared growing experiences with. I loved my teachers, those wonderful nuns who dedicated their entire lives to service to God and His people. The Sisters throughout our country developed the most effective and productive educational system America has ever seen. Hundreds of thousands of Catholic school graduates, operated by these nuns went on to become our country's finest citizens. They became priests, nuns, generals, lawyers, mayors, legislators, governors, congressmen, teachers, doctors, and so forth. Their ability to read, write, serve, solve problems, and make decisions, all garnered from their educations at the hands of these nuns, was the cornerstone of their success. Unfortunately, our Church failed to provide these wonderful orders of sisters the support they needed. Disenchantment set in, fueled by feminists who pushed for more power, brought an end to this great educational system in the sixties and seventies. Today, Catholic schools are manned by lay people, the costs are staggering, and the schools are not as good as they were in my days when they were operated by these terrific women. I will never be able to repay them, but I thank God that they were there, and I pray for them.

I Move On

In my last year at Blessed Sacrament, I began to extend myself into roles that required some leadership aspects. It began in sports where my teammates elected me captain. This did not set too well with a few of the previous leaders who attained their positions because they were the biggest guys and the most prone to enforcing their positions through physical enforcement. While I was an aggressive player in all my sports, fighting was not my style. But one day, after our baseball team elected me captain, John O'Reilly and one of his aides showed up on my street to settle the matter their way. I had been told that they were coming, and I truly considered taking off and hiding. But I decided to stay and see it out. O'Reilly soon appeared and confronted me. I stood my ground and waited for the inevitable. O'Reilly' s friend circled behind me, and I realized what was about to happen. We were on the sidewalk and O'Reilly had backed me up against a hedge. He lunged at me and tried to push me backwards over the hedge and onto the ground where he and his lackey could deliver some nasty blows while I was on the ground. But I evaded his thrust and punched John in the face. The blow did not land on the spot I had hoped for, nor with the force I had hoped, but it did the job. It startled him and brought a little of his blood to the surface. Nothing is as confounding to a bully as the sight of his own blood. John was shocked and then outraged. He rushed me and we both tumbled over the hedge and onto someone's lawn where we rolled over and over trying to exercise mayhem on the other with mixed results. I recall deciding that because I was going to get the worst of this engagement, I would give it all I had and try to inflict as much damage as I could. Neither of us did much and the fight was ended when team members broke it up and we were separated. After a few minutes where each of us assessed the damages, I was startled to hear John compliment me on how I stood my ground. He clearly saw that I had the support of the team, and he was not going to change that. We shook hands and the matter was ended. John O'Reilly and I became good friends, and it was a friendship that lasted many years. We learned to respect each other. But more important for me, I learned something about myself. That was, there are times and situations in life when you must stand up for your beliefs and be ready to fight for them if necessary. It was a good lesson for what was to come many times in my later life.

My school years at Blessed Sacrament came to an end in June 1943. Our teacher, Sr. Mary Edith, was simply a superb teacher. She was an older nun, but very experienced and dedicated to getting the best out of her students. Her specialty was preparing her best students for the scholarship examinations which each Catholic high school in the area conducted. Each school awarded at least two full scholarships and some partial scholarships to the students who scored highest. These exams were conducted at the high school on Saturday mornings in the spring. Sister had collected all of the schools' past exams and she then conducted special preparation classes for the students she felt had the best chance. Unfortunately, these classes were held Saturday mornings, which, of course, conflicted with my sports schedule. I still recall with chagrin, a Saturday morning when I decided to play basketball in the schoolyard and skip Sister's class. Lo and behold, here she came after me with her habit fluttering in the breeze. "Mister Cunneely you get right into that classroom right now," she shouted. Did I ever take a ribbing from the guys, but I hightailed it to the school. Years later I still remember her and what a great person she was, and I remember her with gratitude.

I took about seven tests, and I was awarded five scholarships. I just had a knack for taking those tests. I always scored high in the math section and was able to write essays for just about any subject they asked for. I was awarded a scholarship to Regis High (Jesuit), LaSalle, Xavier (Jesuit), St. Francis Prep, and the one I really wanted, Chaminade High School in Mineola. No one had ever won so many scholarships. Sister Mary Edith was just thrilled and so was my father because that meant free tuition. The reason I wanted Chaminade was because I had played some basketball with and much admired a kid named Joe Mullaney. (Joe later became a star at Holy Cross College and a pro player. He coached Providence University for several years.) Joe and his brother Dave lived in Elmhurst, and they attended and played varsity ball for Chaminade, a school I had never heard of prior. Joe told me about his school. It was a small (four hundred students), all boys, and situated in a rural setting in Nassau County.

Commuting there and back would be a problem, but I nevertheless wanted to go there. It was run by the Society of Mary, the Marianists, and I liked that too. My father drove me out to the school one Saturday morning to take the test. Dad was familiar with Mineola from his work on the LIRR. It was the first time I had ever seen the school. I prayed for a good day, and I never concentrated so hard during a test. Three weeks later I received a short letter telling me I had won. What a joyful day that was for me, and what a turning point in my life. Chaminade High School was and to some extent still is, an important part of my life. And I owed it all to Sr. Mary Edith.

CHAPTER FIVE: THE END OF CHILDHOOD

When I was a child, I spoke as a child,
I felt as a child, I thought as a child.

-St. Paul

Childhood

Childhood is such a precious time in a life. Children come into the world with innocence and an enthusiasm that will never be duplicated later. There is such a complete dependence on the mother, and later the father moves into a pivotal role. An infant needs love and security, every day, all the time. They comes into the world totally self-centered and demanding. And the parents must bring the child from that totally inward state to one of outwardness and selflessness that are the trademarks of truly arrived grownups. It is a daunting task.

I would say that my graduation from Blessed Sacrament School in June 1943 marked the official termination of my childhood. To that point I can state that I enjoyed a mostly happy one. In spite of the split up of my parents, the loss of my mother at age five, the move to my grandmother's home, change of schools, departure of my brother to war service, our low economic status, and so forth, I breezed through my early years a happy, busy, healthy child. Thanks to my father, I was deeply into sports, thanks to the good sisters at Blessed Sacrament I was good student, and thanks to my grandmother, I lived in a secure and loving home. I never had my own room, and seven of us shared one bath.

Until I was in college, we never owned a telephone or a television set. I traveled to and from school on LIRR trains, subways, buses, hitch-hiking sometimes, and, of course, walking. I left early and came home late from various after school activities. We had no luxuries, but I never missed a good, well prepared nutritious meal, and I always slept in a warm comfortable bed. Considering that I was a 'depression baby', my childhood was much better than thousands of others in their childhood at the same time.

I am thankful to all those who took care of me, educated me, advised me, and made sure I stayed on a level path. My great interests were in sports and reading. I had no time for girls. Girls were people I did not understand nor cared much about. In short, they were a nuisance. I did not understand why they were here at all. Naturally, that attitude changed somewhat when adolescence overtook me. I guess that was in my first year in high school. Then it became bothersome. I attended an all-boys' school that did not even permit dances within its walls. So, my association with girls was limited. This was all right, as I never knew what I should say to them or how I was to respond to things they said. My best way to handle this was to avoid contact with them to the largest extent possible.

And so, my childhood evolved into adolescence that brought with it new experiences, new places, new friends, and new insights. It also brought fears—fears that I would not succeed in reaching goals. My scholarship to Chaminade required that I attain Honor Roll status each semester or lose it. The pressure to do well on every test never diminished. But I also wanted to excel in sports and other activities. As such, my high school years were very busy, and pressure filled. The war was still on and worry about the safety of my brother and my cousins was ever present. The future was obscure and uncertain. But I had now left childhood behind. My world had expanded, and I was getting ready for whatever the future might bring.

CHAPTER SIX: HIGH SCHOOL AND COLLEGE YEARS

Fortes in Unitate.

-Chaminade High School Motto

Chaminade High School

Looking back, I believe that my high school years were my happiest in many ways. Despite the pressure of school and athletics and hormone changes, I began to grow socially and intellectually. I found my new classmates, who came from all ends of Long Island, to be stimulating. They were all different in some respects. Some came from wealthy families, and some struggled to pay the tuition and fees. Some brought their lunch in a brown bag; others bought their lunch from the lunchroom menu. Some drove their own cars and most took the bus or hitched. (During my freshman year, while I was still living in Jackson Heights, I walked three blocks to the elevated train station, took the train to Woodside LIRR Station, caught a train to Jamaica, changed to the steam locomotive train to Mineola and then walked a half-mile to the campus. Beginning in my second year, we moved to Great Neck from where I commuted by bus.) I made lifelong friends at Chaminade because there were so many great guys who went there. We all shared the anxieties, disappointments, and pressures of high school years at a school run by the Marianists, dedicated to academic and athletic excellence.

When I first came to Chaminade, I knew no one. As I lived furthest from the school than anyone else, every one of my classmates was a stranger. All of them marveled that I came from so far away, every day. But I was ready for a change and a wider horizon. I genuinely enjoyed meeting these kids, learning as much about them as I could, and making as many friends as I could. Though I was certainly not an extrovert, in fact I was more of a silent, inward type; nevertheless, I tried to make friends with everyone and to enjoy the individuality of each one. I joined no cliques and though I was perceived as a "brain" and a budding basketball star, everyone was my friend.

Chaminade was more than I expected. In 1944, it had approximately four hundred students, all boys. The teachers were mostly Brothers of the Society of Mary, the Marianists. They were tough, dedicated teachers. They knew their stuff and demanded complete student compliance. It was a no-nonsense institution. Students were required to wear a coat and tie and to be on time. When the teacher entered the room, students stood up. Each class began with a prayer. Mass was held on Friday and mandatory on the first Friday. Fifty years after my graduation, I returned to Chaminade for a reunion. I was surprised favorably to see Chaminade now had over 1500 students, all boys. Students had to wear a coat and tie, and each class began with a prayer. The school had a long waiting list. Long Island parents recognized that here was a school that had not watered down its standards and retained its revered traditions. I was very proud. Chaminade over the years had turned out countless graduates who attained leadership positions in government, military, entertainment, the Church, and business. Most of them attribute their later success to their years at Chaminade, where they received a terrific education with a strong Catholic flavor, and a special devotion to the Blessed Mother.

When I graduated from Chaminade in June 1947, I was no longer the same green kid that had entered four years ago. Despite the pressures, I was able to retain my academic scholarship by making the Honor Roll each semester. I also participated in sports, especially in basketball where I played all four years. In my senior year, I was team captain. I also played other sports, wrote for the school paper and yearbook, and participated in debate. I was runner up for class valedictorian and graduated with the highest-grade average. I was given an award as the student with the highest grades in math and English and given the honor of being the athlete with the best academic record. This made my dad and family proud. They had sustained me through a tough time in my life, and because of them, I was ready to move on. I would be the first Cunneely to attend college.

Looking back at my high school days, I realize what a lasting influence those years had on me. They were my teachers first. After years of having only women teachers, at Chaminade there were only male teachers, most were Brothers or priests of the Society of Mary. They were real guys who were very smart, dedicated to their students, and devoted to the Mother of Jesus. Many of them joined us on the playing field or basketball court after class where they let us know that they participated in all facets of life with equal enthusiasm. They were also our counselors and friends. I especially remember Bro. Louis Nath, S.M., Fr. Tom Stanley, S.M., Bro. Gerry Morris, S.M., and Bro. John Darby, S.M. I also remember with fondness my coach Eddie Flynn, who got me started in coaching, a profession I was aiming at but only performed part time in later years. I also formed lifetime friendships with several classmates. Although we went our separate ways, we joined together whenever we could, called each other often, and remained friend to the end. I admired all of them. They were such good men, lived exemplary lives, and remained true to the beliefs they formed in their Chaminade days.

In my senior year at Chaminade, I began to think seriously of joining the Marianist order. I was counseled in this by Bro. Lou Nath. It culminated in an offer of a scholarship to Dayton University, a Marianist college, where I would play basketball and take up religious studies. However, after much thought and reflection I declined. Instead, I enrolled at Fordham University, a Jesuit school in the Bronx. It was a momentous decision for me and clearly changed the direction my life would take. It is a decision that I think about often, but never regretted.

800 See Diplomas Awarded At Chaminade

An audience of more than 800 relatives and friends saw the graduation exercises at Chaminade High School, Mineola, last night when 68 seniors were awarded diplomas. The Rev. William J. Gately, pastor of St. Boniface Martyr Church, Sea Cliff, addressed the graduates.

Denis Francis Mullane, Floral Park, valedictorian, won top honors by obtaining the gold medal offered by the Chaminade Mothers Club for excellence in the study of religion; honor plaques for excellence in German and speech work, a gold pin by qualifying for the Chaminade Honor Society for four years; the "Most Representative Chaminade Man" alumni award, and the Long Island Daily Press award for scholarship. Mullane was also the sodality prefect and president of the Chaminade Speech Society.

Anthony Edward Scudellari, Hempstead, salutatorian, merited an honor plaque for excellence in Spanish, and a gold pin for four years as a member of the Chaminade Honor Society.

Christopher Patrick Sweeny, Freeport, class president, gained highest honors for excellence in dramatics and a plaque by distinguishing himself as the best student athlete.

Eugene Paul Cunneely, Great Neck, one of the four-year scholarship winners upon his entrance to Chaminade in 1943, gained a fourfold award by meriting the honor plaque for excellence in English and Mathematics, a gold pin for four years in the Chaminade Honor Society, and as captain of the varsity basketball team, the Long Island Daily Press award given to the student holding a monogram in a major sport and attaining the highest scholastic average.

DENIS FRANCIS MULLANE

EUGENE PAUL CUNNEELY

William Henry Caulfield, Glendale, the other four-year scholarship winner in 1943, also copped double honors by attaining the award of excellence in social studies and Latin. The plaque for excellence in science was won by Jacob William Zimmerman of St. James.

Edward Joseph Boyle, Freeport, won the American Legion trophy award presented by the William Patterson Post, Mineola.

Gold medals for four years' service in the Chaminade Band were awarded to James G. Avenius, Garden City; Joseph L. Costa Jr., Hollis; Donald C. Ryan, Rockville Centre; Francis P. Mullen, Floral Park; William F. Hirchmann, Jr., Rockville Centre; John E. Jarvies, Queens Village; William A. Keenan, New Hyde Park; Francis P. Kelleher Jr., West Hempstead; Robert W. Levesque, East Williston; Francis B. Melvin, Hempstead; Rudolph D. Parsons, Roosevelt, and Francis V. Zepf, Massapequa.

The Rev. Brother John T. Darby, S. M., principal of Chaminade, presided. Citations and presentation of the graduates were made by Brother Clarence H. Walters, S. M., assistant principal. The Rev. Bertrand Clemens, S. M., school chaplain, gave the invocation.

The Chaminade Concert Band under the baton of Angelo Feraindo provided musical interludes.

The Jesuits at Fordham University

My college years were totally different from my high school ones. The pressures were lifted from me, and I decided to take things a little easier and to begin doing the things that I wanted to do but I could not; that included studies. Fordham was an excellent school operated in the true Jesuit tradition. Whatever your major was, you had to take a large dose of Philosophy, and in those times that meant Scholastic philosophy as generated from the great mind of St. Thomas Aquinas. The Jesuits were excellent teachers, masters of their subject areas, but each different from another in style and attitude as day and night. They were a delight and a frustration. Fordham was an all-male institution, with a great reputation for academic rigor and its football teams. Its purpose was to graduate men who were men for all seasons. They would be Fordham men.

From the Fordham Undergraduate's Handbook:

The Fordham man is always a gentleman. He is kind, he is courteous, he is above all, thoughtful of others rather than himself. He is ever mindful of Cardinal Newman's lasting definition, "A gentleman is one who never hurts another's feelings," and hence he is merciful toward the weak and unfortunate, just as he is merciful toward the absurd.

A Fordham man is loyal; he is loyal to his teams, he is loyal to his fellows, he is loyal to his Alma Mater. Loyalty is a distinctive note of character, and character is an essential requisite of gentlemanliness. In turn, unless he is a gentleman, no undergraduate is truly a "Fordham Man."

He is careful of his actions both at college and when without the "hallowed gates," and his speech is ever guided by an upright mind. He is loyal, he is happy, he is a good fellow, but remember: The Fordham Man is always a Gentleman.

These are some high-minded words, and they represented a strong challenge to us freshmen. They presented an ideal that few of us were up to at that time. I would be the last to confirm that many of us met that challenge during our college years. Several did, but more of us came closer to attaining those goals in our after-college days. But looking back at those years, I better understood what it was the Jesuits were aiming for through their educational process. They clearly wanted us to become more like Jesus. They were aiming very high. I had great Jesuit teachers at Fordham including the renowned Ignatius Cox, S.J. When you attended Fordham in those days, regardless of your major area of study, all students really majored in Scholastic philosophy. Based largely on the writings of St. Thomas Aquinas, our philosophical studies delved into the existence of God, the creation and nature of the universe, man and his relationship to the Creator, the purpose and meaning of life, critical thinking, and how all of the above should affect our daily life, our moral behavior, and our continuous search for reunification with the Creator. This was pretty heavy stuff. It caused me and my fellow students to think about a lot of things we never before contemplated. These studies, imposed by the Jesuits, opened a whole new arena of reflection and wonderment, and made a great difference on how I looked at my life and where I would take it.

I got from my Jesuit years many attributes I needed to proceed into the real world with any hope of success. These included the ability to think logically when approaching critical decisions. Also, there was a reliance on the Creator to help me meet life's trials and tribulations. Another was admiration for leaders who employ all of their God given talents and intellectual powers to overcome adversity. I also fell back on the Handbook's special emphasis on loyalty as a fundamental quality that one finds in a gentleman wherever he is, whatever he is doing. In my later years in the military, I soon realized that loyalty was essential to the successful completion of the very difficult and dangerous tasks that we enjoined.

With less personal pressures and a more relaxed lifestyle, I began to enjoy my college years. I continued my athletic endeavors in both freshman and JV basketball, but I began to widen myself in other ways. I began to enjoy the Broadway Theater, squandering what little money I had on seeing plays and musicals that were readily available a short subway ride to Manhattan. I even attended a performance at the Metropolitan Opera to see what that was all about. I was impressed. As Fordham was a short subway ride to the Yankee Stadium, I took advantage of that and the low bleacher seat prices to watch about 20 to 30 games a year. You knew where to find me when the Boston Red Sox were in town. I still have great memories of watching some of the great stars of that era including Bob Feller, Ted Williams, Allie Reynolds, Phil Rizzuto, Bill Dickey, Tommy Henrich, Lefty Joe Page, and my all-time favorite, Joe DiMaggio. I also broadened my reading, perhaps at the expense of my grades. I began to read the great novelists such as Hemingway, Fitzgerald, Rand, Faulkner, et al. Politics now became of interest to me as we came out of World War II and began to resolve the problems of post war Europe, Japan, and the reentrance of our veterans into the civilian work force. We also saw the beginning of what became known as the Cold War. In summary, life was becoming more interesting and challenging, and I began to realize that when my school days ended, and that was coming swiftly, I had lots of big decisions to make.

During my college years I began to think about what I would do after school days ended. At the beginning of my junior year, I decided to major in English. It was a hard choice between English and math, but I decided that I wanted to teach high school English and coach basketball, preferably at Chaminade. I received welcoming responses from my former Marianist teachers and administrators. I offered my services to Ed Flynn, Chaminade's varsity coach and I began to help coach the JV squad and scout opposing teams for the varsity. I truly enjoyed this, and I became more proficient with experience. I began to feel more comfortable in a leadership role. In my junior and senior years, I also played on many teams in Church leagues, which were highly competitive and also on some semipro teams. While not as big as I wished, I was a good shooter and an excellent defensive player thanks to the mentoring of my Fordham coach, Bob Mulvihill. Two of the teams I played for were The Fordham All Stars and van Breda Kolff's All Stars. These teams would play games at different locations around New York and New Jersey against the local team. These games were played in church halls or community centers where the home team had a significant advantage. But I don't recall our ever losing, especially with Butch van Breda Kolff who was a great star at Princeton and the New York Knicks. Butch was a big strong player, unstoppable near the basket, and the most competitive player I ever knew. Many of these games ended in a wild melee. Our plan was to coast through the first half staying even or just a few points behind. In the second half we then fed the ball to Butch who would carry us to victory. Butch would take the largest share of the revenue (normally @$100 and the rest of us about $10). However, he always bought the beers. The impressive thing about it was Butch's capability to down about six beers before the game without any noticeable impact on his stamina and performance. Those were great days. I played with and against some of the best basketball players in the country. New York was the center for basketball turning out great players for the colleges and eventually the pro leagues that were starting up. Some of those I played with and against included Joe Mullaney, Bob Cousy, Connie Simmons, Dick and Al McGuire, Sam Mele, and Ernie Vandeweghe.

Work

At an early age I learned the meaning and value of work. We were a poor family and while we never missed a meal, there was never any loose money available for luxuries or vacations. When I reached the age of 15, I began working during the summer school vacations. My first job was at Rockaway Beach for the Parks Department picking up debris off the beach with a stick that had a nail at the end of it for snagging papers and other garbage, which I then deposited in a bag. This was a terrible job as I was out in the hot sun all day and the foreman was always watching us to be sure we were not sitting down on the job. Later, I was fortunate enough to be assigned to the First Aid station which was located under the boardwalk and out of the sun. There I attended to small cuts and other wounds, frequently sunburns, that beach attendees came in with. One summer, I got a job in Manhattan with the State Department of Labor. This required me to commute to Manhattan every morning by LIRR and subway and return at night. In the New York heat and humidity, this added to my discomfort and unhappiness with work. I also got a job one summer, through my brother's influence, with Woman's Home Companion, a popular magazine for ladies in the 40s and 50s. But my best break came when I got a job with the U.S. Post Office. I began to work summers and during the Christmas rush delivering mail twice a day. I earned one dollar an hour. But then I got another break. I was able to demonstrate the ability to memorize postal route by streets. Thus, I was given inside work sorting incoming mail into boxes which represented the delivery route for each postman. Now I could work inside and did not have to lug around a heavy bag of mail that made my shoulder ache each night. Later I began to work nights which upped my pay to $1.25 an hour. The postal foreman took a liking to me (we were both Irish) and often let me have overtime. I began to make some good money. Working summers and holidays at the post office through high school and college helped pay my tuition and other expenses through Fordham. More than that, my work years taught me valuable lessons regarding the relationship between work and money. You certainly take care of your money and spend it more prudently when you had to personally sacrifice so much to earn it.

I Make an Educational Choice That Changed My Life

When I registered for my freshman year at Fordham, I made a choice that would have a major impact on my life. In 1947, our country was emerging from six years of World War II. The nation was trying to return to some semblance of normalcy. Soldiers and sailors coming back from years of service were attempting to throw off the memories of their years away from home, family, wives, sweethearts, etc. They were planning to make their future lives more of what they had originally dreamed about. Thousands used the new G.I. Bill to attend college. Many were my classmates at Fordham. Admiration for these men and the military as a whole was widespread. I made a choice to take ROTC as one of my science requirements. Not only did this choice permit me to avoid taking Chemistry/Physics, but it also paid $29 a month in the last two years. As nutty as that may seem today, $29 was a lot of money in those days for a struggling college student. Uncle Sam made a good bargain for itself. I never dreamed at the time I registered for ROTC, that that choice would have such an impact on my future. You never know, do you?

I truly believe that it would be impossible today to receive the quality education that I had. It's just not being offered, and the dumbing down of college curricula is very hurtful to the young people of today. In each school I attended, I was challenged to study, memorize, and be tested on what I was supposed to know. The skill and dedication of my teachers were unequaled. The nuns in grade school were terrific teachers always pushing us to our full potential, and of course, it was the Sisters who taught me the basics of my Catholic faith and prepared us for the Sacraments that are so vital for a strong spiritual life. The Marianists brought their dedication to the Virgin Mother right into the classroom. And the Jesuits made us think logically and reach decisions based on firm principles. At Fordham, although I majored in English, I really studied Philosophy.

Here are some of the subjects I was taught there:

Logic	Critical thinking, how to come to the right conclusion through reasoning
Epistemology	Study of the origin, nature, methods and limits of knowledge
Cosmology	Theory of the nature and principles of the universe
Metaphysics	Study of the origins and nature of reality and the origins and structure of the world
Rational Psychology	Study of the mind as a tool of reason
Natural Theology	The nature of God as we know it from His revelation
General Ethics	Study of natural law and innate desires to do what is right
Special Ethics	Application of ethical behavior in the real world e.g., business

I'm not sure that you can find any of these subjects in today's college curricula. But they opened up a new world for me because they addressed concepts that had been in the back of my mind and that few people ever brought up. It is stated that no man is truly educated who has not majored in Philosophy because no other studies comes to grips with the most vital ideas that the wisest of our forefathers had concluded about the origins and nature of the world; why we are here and what should be our goals. My philosophical studies at Fordham under the great Jesuits had a lifelong impact on my life that still governs my thinking and my decisions. Thank God for that.

Graduation

In June 1951, I ended my college days at Fordham University. It had been a good four years and looking back, I believe I had used them wisely. While I did not apply myself to my studies as hard as I had in high school, I had no pressure to do so as I was not under an academic scholarship. I, nevertheless, had above average grades and matriculated with low honors in most of my classes. During my college years I intentionally tried to broaden my horizons and began to open myself to many areas that I had before neglected. I mentioned that I now became more interested in the arts, drama and music. While I still continued my love for sports, I now attempted to learn more about politics and world events. I began to understand that I would soon be entering the 'real world' about which, up to this point, I knew little about. I now more clearly recognized that what I had been given in the classroom, while excellent in many ways, was also limited. It did not always coincide with what I would find outside of the ivy-covered walls. I wanted to be ready. I also realized how much my life was going to change once I left college. I would not only be leaving school, but I would also be leaving my home. Probably for good! Would I be able to handle that? My plans to become a teacher and a basketball coach began to fade and that left me feel very insecure. Manhood was coming upon me at a fast clip. Ready or not, the time to stand up and make my own decisions and live with them could no longer be put off.

Graduation day was hot and sunny. It was held outside. My dad was there and so was Alice Marie who took off from work to attend. I proudly received my diploma awarding me a Bachelor of Arts degree. I was also sworn in as a Second Lieutenant in the USAF Reserve with orders to report for active duty in July. I never had my dream fulfilled of coaching basketball at Chaminade High School. The Korean War was upon us.

CHAPTER SEVEN: ALICE MARIE CURRY

Faint heart never won fair maiden.

-Anonymous

The First Woman in My Life... and the Last.

During everyone's life there come decision times. Sometimes you know that you are about to make a momentous decision that will have great consequences. But most of the time you don't. You just make them and go on unsuspecting. My decision to join ROTC at Fordham was one of those. I certainly had no intention of pursuing a military career.

Far from it. But that is exactly how it turned out. We had just come out of five long years of a brutal and costly war. Everyone was sick of war and hoped to never see another in their lifetimes. Still, all of us revered and respected our servicemen. While we did not yet know even a part of their heroism and bravery, we knew that what they had accomplished was of huge historic significance. Men in uniform were loved and respected. To wear the uniform of our country was something to be envied. And while I had no ambition to join the active military, becoming a reservist and wearing the ROTC uniform was very acceptable. So, I applied for ROTC with absolutely no idea of what that decision would mean to me in the years to come. So, life plays its tricks on us in subtle ways that leave us surprised and unsuspecting.

Throughout my high school and college years I had a very limited interchange with females. Chaminade, an all-male school, did not even permit dances on the premises. I knew very few girls and never dated any of them. I also had no money to spend on such things. I was busy with school and sports and that was all I needed. I did not attend my high school Senior Prom as I had no money for such a thing, and I knew no one to ask.

Fordham was also an all-male college when I attended. Any girls I met were from outside the school scene. Living in Great Neck, Long Island above the LIRR Station meant that I did not live on a street or in a neighborhood where there might be some girls close by. I left early on school days, taking the 6:30 am train to Penn Station, and then the 8th Avenue Subway to Fordham Road and a half-mile walk to the campus. When playing sports, I returned late at night. My homework assignments were often completed on the subway or LIRR. It was a closed life, regimented and without much freedom to expand my horizons.

Thanks to my father's promotion to Station Agent and our living in quarters provided by the company with all utilities paid, our family now had money enough to spend summers at Rockaway Beach. This was great for my grandmother especially. We rented a bungalow about a block from the beach and my father, Aunt Anna, and I commuted each day, my father by car, and Anna and I by LIRR. We thus avoided the stifling hot humidity of New York during the worst months. Rockaway in those days was a great gathering place for the masses that tried to escape the heat, particularly on the weekends. I loved Rockaway Beach, and it was there that I was able to make new friends whom I would have never had met otherwise. One person I met there was a girl named Alice Marie Curry. It did not seem like a big deal at the time, but again I was wrong. It was a huge deal and one that changed my life completely.

Here's how I remember it. One of my friends at Rockaway was Rich Leahy. Rich lived in Manhattan, and I only saw him in the summers at Rockaway. We enjoyed each other's company, but Rich was more advanced than I on the subject of females. Rich was always after me to find the company of girls, and he seemed to find them much better than I. He told me that he knew these two girls who were really neat looking, and he wanted me to go with him on a double date. I was not enthused. I did not know these girls, had no money, and felt very uncomfortable in such a situation. Rich kept after me. One day we encountered them near the boardwalk. Rich introduced me. They were cousins and lived in the Bronx. One, Marge, was quite short, perhaps five foot one or two at most. The other, Alice, was taller and quite nice in many ways. She seemed friendly, easy to talk to, and had a terrific figure. When we left them, Rich, again nagged me to double date them. Rich was tall and slim, about six feet, one. I relented but told him that I would only agree if I got the tall one. Surprisingly, Rich agreed and a few days later we met the girls and took them to 103rd Street where all the Irish dance halls were. I truly don't remember much of that first date with Alice Marie Curry, but I must have enjoyed it because I did begin to call her and when possible, asked her out. I had just begun studies at Fordham, and she was working downtown as a secretary, so we did not have any torrid romance.

We lived far apart; I had no car and very little money for dating. I was also tied up with sports, school, and part time work to pay for my college. When we dated it was often to attend a ballgame or she came to see me play. Occasionally, I had enough scratch for a movie and a beer after. A favorite date was to walk from her home to Yankee Stadium for a night game and a few beers after at the Concourse Plaza Hotel. Though we never were steady, Alice was about the only girl I ever dated from my college years onward.

But there was no thought of marriage, at least on my part. I was just not ready mentally, financially, or otherwise to contemplate such a lasting relationship. Life was still very complicated for me, and the future uncertain. And that uncertainty exploded, as I was about to enter my senior year in June 1950, the Korean War commenced. Because I was draftable and a member of ROTC, it now became clear that my next few years would be in uniform—hardly a time to think about marriage.

Alice helped me in many ways during college. As she worked for Harper's Publishing, Alice frequently provided me textbooks free of charge. And once in a while, she came up with tickets to a Broadway play. She was my date at Fordham's Senior Prom, the first prom I ever went to. And Alice attended my graduation from college where I was commissioned a Second Lieutenant in the USAF. I had told her that rather than beginning my teaching career, I would be entering active duty right away. I also told her that I would be going to flying school in October. I don't recall her reaction to that news, but it would not change anything anyway. But I'm sure a lot of things must have raced through her mind. The future for both of us was completely up in the air (no pun intended). We made no promises to each other. We were neither engaged nor even going steady. No strings were attached to either of us. We would have to see what the future would hold for us and hope it would be good.

Alice wrote to me often while I was on active duty, and I looked forward to her letters. I kept a photo of her in my wallet. When I was home on leave, we saw each other, but flight school loomed ahead and after that I was certain there would be an assignment to Korea. The next few years were not conducive to any lifelong decisions for us, and besides, I was not at all sure that Alice was that interested in me. I knew she was dating others and I certainly had no reason to object to that. I was into my new career, learning all that I could about the Air Force and flying. This was not a time for distractions, and the realization that my life might be a short one was ever present. I was not a good candidate for marriage. That would have to wait.

You will read much more about Alice Marie Curry later.

PART II:
I LEAVE HOME AND BEGIN MY CAREER IN THE AIR FORCE

CHAPTER EIGHT:
I JOIN THE UNITED STATES AIR FORCE

Off we go into the wild blue yonder...

-The U.S. Air Force Song

The same day that I graduated from Fordham University, I was commissioned as an officer in the USAF Reserve. I also received my orders calling me to active duty. I was very disappointed that I would not go directly to pilot school. Instead, I was assigned to the 3664th Basic Training Squadron, Sampson AFB, Geneva, New York. I was decked out in my new blue uniform with a gold bar on each shoulder. I received my first salute and presented the saluter with a crisp dollar bill. My father was present and so was my future wife, Alice Marie. It was great day because I had reached a point where I was now setting out on my own. I would now be making my own personal decisions, living away from home, and seeing what I was truly made of. The specter of learning to fly was in front of me and I knew that that would be the greatest challenge I had so far faced. That summer, I packed my B-4 bag and left for my first assignment. Dad arranged my transportation. I had a Pullman berth on the night train leaving Grand Central Station. I arrived at Geneva, NY the next morning and took a bus to the base. It was a hot, humid day and I walked about a quarter mile from the bus stop to the base headquarters. There I showed my orders and was assigned a room in the Bachelor Officer Quarters (BOQ). It was the first time in my life that I ever had a bedroom of my own. That afternoon I reported for duty with my squadron. I soon realized how little I knew about the military and how much I had to learn in a hurry. My confidence was shaken.

Sampson AFB was a former U.S. Navy installation that the Air Force now used to take on the load of training newly enlisted civilians in how to become military in about eight weeks. The base was situated on Lake Seneca, one of the Finger Lakes and in one of the most scenic areas of New York State. The Officer's Club was right on the lake front and in many ways, it was a pleasant place to be stationed. But there were no airplanes at Sampson and no runway or hangars or anything else that reminded you that you were in a fighting Air Force. I found that depressing and I was really anxious to receive my orders that would send me to flying school. The sooner, the better!

I was assigned as Assistant Adjutant, with additional duties as Insurance Officer, Historical Officer, Personnel Affairs Officer, Recorder Unit Fund, Council Squadron Records Officer, and Squadron Supply Officer. I was also authorized to sign the Morning Report. In short, I was absolutely in over my head. Most of these duties were completely new to me and I possessed little to no knowledge of what they were all about. All I could think about was getting through the next few months and getting off to flying school. The Commander, Major Houk, was understanding and so was his Adjutant, Captain Syl Cleary. The mission of the squadron was to welcome new Air Force recruits, bed them down, and train them. In a few months they would learn how to dress and act as an Air Force enlisted man. They were given haircuts, clothing, bed clothes, a cot, a footlocker, and a drill sergeant who ruled their lives completely 24/7. They would be taught how to march, salute, make a bed, clean latrines, do KP duties, fire an M-1, and become physically fit. My duties in the Squadron were to oversee the payroll, records, and training of these raw recruits. As I was just a smidgen ahead of them myself, I was in no position to act as an expert in military matters. But I did my best and tried to learn as much as I could about what the military was all about and how it operates. The blind leading the blind is apropos here. By the time I left Sampson, I did learn much and what I learned came in useful in the years ahead. But in the summer of 1951, I had no idea what lay ahead. Sampson was an Air Force base that had no runway and no airplanes. If I had to serve in the Air Force, I wanted to fly AF planes. When my orders came through for pilot school, I was truly a happy man. I was ready and willing to see if I had what it takes to become an AF pilot. It would be the biggest challenge of my life.

In September 1951, I received Special Order 217, requiring I report to the 3306 Training Squadron Contract Flying, Bainbridge Municipal

Airport, Georgia, on October 2. My class would be 52-G. I would now be leaving my life as a "ground pounder" and begin my quest for those precious AF wings. Now, I would see if I had the right stuff or not. I could not have been happier.

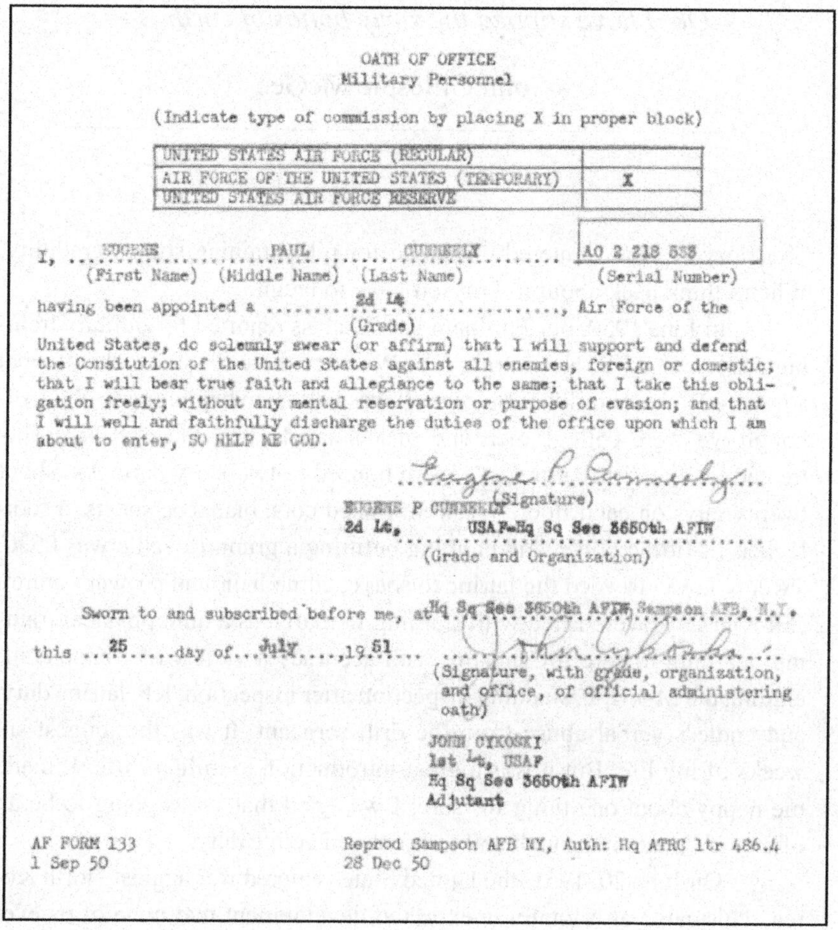
OATH OF OFFICE
Military Personnel

(Indicate type of commission by placing X in proper block)

UNITED STATES AIR FORCE (REGULAR)	
AIR FORCE OF THE UNITED STATES (TEMPORARY)	X
UNITED STATES AIR FORCE RESERVE	

I, EUGENE PAUL CUNNEELY AO 2 218 538
(First Name) (Middle Name) (Last Name) (Serial Number)

having been appointed a 2d Lt, Air Force of the
(Grade)
United States, do solemnly swear (or affirm) that I will support and defend the Consitution of the United States against all enemies, foreign or domestic; that I will bear true faith and allegiance to the same; that I take this obligation freely; without any mental reservation or purpose of evasion; and that I will well and faithfully discharge the duties of the office upon which I am about to enter, SO HELP ME GOD.

Eugene P. Cunneely
(Signature)
EUGENE P CUNNEELY
2d Lt, USAF-Hq Sq Sec 3650th AFIW
(Grade and Organization)

Sworn to and subscribed before me, at Hq Sq Sec 3650th AFIW Sampson AFB, N.Y.

this 25 day of July, 1951

(Signature, with grade, organization, and office, of official administering oath)

JOHN CYKOSKI
1st Lt, USAF
Hq Sq Sec 3650th AFIW
Adjutant

AF FORM 133
1 Sep 50

Reprod Sampson AFB NY, Auth: Hq ATRC ltr 486.4
28 Dec 50

CHAPTER NINE: AIRBORNE

Oh, I have slipped the surly bonds of earth...

-John Gillespie McGee

The story of how I entered my name for pilot training strains credulity. When I think back about it, I myself have to laugh.

In June 1950, our Fordham ROTC class reported for military training at Stewart AFB, Newburgh, NY. We were there for six weeks of what amounted to basic military training. It was hell from beginning to end. Although we were college men and officer candidates, the drill sergeants treated us like dirt recruits. We were housed in two story barracks, about twenty guys on each floor. We were issued cots, blankets, sheets, a footlocker, ill-fitting boots, and clothing befitting a grunt. Reveille was 0530. Twenty guys crowded the latrine to shave, comb hair, and answer nature's call. Our days were filled with learning to march as a unit, physical training, learning to care for and fire with accuracy a variety of weapons including the M-1 rifle, standing inspection after inspection, KP, latrine duty, and endless verbal abuse from the drill sergeant. It was the longest six weeks of my life. But it was a great introduction to military life. It made me happy about one thing for sure; I was glad that I was going to be an officer if I made it through and ever entered active duty.

On June 20, 1951, the United States entered war against North Korea. This came as a totally unexpected development that none of us ever anticipated. Especially upset were the ROTC guys who were veterans of WWII. These guys had just come through the big war alive and were now about to begin their civilian lives. The last thing they wanted was a return to active duty and another combat tour. Many of them packed their duffel bags and began heading for the gate. The rush was halted by the base commander, who assured these guys that active duty was not in the cards for them, as they had served their time in the last war. He told them that it would be guys (like me) who had never served that would be going into this war. That seemed to have reassured them, but it certainly did not make me feel very good.

On returning to Fordham for my senior year, it became clearer that instead of teaching and coaching at my old high school after graduation, it would be active duty with the AF that lay ahead for me. My well thought out plans had been changed. And even stranger things began to happen.

One of our AF ROTC instructors was a pilot. He was a Captain and wore those silver wings above his left breast pocket. I admit that those wings intrigued me. Somehow, I looked on them as something special that placed Captain Scott a rung above the other officers who did not wear them. I began to think how great it would be that when I put on my uniform next summer, I could also wear those wings too. Scott began to urge me to put in for flying school. At first, I thought he was joking, but he was persistent. I had never been in an airplane in my life until I flew in the C-47 at ROTC camp. I was an English major and truly had never given a thought to flying because I believed that that was simply way beyond realism. In any event I was cajoled into taking the pilot's test. It was one of those three-hour exams that included math, verbal skills, science, physics, engineering, and aerodynamics. I truly had to guess at half the questions that were mostly multiple-choice. When finished, I left totally convinced that that was the end of that. Nope—a few weeks later I received a letter from the AF congratulating me on passing the pilot's exam. I was now to report to Mitchel Field on Long Island to take tests on reflexes, visual acuity, ability to recognize and respond to various positions related to climbing, diving, and inverted flying. This battery of tests came very close to making me nauseous. I was sure that the results of this experience would doom any chances I may have had to fly. But I was wrong again. Three weeks later I received an AF letter congratulating me on passing those tests and ordering me to report for a flying physical exam. It was then I surmised that with the Korean War in full swing, much of it being fought in the air over North Korea, the AF's needs were so great that they intentionally overlooked the obvious defects of pilot candidates like me. A few months later, I was informed that I was officially accepted for flying school upon my entering active duty in the summer of 1951.

When I told my father that I was going to go to pilot's school, his response was terse: “Are you out of your mind?” It was a good question, and I had no answer.

So, that is how I took on the challenge of becoming an AF pilot. It began almost as a whim, but due to my phenomenal success in the tests, it all became a serious piece of business and would result in my undertaking the most serious challenge I had faced to date.

And so, after a flight from New York to Savannah, Georgia, I took the train to Bainbridge where I would begin my flying career. The train ride was terrible. The train rambled through southern Georgia stopping frequently at places where I could not believe anyone lived. I was wearing my blues and it was insufferably hot and humid. By time we arrived at Bainbridge, I was soaking wet. On top of that, at the Bainbridge RR station, no one seemed to know anything about any AF base in the vicinity. I was totally exasperated. Finally, someone thought they had seen some activity at the old airfield north of town. So, I took a cab out to this place and after much exploring found some old WWII buildings still intact. Thank God inside of one of them was an enlisted man sitting at a desk reading a comic book. He put it down just enough to take my orders and assign me a room. The room was in a newly constructed single-story building, made of stone blocks, concrete floors and no hot water. The room was furnished to accommodate four men. It had four steels cots and four chests of drawers. The common bath was at the end of the hall. There was no heat and no air conditioning. I thought that this was hardly appropriate quarters for AF officers who were to be the next crop of aerial fighters. I may have been right, but that did not change anything. I lived there for the next six months. It took two months before we received hot water and heating. There was no mess hall or officer's club. Worst of all, there was not a single airplane on the whole base. I could not have been more flummoxed. I had spent long, awful, boring months at Sampson waiting to get to flying school, only to have arrived on a base with no airplanes, no instructors, and barely livable conditions. Well, in the rush to turn out more pilots in a hurry to meet the needs of Korea, the AF contracted Southern Airways to provide pilot training at Bainbridge Air Base, GA. The trouble was Bainbridge AB was in no condition logistically to accept the personnel and planes to begin the training program.

What happened was the cadets, student officers, and base personnel began arriving, but there was nothing for them to do. In fact, we all sat around for nearly a month before the planes and instructors and maintenance people arrived. Worse, there were no eating facilities on base, as the mess hall had not been completed. We had to drive into Bainbridge to eat, although the local Elks Club allowed the officers to dine in their facility, but that was open only in the evenings. To kill time, we student officers took out platoons of cadets and gave them marching and drill exercises. This did not sit well with either the student officers or the cadets. It seemed like an eternity before our planes arrived on base and the business of learning to fly began.

The aircraft were T-6 Texans. T stood for trainer. These aircraft went back to WWII and had recently been pulled out of mothballs and overhauled. They were two-seaters, powered by an eight-cylinder Wright engine. My first impression was they were bigger than I thought they would be. They were painted yellow, and noisy, with a standard retractable landing gear that just begged student pilots to ground loop them on landing. My confidence began to erode.

Our instructors were all civilians; most were former military pilots. But I was shocked when I met my instructor. His name was John Hoover, and he was younger than I. My first thought was, is this the guy whom I am going to rely on to keep me from killing myself, and he looks as though he doesn't know much more than I do about flying airplanes. Hoover was a slim young blond man with a ready smile and a great sense of humor. He had never served in the military, but was born and raised in Lock Haven, PA, home of the Piper Cub Aircraft Company. From an early age he was destined to become a pilot. He had soloed in his teens and had rolled up thousands of hours in civilian aircraft. When Southern Airways advertised for instructors he applied, and after a grueling training program in the T-6, he became a certified instructor. I would be his first student. Both of us would be challenged by the other's inexperience.

At Bainbridge, I roomed with two other second lieutenants. Bob Stapleton was a Texan and a graduate of Baylor University. He was a soft-spoken man, a strict Baptist, a former Marine, and a great lover of western music. The guitarist Chet Atkins was his favorite and he played his records daily. Charles Bo Dudley was an Alabama native, and a graduate of Georgia Tech. Bo was the first true southerner I had ever met and if there was ever a classic redneck, Bo was it. He referred to me as the Yankee, and often as "the dumbest Yankee he had ever met". Bo had no use for colored people and was a strict segregationist. He was very disciplined and seemed cut out for military life. I strived to get along with Bo because his knowledge of engineering helped me get through that course. I employed shameless flattery to obtain his help with engineering assignments and test preparation. But Bob and Bo did not fly in my flight. Each student officer was assigned to a flight with three cadets. In my flight there were three cadets who were taking additional courses toward their officer training because all pilots had to be officers. So as an officer, I had it relatively easy in ground training taking only courses dealing with flying. These courses included navigation, engineering, principals of flight, and weather and radio communications (including Morse code). Our schedule was to be in ground school half day and fly the other half, alternating mornings, and afternoons on a weekly basis.

At last, my flying career was underway. On an early morning in October 1951, Hoover placed me in the back seat of the T-6. The first thing that struck me was that from the backseat you could not see anything forward. The plane was tilted downward from front to back and looking forward all you could see was the back of the instructor's head. On the intercom, Hoover told me what was going on and I could monitor his transmissions to the tower. These I would have to learn because in the next flight I would be in the front seat. Hoover started the engine and with the canopy open the noise was overwhelming. We taxied out and Hoover did all the necessary pre-takeoff engine and control surface checks. Then we took the runway and began our takeoff roll. My heart never beat more rapidly. I could feel the aircraft swerve and then straighten as our speed increased and then we broke ground. My heart leaped as I saw the runway fall away and I realized that we were now airborne. There was no turning back now and while this was definitely a thrilling event, I confess that I wondered to myself, could I really do this alone without something happening that I would not sense quickly enough or do something completely wrong that would end in disaster. Hoover, who was talking throughout and describing what he was doing and what was taking place, clearly was in total command of the situation and it was absolutely clear that he was a skilled and experienced pilot. That was reassuring and from that point on I had the utmost confidence in him despite his youthful looks. It took a while, but I slowly began to truly enjoy this flight. I had many concerns about becoming a pilot, and truly no idea that even I could, but at least for now, I became exhilarated with flying above the earth. Just looking down on the passing landscape—seeing miniature cars, houses, roads, farms, towns, lakes and people—wondering if they realized that I was up there above them, free as a bird, carried on invisible winds, and having the time of my life. Hoover then let me handle the controls. I tried some turns and discovered that you had to use rudder with your ailerons to get a coordinated turn. We did a few dives and climbs and rolls. I felt some G-forces on my body. This was an experience that I had never really imagined. Come what may, I decided then and there that I would succeed in this, the most challenging test of my life. I would not permit them to wash me out. I will be a pilot—I will earn those wings.

The Texan

The T-6 Texan was the first airplane that I flew. The thrill of taking off in a military airplane that I was actually controlling was an experience I will never forget. Of course, my instructor John Hoover was there to ensure that I did not kill us both. I could feel his hand on the controls, and he made certain that I always had sufficient airspeed to sustain flight, especially at low altitudes. Before my first flight I had pored over the Tech Orders on the T-6. This manual told you all you had to know about the aircraft, including starting, flight procedures, stall speeds, and emergency procedures. Every pilot had to know the Tech Order almost by heart. The emergency procedures you did have to know by heart and your instructor continually asked you to recite them by heart. In addition, I had spent many hours sitting in the cockpit, on the ground. I came to know where every switch was located, every fuse, circuit breaker, and light. I would sit in the cockpit and have another student test me on locating, blindfolded, every item in the cockpit of the T-6. At night, in the dark, you had to be capable of finding the exact item you needed to remedy a bad situation that every pilot had to be prepared for. So, while I was truly exhilarated by getting the T-6 into the air and putting it through its paces, the proximity of something dangerous and life threatening was always present. There is an old saying about aviation that still pertains: "Flying is thousands of hours of boredom, interrupted by moments of sheer terror." During my years of flying some of the latest and fastest fighter aircraft in the world, that saying proved itself over again many, many times.

By today's standards the T-6 was an ancient machine. After World War II, T-6s were modified to become AT-6Gs for use in pilot training. The idea was that if you could fly the T-6, you could fly anything. I found that to be true. It was the most difficult plane I ever checked-out in. North American manufactured the airplane. The Navy used them and called them SNJ's. Some T-6s were used in the Korean War as Mosquitos, which were flown with Forward Air Controllers (FACs) who spotted enemy targets on the ground, called in fighters to attack, and even marked the targets with smoke flares. I flew against ground targets in Korea under the control of FACs flying T-6s. When hitting those targets, I always thanked the Lord that I was in my fast-moving Sabre and not in that slow moving T-6. I had great admiration for the FACs and their skill and courage in flying in so close to enemy fire in order to help our ground troops. I did not envy their job one bit. Without them, we could never have found enemy forces that were camouflaged or well hidden under foliage or in buried fortifications. The T-6 was a two-seater with duplicate control in the rear cockpit. Normally you would take off and land with the canopy open. That always provided the allusion that you were flying one of the older open cockpit planes of an earlier era. Due to the high nose angle, the pilot had to zigzag as he taxied out in order to see what was ahead of him. Failure to do so often ended in a taxiway collision that was heavily frowned upon and punished. All in all, the T-6 was a great training plane because it required close attention all the time, was highly maneuverable, honest in stalls and spins, and made every landing and takeoff something that could not be taken lightly.

Actually, flying an Air Force aircraft excited me more than anything I had ever done in my life so far. It was true that I was just a fledgling airman and there was no guarantee that I would make it through training alive or successfully, but I came to the realization that this was something that challenged me to the utmost and which I was determined to become good at. On the night before our training began, all of us were called to the base gym where we were addressed by the Commander, Lt. Col. Dawson. After welcoming us, he then proceeded to tell us that within a month, 25 percent of us would be one... washed out; and that within another month, 10 to 15 percent more would be gone. We were stunned. That was not what we wanted to hear. I did not sign up for this program, taking all those tests, and then coming down to this hellhole they called Bainbridge, GA to get kicked out. I made up my mind, in no way are they going to wash me out. I wanted those wings; that's why I signed up and put myself through all of this, and I am going to get them. And that was my attitude for the next year. But to get there, I had to master the T-6. And that's what I set out to do.

Racism in the South

My days and nights in Bainbridge, Georgia flew by. From Monday through Friday, it was half day on the flight line and half in Ground School. Weekends were another thing. I had no car so I could not escape the local area. Bainbridge gave me my first actual experience with racial segregation. Black people did not attend the same schools as white, and restrooms, water fountains, and other amenities were separately assigned and designated. In the movie house, they could only sit in the balcony, and they had to sit in the rear of the local bus. They also had their own eateries. When I walked down the sidewalk, a black person would step aside, sometimes into the street, to allow me room to pass. Nothing in my past in New York approached such blatant racism in practice. We had no black airmen assigned at Bainbridge, and I could understand why. It would have been hell for that person. On weekends I mostly played ball in the gym or went to the movies. To provide Sunday Mass for Catholics on base, we had to fly a T 6 up to Albany, GA and bring a priest back by air. South Georgia was not a friendly place for Catholics.

CHAPTER TEN: BECOMING AN AVIATOR

Flying is hours and hours of sheer boredom, interrupted by moments of heart stopping terror!

For the next six months, I flew the T-6 almost daily. Hoover spent the next few weeks of my training, concentrating on takeoffs and landings. My pre-takeoff, takeoff and pattern work were deemed satisfactory, and my confidence grew. However, my landings were inconsistent. Some were good and some not. We were taught to make three-point landings, that is, you rounded out as close to the runway as you could gage it and then let the aircraft stall and touch down on the two front wheels and the tail wheel simultaneously. This would be a great landing if you rounded out a foot or so above the runway. But sometimes I put the aircraft into a three-point attitude about ten feet above the runway. With airspeed decreasing rapidly, the result would be a ton of aircraft falling ten feet out of the sky onto the runway creating a terrible jolt on the landing gear. Even if the gear did not fold up under this force, the pilot would have to log a hard landing and have the landing gear and the struts inspected. Fortunately, John Hoover usually rescued me from this disaster just prior to the aircraft's smashing into the asphalt. Slowly I began to resist the urge to bring the nose up too early on my landing approaches and my landings became more consistently safe and sound.

Each morning, Hoover would brief the four of us on what we would be doing that day. Because so many of us were still working on landings and takeoffs, Hoover would take the first student and they would fly to an auxiliary field, Stephen Decater Field, where they would enter the pattern and practice landings and takeoff. The other three students would get on a bus and be driven to the field and wait their turn to fly. When the first student completed his flight, he would get out and, with the engine still running, the next student would climb up and take his place for his flight. The last flight would land back at Bainbridge. On a cold clear day in early December, I was the last student of the day. I took off with Hoover in the backseat and we just reentered the pattern and began shooting landings. My first two were pretty good, but the third a little bumpy. Hoover directed that I taxi into the tarmac area and stop. He never told me what he was doing, but the next thing he was out of the cockpit and on the wing. He shouted at me to take it off and shoot three landings. He would be in the tower and would be available if I had any problems. I was shocked. Frankly, after that last touchdown I did not feel ready to solo. Hoover reassured me that I was. So off I went without my security blanket in the backseat. I was quite nervous; however, this was what I wanted wasn't it? I concentrated on my checklist and on making sharp radio transmissions. I took off and re-entered the landing pattern. I did my GUMP check, GUMP meaning GAS, UNDERCARRIAGE (landing gear), MIXTURE, and PROPELLER. I turned off the base leg onto final approach, rechecked my gear down lights, received clearance to land from the tower and cross checked my airspeed and kept the nose lined up with the runway. Then as I came over the approach lights, I began my round off. *Not too early now. Take it slowly. Be sure you get down as low as you can, but don't let the aircraft hit too soon.* Slowly, I pulled back on the throttle, the aircraft descended closer to the runway. *Will it ever touchdown?* I wondered. Then it did, gently on all three wheels—the best I had ever done! My confidence rose to new heights. I can do this. Who needs an instructor? Not me. Two more takeoffs and landings and I taxied back to the parking area. Hoover came over and climbed into the back cockpit. His praises for my performance met warm ears. I was on a real high. I had done it. Well, I knew I could, but there's always a little doubt until you actually pull it off. I had completed the first major hurdle on my long road to winning those wings. It was a good day.

The 50 Hour Check

The next huge hurdle for us flying students after soloing was the 50-hour check. This check ride was administered by one of the military pilots assigned to the base. And these were the toughest to pass. The flight consisted of the preflight check, the startup and taxi out, the takeoff and then extensive air work which included acrobatics, climaxed by a three-turn spin. On the return to base, you could expect the check pilot to pull the throttle back to idle and require the student to simulate a forced landing almost to touchdown. Hoover felt that I needed more practice on my acrobatics and the spin before he would release me for the check ride. I confess I did have problems with the spin and on every solo flight I had, I practiced them even though they terrified me. But something went askew. I was sitting in the students' waiting room when Capt. Kent, one of the military check pilots, called my name and told me to meet him in the briefing room for a 50-hour check ride. I was truly shocked. I was sure that Hoover wanted to give me a few more practice rides. I asked Kent if he was sure I was on the list. He was dead certain, and he was not one I wanted to trifle with. So out we went. Everything went smoothly, but I was still worried about my acrobatics. But it was my day: my rolls were smooth, and the loop was the best I had ever done with the nose coming over the top and down right on the line. Now the spin. Kent asked for a three-turn spin. I brought the nose up, retarded the throttle, and shoved forward on the right rudder. The Texan rolled to the right, and whip lashed into a spin. The nose of the aircraft revolved to the right. G forces shoved my body against the left bulkhead, and I watched the earth fling itself in a violent circle around the nose of our helpless plan. All of the flight instruments were whirling crazily back and forth up and down. One, two, three, now recover. I shoved forward on the left rudder and flung the stick forward till it was up against the dashboard. Almost immediately, the rotation slowed stopped. The nose of the aircraft was pointed straight downward toward the earth and the airspeed began to increase rapidly. When the airspeed reached a level for safe flight, I began to bring the nose back up to level flight. It was the best spin recovery I had ever done. On the way back to base, Kent took the throttle and called out that we just had engine failure. I was ready. I called out the emergency procedures for an engine failure. At the same time, I turned toward a pasture that I believed I could glide to. Between our plane and the field was a small river and a line of trees. But my glide easily took me over those and permitted a touchdown that would have been in the center of a freshly plowed cornfield. (We only made a low approach, of course.) My landing back at the base was not perfect but satisfactory. I had passed my 50-hour check with one of the toughest check pilots in the group. Hoover was shocked when be found out that I had been erroneously called out for the check. But he smiled and laughed heartily when he learned that I had passed with flying colors. With that check ride behind me, I felt more confident that I would get through my Primary Training in the T-6. After that, who knows what would happen.

At Christmas time 1951, a charter flight was set up from Tallahassee to Newark, NJ. A goodly number paid over $100 to get back home for the holidays. During the Christmas vacation, I spent time at Great Neck with Dad, Grandma, my brother and his wife Pat, and the family. It was a most needed getaway. I also saw Alice Marie Curry a couple of times including New Year's Eve when, unfortunately, I had to return to Newark Airport for the flight back. Arriving back in Georgia, I was tired and a little down. But it was back to work on my way to getting Wings.

Night Flight

Re: night flying: Remember that the airplane doesn't know it is dark.

From January 1952 to May, I continued to fly the T-6 and improve my skills in making smooth landings, instrument flying, (needle, ball, and airspeed), and acrobatics. The biggest challenge that came up was night flying. Up to this point we flew mostly in clear weather and in the daylight when the horizon was in constant view. But flying at night was a different story and many of us students were fearful whether we could hack it. On my first night flight, Hoover taxied out and took off. But sitting up front, I saw fire shooting out from the engine vents right outside the cockpit where I was sitting. Alas, no one mentioned to me that this was common. It happened on all takeoffs, but in the daytime, you could not see it. When we called out to Hoover that we were on fire, he just laughed and told me to pay no attention to it. When my heartbeat finally returned to normal, I began to adjust to seeing how differently things looked at night versus the day. I found that without a clear view of the horizon, I had to peek at my attitude instruments a lot more to fight off vertigo. I also began to recognize things on the ground by streetlights, car lights, and reflections. I also found that you could be easily fooled at night. Bright stars looked like other aircraft or even UFO' s. You also had to be more aware of where you were as frequently you lost sight of geographic points that you can recognize in the day, but not so at night. Most of all, you had to use your eyesight differently at night. Keeping your cockpit lights dimmed and not staring at bright lights on the ground made it easier to see what you wanted to see. And then, of course, there was the matter of landings. Approaching the runway at night, you had to be very attentive to airspeed and round off. You also had to have your head on a swivel in order not to collide with other aircraft in the pattern. The fear of mid-air collision was very great with me throughout my flying career as it usually ended in a fatal crash at an altitude where parachuting to safety was not probable.

My night flights were okay, and I soloed in short order. However, night flying washed out several student pilots. One, who had a lot of previous flying experience actually quit after his first flight. "You can't get me to go up there at night," he said. "No one is supposed to fly at night, and that's it." And he meant it. He was gone a few days later. But, while night flying brought butterflies to my stomach in those early days, later I truly loved night flying. Having flown many years in Air Defense Command, where we sat alert in 24-hour shifts, and were proficient in all-weather flying, night flying was just a way of life, and that included night formation flying and tanker hookups.

Goodbye to Georgia

My days in the T-6 ended in May. We awaited our next assignment. Each instructor recommended what type of aircraft each of his students should fly. To my chagrin, Hoover told me he recommended that I go to multi engine aircraft. (B-26s). Hoover felt that I would be best suited to flying large aircraft rather than fighters. I was disappointed. I truly felt that in my last three months in the T-6, I had shown that I could handle a fighter type plane. But when the orders came out, I was programmed to Bryan Field, Texas to fly the T-28/F-80. The F-80 was the first jet fighter the USAF had and was being used in Korea. I was pleased, feeling that the Air Force was smarter than my instructor about my abilities. Little did I realize that I and many others were being sent to fighter school because the AF needed us badly in Korea. When I told Hoover what my orders were, he just laughed. He told me that I would do all right. We said goodbye and I thanked him for the many times he saved me from destruction. I never saw him again. Goodbye Georgia, hello Texas.

Deep in the Heart of Texas

When I arrived at Bryan AFB, Texas, I found myself on an aging World War II base with buildings that still belonged to that era. It was the middle of June and temperatures were hot and humid. I was assigned a room in an old barracks that had been partitioned for student officers. It was not a totally complete room, as the walls did not extend from floor to ceiling. Besides a bed and a chest of drawers, there were no other amenities. No air conditioning and the latrine was at the end of the hall. The airfield had two parallel runways; one was used for landings and takeoffs by the T-28s, the other by the T-33s. But this led to many close calls when students overshot the runway they were aiming for and flew across the flight path of the other runway. Each day we witnessed near collisions and frantic radio calls from the tower, mobile control, instructors, and frantic student pilots who saw their lives pass before their eyes. My instructor was an ex-Navy pilot named Jim Ashmore, and he was a good one. He had lots of military experience and brought a fighter pilot's viewpoint to his work. He saved me many times from killing us in this wild traffic pattern, but I kept working on it and got better every day.

The North American T-28 (Trojan) was a super airplane. It had a tricycle gear that relieved us of our fears of ground looping on landings. With a bubble canopy you could see straight ahead taxiing and taking off. But the greatest challenge was the landing pattern. In the T-6, we entered the landing pattern on the downwind leg, performed our pre-landing checks, turned on to the base leg, rechecked everything, made our radio calls, and then turned onto the final approach. You had plenty of time to get everything done and rechecked. In the T-28, we were taught to fly a fighter pattern that consisted of making your initial approach lined up with the runway about one to two miles from the field. Over the end of the runway, you banked sharply (90-degree bank), retarded the throttle, dropped the gear, set the flaps and while continuously turning 360 degrees and lowering the nose, you made your landing radio calls, lined up with the runway and landed. During all of this you had to keep double-checking your airspeed and by no means overshooting the runway. We worked on this pattern over and over on every flight until the instructor was confident that you were in full control of a very fast-moving situation. Also, he made sure that if things did not work out, you recognized that quickly, and aborted the landing and went around to try again. I became more proficient at landing the T-28 and that reinforced my self-assurance that I did have what it took to be a fighter pilot.

Many of my fellow student pilots did not want to continue flying the T-28. Like me, their basic instructors recommended that they go on to multiengine aircraft. When we arrived at Bryan Field, the Wing Commander, Col. McGee addressed this issue. He told the student body that they would love the T-28 and they should have an open mind. If after two weeks of flying the T-28 they wanted to transfer to B-26s, he would arrange it. That seemed reasonable, but after two weeks, several students asked to be transferred. McGee was livid and absolutely refused to accommodate them. Instead, he told them they would have to be personally flunked by him on a check ride before they would be released and then they probably would be washed out. That was the first time that I had witnessed a breach of honesty by an officer. I decided that I would continue flying the T-28, as the more I flew it the better I liked it, and my self-confidence confidence grew.

My days in Texas flew by. And I began to enjoy my off-duty time more than before. When I arrived in Bryan, I opened a checking account with the local bank. The president was Travis Bryan and he had built a reputation as a banker to AF officers. He told me (and all others) that I could write a check on toilet tissue in some foreign place, and his bank would honor it. A few weeks later I bought my first car. I bought it from the Chevrolet dealer in Houston and Travis Bryan made me the loan. It was a four-door sedan and I believe it cost me about $1,600. At last, I had my own transportation and that permitted me to spend weekends in Houston or Dallas, anywhere but in Bryan. I could now eat downtown or check in at a hotel with a pool on a weekend. Life got a lot nicer, and I felt that another milestone had been reached. My flying grades were good, and I was looking forward to becoming a jet pilot. Then, out of the blue a sudden unexpected change took place. I, and several other student officers, received orders to report to Pinecastle AFB, Orlando, Florida (later named McCoy AFB) in August for advanced training in the T-33. We were to form an all-student officer class there. I was elated. In August 1952, I drove my new 1952 Chevrolet from Texas to Florida and reported in for jet training in Class 52-G. I felt ready to become a jet pilot.

Pinecastle AFB, Orlando, Florida

Pinecastle was a new base. It had fairly new buildings including a BOQ where I was assigned a nicely furnished room and a private bath. It was close to the mess hall where I ate my breakfasts before reporting to the flight line. The flight line was a short drive, and we had a great operations building with new briefing rooms. Strategic Air Command owned the base, and a wing of B-47s took up a large section of the tarmac. Also on the base were F-84s. Pilots flying these were newly graduated who were going through combat training in preparation for assignments to Korea. Our training squadron consisted of approximately thirty-seven student officers who were picked from the different basic training bases (just completed T-28s) and were brought in for jet training. It was great situation. No more cadets and the instructors were all Korea combat returnees. On our first day on the flight line, we were taken out to the tarmac to survey the parking area and the routes to the taxiways and runway, etc. All of us were excited about transitioning to jets. As we were standing outside the ops building, a flight of F-84s took off on a gunnery mission.

As the second element broke ground, number four rolled over and crashed off the end of the runway. A huge fireball emptied, and black smoke billowed into the sky. No chute was observed, and it was clear that we had just witnessed a fatal air crash. That put all of us in a state of shock. As fire apparatus rushed out to the scene, we were ushered back into the ops building. Our instructors had a difficult time settling us down and explaining in a manner they felt was most appropriate that such accidents do happen, but not that often and that we should not dwell on this incident. Their words were not immediately soothing to our frayed nerves. This accident on our first day of jet training really shook us up and I'm sure everyone of was affected by it. Individually, we all had to deal with it.

I was assigned to Nuthouse Flight and my instructor was Capt. Bill Wyatt. Wyatt was no more than five feet, four in height and had to sit on a large cushion so he could see from the backseat. He was a real character and a delight to be around, but I did not rate him high as an instructor. His briefings were short and incomplete. He made things up in the air and you rarely knew what was coming up next. Wyatt had a girlfriend, who worked in a hotel at Daytona Beach. Every flight ended with a low-level buzz over the beach right past the beach side of this hotel. While these buzz jobs were fun and exciting, I never understood why the folks at Daytona did not complain about it through channels.

Wyatt had another flaw that got me into trouble. To speed things up on our first flights, when the student jumped into the cockpit, Wyatt would taxi the aircraft out to the runway while the student was strapping in and getting ready. Sometimes he would even begin the takeoff. When I was sent out on my solo flight, I found out that I did not know how to taxi the aircraft. The T-33 did not have nose wheel steering and you steered using the brakes. But you had to keep the hydraulic brakes pumped up for them to be effective. I did not know how to do this because I had never practiced it. I barely made it on to the runway, but I could not point the nose wheel straight down the runway and takeoff. After several embarrassing moments on the runway going in circles, I had to abort my first solo takeoff. The Section Commander, Major Desutter, was furious, called me into his office and really did a chew job. I did not want to rat on Wyatt, but when I saw Wyatt, I was really angry because he had never let me taxi the aircraft and takeoff. We got that fixed quickly and I had no further troubles. But I really took a lot of ribbing about that incident. I also got my first pink slip signifying an unsatisfactory flight.

The T-33

The Lockheed T-33 was a stretched version of the F-80 and became the world's standard for advanced jet training for more than 20 years. It was a marvelous aircraft, easy to fly and land, and yet demanding enough to be useful as a trainer. Initially, most of us were anxious about the conversion to jets. Up to this point we had all began our flying career in piston driven airplanes with a propeller providing the thrust. But jets changed all this. They used a different fuel (JP 4), which was kerosene based. Once started, they needed no warmup time and pre-takeoff checks were minimal. Thrust came from air being pulled through the intake vents, compressed, and then blown out the rear at such a high speed that the aircraft was pushed forward. As jet engines used more fuel, they became more efficient at high altitude where the air was less dense. But the biggest change we saw in jets versus piston driven aircraft was speed. Things just happened more quickly in jets from takeoff to landing, and the pilot had to be ready for that factor. There was much less time to react, and you had to project yourself out ahead in every phase of flying. I have always thought that the major difference separating fighter pilots from multi-engine pilots was the capability to think ahead of the aircraft, anticipate your next moves, and perform flight operations in a more responsive manner. So, the leap from T-28s to T-33s was a large one and was designed to separate the men from the boys. There was so much less room for error. The F-80 was America's first jet fighter. It was still in use in Korea even though the F-86 Sabre Jets were now in action. Little did I know in the summer of 1952, that I would log thousands of hours in the T-33/F-80 and actually fly 30 combat missions in the F-86.

These were the happiest days I experienced during my year of flight training. I loved the flying, and I loved Orlando. Now with my own transportation, I was able to travel around when off duty and see Florida. Orlando was a beautiful city in 1952. It was a small town, and the people there respected us and were friendly to the airmen. Around the town were many small lakes with lovely homes on them. It had several good places to eat, and I often spent time at the Orlando Naval Base Club. Some weekends I drove to Daytona Beach where you could drive your car on the beach. Life was pretty good and the heavy pressure of my days at Bainbridge and Bryan were behind me. I loved flying over Florida, except all of us were worried about the huge areas of swamp that was below. Having to go down in the Everglades was not conducive to a good day. In fact, in the 50's, Florida was more swamp than anything else, crowded with gators, snakes, and other nasty things. Another thing we dealt with was the afternoon thunderstorms. Almost every afternoon huge buildups occurred that frightened the cadre and frantic calls went out calling all birds back to the base. While this was done to save us from being caught in storms, we were not ready for, it also resulted in thirty some odd aircraft feverishly racing back and jamming up the approaches to the field and the landing pattern, which was just as dangerous. That was why I preferred to be scheduled for morning flights. My class work also became easier, especially in engineering as the T-33 was much simpler engineering-wise than the piston driven planes we had flown. My days at Orlando raced by. In early October, we received our orders. Mine were to report to Tyndall AFB for transition to the F-94 all-weather interceptor and then on to Alaskan Air Command. I was truly disappointed. I had been hoping to go to Korea in one of the fighters we were employing there, especially the F-86. It appeared that that would not happen. We were scheduled to be awarded our Wings on October 25.

Then another unexpected event occurred. Two days before graduation, I was notified that my orders had been cancelled. I was told to standby for further word. On Saturday morning, October 25, class 52-G was assembled in the Academic Building at 0900 and there we had our Wings pinned on by Col. J.E. Roberts. I had done what I so fervently wanted to do. I could wear those cherished wings above my left breast pocket that was an empty space beforehand. Of course, the wings I received that day were without a star or a wreath. These wings were worn by fledglings, who still had a lot to learn about aviation and its many unexpected ins and outs. However, those wings represented a successful completion of a dream I had to become a military pilot in the USAF and for which I worked hard day and night for over a year. And so, that moment was a great accomplishment for me because I certainly was not at all positive that I could do it.

But I still had no orders. After saying goodbye to my fellow officers, I sat in the orderly room awaiting word for several hours. My car was packed, and I was all set to head for New York and a short leave before reporting in to wherever it was that I was sent. That afternoon, the airman told me that there was an opening for an assignment to Korea in F-86s with TDY at Nellis AFB, Las Vegas, Nevada. The one problem was I had to wait another month for this assignment, reporting in December. Despite the delay, I happily accepted that assignment. I wanted Korea and F-86s all along. One more month would not hurt too much. I got my papers and headed for the car. A new adventure would begin for me in December, but for now I had my wings and my orders and a nice vacation ahead. My student pilot days were over.

CHAPTER ELEVEN: I BECOME A FIGHTER PILOT

Every man a tiger!

-Colonel Clay Tice
Commander, 3995 Flying Training Wing (Fighter)
Nellis AFB, Nevada

After a year of intense training learning to fly without killing myself and others, I headed home for five weeks of leave. It worried me that I would not be able to fly for such a long period of time, but I wanted to spend time with my family before going off to combat training and Korea. It was especially good to spend time with my grandmother who had raised me from the age of five. I realized that as I was destined to war and my beautiful grandmother was well into her eighties, there was a chance I would not see her again. I spend time with my dad, and we went to some hockey games. Most of my college buddies were away in the service somewhere so I did not make much contact with them. I did find an old friend who was with the National Guard at Floyd Bennett Field and so I did get in a few T-6 flights with him. I also had several dates with Alice Marie. It was good to see her again and know that she was still unattached. She promised to continue to write. It was a good leave, but the problem was I had to leave before Christmas to report in at Nellis. It was a sad day when I got into my Chevy and took off for Nevada. It took me five days as I headed south to Texas and then drove Route 66. I arrived at Nellis on December 28. It was a long, melancholy trip. I was anxious to start combat training, but leaving home, family, and friends again, was a bitter pill. It was particularly bitter that I was driving across the wastelands of Texas alone on Christmas Day.

After leaving home, my thoughts turned to what was ahead. First, I would be flying the F-80, the Air Force's first jet fighter. Then I would transition to the F-86. Both of these aircraft were in combat in Korea. The war was entering its third year and had become pretty much a stalemate. America's fighter pilots had become national heroes. With the F-86 the ratio of MiG-15' s shot down to our losses had escalated to about 12 to 1. Names like Jabara, Kincheloe, McConnell, Thyng, and Davis would become household names. While I did not expect to reach those heights, I did hope that I would perform bravely and up to Air Force standards.

When I reached Las Vegas, I checked into the base and was assigned a BOQ room close to the Officer's Club. Compared to the bases I had been assigned previously, Nellis was a vastly different atmosphere. It was a real fighter base, and you never could mistake it for anything else. At that time, it was the busiest airport in the world. From dawn to late at night, the roar of jet engines never ceased. With two parallel runways adjacent to Sunrise Mountain, about nine miles north of the city, it seemed that fighters were constantly taking off and landing from sunrise to sunset, and beyond. The entire culture of the base was geared to fighter pilot training, tactics, gunnery scores, inter-squadron simulated aerial combat, and endless braggadocio. The Officer's Club was raucous and filled with loud laughter, singing of ribald ditties, and alcoholic binging. The flying suit was the uniform of the day. I could see that my three and a half months at Nellis were going to be like nothing I had experienced before. It was clear that what was to come in my combat training would be the severest test of my flying career. The motto of the commander, Col. Tice, was, "Every man a tiger." In other words, every fighter pilot, was to be smart, aggressive, daring, and determined to survive and win, just as that most lethal and beautiful of felines, the TIGER. And that was what they were going to make of me and fast.

My First Day—My First Pink Slip

My months at Nellis came fast and furious. I was assigned to the 3599th Flying Training Squadron-Cobra Squadron. My first day of flying at Nellis still burns in my memory. I was briefed on the local area and told to take off and cruise the training areas for about an hour, seeing all the gunnery and bombing ranges as well as the landmarks for navigation. I was assigned an F-80. After a walk around inspection, I strapped in. But the problem was I did not know how to start it. This cockpit was different than those I had flown at Orlando. I was embarrassed to ask the crew chief to show me how to start the engine. I cautiously worked my way out to the runway and off I went. The weather in Las Vegas was most usually clear with unlimited visibility, which made it a great location for combat training. After an hour or so, I returned to base, entered the traffic pattern, which was very crowded and landed. It was my first jet flight in almost two months. But as soon as I entered the ops building, I was told to report to the flight briefing room for my next flight. The instructor, Major Kenny, had already completed most of the briefing to the other two student pilots. It was to be another orientation flight with Major Kenny leading a flight of four. I would be the number four man. Out we went. Again, I had to ask the crew chief to help me start the aircraft as I could not recall all the steps the first chief had shown me. Checking in on the squadron frequency we then taxied out and took the runway. We took off in formation with about a ten second interval between the first and second element. Bob McDermott was the number three man, and I flew off his wing. Right after takeoff Bob began his cutoff turn and brought us into close formation with the flight leader. But then our troubles began. Bob overshot a little and when he backed off the throttle, we both fell behind the leader and had to try and close again. Each time Bob moved the throttle, I had to compensate quickly and then I began to overshoot and undershoot. I always thought I was a pretty good formation pilot, but somehow neither my element leader nor I could get in sync with the flight leader. This brought forth some pretty nasty calls from Kenny. I could not understand why I could not smooth out my airspeed corrections. This F-80 lagged when you pushed the throttle either forward or back. After about 15 minutes of climb out with large turns, Kenny signaled that we should fall into a tactical formation. Here we separated from each other as elements and increased spacing between aircraft. This formation would be used in combat situations in order for each pilot, other than the lead to check each other's six o'clock for potential enemy attacks. But again, the three of us had difficulty in holding position. Kenny was furious and he let us know it over the radio. The climax of this flight came when Kenny signaled us to take up a trail formation, i.e., the number two directly behind the leader, number three behind number two, and number four (me) directly behind number three. Kenny then began a series of barrel rolls, followed by several high G turns and then two loops in succession. As some of these maneuvers were four G's and more, I was constantly blacking out, coming to, and then blacking out again. I had great difficulty keeping number three in sight. I recall at one point seeing the flight leader and number two climbing at my 12 o'clock while I was still diving toward the desert floor. I was certain that I would not survive this flight. Finally, Kenny wagged his wings for us to join up and we returned to base. Again, the three of us had much difficulty in staying in close formation. As I was number four, I was constantly too far forward then too far behind. It was like being on the end of a whip. When we landed, I was absolutely the happiest man alive. I had survived, but barely. Kenny's debriefing took about three minutes. He told us that that was the worst flight he had ever led. He stated that we were the sorriest bunch of pilots he had ever seen and that we would not last ten minutes in combat. He then presented all of us with pink slips indicating an unsatisfactory flight. What a way to start my combat training!

Afterward, the three of us talked about what had taken place. All of us agreed that the fuel controls on our F-80's would not respond the way we expected. But why? A little research uncovered that these older versions had a different fuel control system than those we flew in flight school. (Which was why I did not know how to start it.) They were much slower to respond to throttle movements and you had to compensate for that by managing the throttle with anticipation and not overcompensate. Kenny, of course, did not advise us of this before our formation flight, and we decided that he had purposely neglected to do so. It was quite a day. It completely shook my confidence, and I began to wonder if would really be able to hack this program. Maybe I was not the hotshot I thought I was. It was up to me to regroup and just work harder at showing that I learn quickly and then adjust. The course was intense, and I flew at least one mission daily, and sometime two. When we flew in the mornings, we had ground school in the afternoons and vice versa. I liked the morning flight schedule better even though it meant 0600 briefings. I was up at 0500, hit the mess hall for breakfast and then to the flight line. The missions were short, usually an hour. Formation takeoffs, then a trip to the gunnery or bombing ranges, then reform and return to base. In the "every man a tiger" arena, the rule was that any formation could jump (attack) another, at any time after takeoff, and anywhere. This always created an attitude of caution and awareness. You always kept your head on a swivel, searching out for a possible sneak attack from above and behind. When your formation was jumped by another, or your flight attacked another, it was pure mayhem. The goal was to get behind (the six o'clock) of another flight and shoot some gunnery film footage showing that you had a kill. These surprise attacks ended up in a wild gaggle of fighters turning and diving to get advantage. The radio waves were filled with pilots screaming at their wingmen or leaders, calling out "enemy" aircraft, and calling for tighter turns or dives to either get in position for a kill or escaping from an enemy attacker getting you in his sights. These gaggles were filled with tight, gut busting, black out, 5G turns and pullouts. As all the aircraft in the gaggle were the same, you had to be careful that you did not "shoot down" one of your own flights. These fights could start at 25,000 feet and end up on the deck. One "aerial battle" I recall was initiated at high altitude and gradually ended at about 500 feet. The trouble was we were over the Las Vegas Strip. I looked out and saw the Sands Hotel in my sights. We put on some air show for the gambling visitors. Clearly, the mission of the 3995 FTIV was to prepare us for combat and to do it as realistically as possible.

Mach One—I Fly Faster Than Sound

As you might guess, we had many accidents at Nellis. Our average was about one a week. They were usually fatal. At high speeds and low altitudes, a mistake was critical and rarely forgiven. On one of my air-to-ground missions, my element leader had target fixation and flew into the target. On another mission, one of my fellow students collided with an F-86, one pilot ejected safely, the other died. Whenever there was an accident and a fatality, everyone was affected mentally and otherwise. Many said that if you survive Nellis, you'd survive Korea. I hoped that was true, but could I survive Nellis? Clearly, this was the most challenging period of my life. There was so much at stake, and I had so much to learn in a hurry. Gradually, my fighter flying was improving, but I still made many mistakes that, fortunately, did not end in disaster. In early February, we checked out in the F-86 Sabres. As there were no two-seaters in the F-86, student pilots just read the Technical Orders (TO's), memorized the emergency procedures, sat in the cockpit, and learned where every switch was, and listened carefully to the instructors. Then you went out and took off, flew around the area, and returned and landed. My first flight in the F-86 took place on January 31, 1953. I found the F-86 was an easy plane to checkout in and I truly loved every minute I flew it. It was capable of Mach one flight. You did this by taking it up to altitude, rolling it over and diving toward the ground at full throttle. As it approached the speed of sound, the plane began to shake. All the cockpit needles wobbled and the stick shook. Then, just when you thought the plane was coming apart, everything smoothed out. The Mach needle read 1.1. I just flew faster than sound. Then I pulled back the throttle and eased the nose back to the horizon—I had done it just like Chuck Yeager had! It was another milestone in my career. (Years later, I would push my F-106 to Mach two). On another day, I took my F-86 and flew over to the Grand Canyon and took it down into the canyon tracing the Colorado Rivers' many curves. It was a glorious flight over one of the world's great wonders. Of course, at that time there was very little air traffic to worry about.

My days at Nellis AFB and Las Vegas were busy and trying. But there were times for rest and relaxation and for tasting the delights of entertaining Las Vegas, the gambling capital of the USA. With my car I was able to get into town on the weekends and also in the evenings. In fact, in the last two months of my training, I usually went to bed in the afternoon after ground school or my last flight. Then I got up and dressed and went to the Strip. My favorites were the Sands and the Desert Inn. In fact, the bartenders came to know me at those places, and I truly enjoyed the shows and the food. Through some friends at the casinos, I got to meet the Mills Brothers, who were headliners on the Strip and Kay Starr who came out to Nellis and put on a great show for the troops. As the war was still going on, many folks were kind enough to buy our drinks or invite us into the shows. I also tried my luck at the craps tables and at blackjack (21). My luck was about as good as the average sucker—some nights I was a winner, and some nights I was a loser. I recall one night when I started off in a great rush but ended up losing it all and then a sum amounting to my month's pay. Of course, that wasn't an overwhelming amount, but it taught me a valuable lesson about gambling. I usually stayed on the Strip until the early hours, returning to the base in time to suit up and go flying. It's hard for me to believe I actually did that, but then I was still young. Some of the guys got to meet the showgirls who would join us between shows. One good friend, Bill Hummer, began a serious relationship with one of the girls at the Desert Inn. In fact, they married just before he headed for

Korea. Many of us thought that that was a bad idea. It did not turn out well as Bill was killed in Korea. All in all, my experiences at Nellis were beyond what I had ever dreamed could happen to me. In little over a year, my life had changed dramatically. I was now a military pilot and preparing to enter combat. I was still single and totally independent. While I enjoyed flying the latest fighter aircraft, I could not forget that there was a price that had to be paid. My future was an open book and I had to be ready mentally and spiritually for what may come.

Bill McCristal

One of the traditions at Nellis was the lowering of the flag each day at 4:30. The base loudspeaker would resound with the National Anthem and everyone on base stopped, faced the flag, and saluted as the flag was lowered. At about the same time the flag reached the airmen's hands, a fighter aircraft would zoom right over the top of the flagpole. It was a fitting end to the workday on a fighter base. The flyby over the flagpole was usually performed by one of the instructor pilots. All of us were zeroed in on how close to the top of the pole the aircraft flew. One of the most popular and outlandish instructors we had on base was Capt. Bill McCristal. McCristal had already made a name for himself with his funny antics, but on this particular occasion he outdid himself. McCristal was obviously determined to outdo all others in how near to the top of the flagpole he could maneuver his F-80. And he did. To the shock and amazement of everyone, McCristal's fighter clipped the top of the pole and the brass ball that sat on top was shattered. The top of the pole flew off in splinters. You can imagine the repercussions of this incident and from then on Col. Tice made sure that the flyover was a flyover and not a fly into.

Our relations with the instructors were not always happy. Though they were all great pilots and combat veterans, not all were good teachers. I got my share of pink slips when I was unable to demonstrate the degree of aggressiveness or accuracy that was expected. I tried to learn from my mistakes, and I always listened carefully when the older, more experienced pilots talked about their missions. There was always more to learn, and the one thing I tried to avoid was letting my confidence become false bravado. My skip bombing and air to ground gunnery scores were improving daily. The one area I had difficulty with was air to air gunnery. My scores there were below average and nothing I did, seemed to help. In air to air, we fired at a large fiber target (called the rag) towed by a B-26. Each plane in the formation had its 50 caliber bullets tipped with paint that would show on the target when it penetrated. After the flight, the tow target was dropped to the ground and then an airman counted the bullets holes for each color. My scores were usually the lowest. Several times I spoke to the instructors and had them look at my gunnery film. They always said that I was tracking the target well and that I should be scoring well and to keep at it. But nothing worked for me. One day I took down one of the gunnery films shot by one of the instructors. When I looked at the instructor's film, I saw that these guys were closing in on the target and firing well beyond the point where, for safety reasons, I stopped shooting and broke off. I was astounded at my stupidity. The instructors would not tell me to press in closer before firing because that was against the safety rules. But now I knew what the secret was. From then on, when I reached the point where I previously began firing, I counted a few more seconds and waited until my sight picture was filled with the target. I then fired and pulled up sharply over the rag, scaring the hell out of myself. But, boy did my scores get better!

By the end of my combat training, I was feeling comfortable and confident that I could be a fighter pilot. I now realized that every flight had to be prepared for and flown at a high intensity level. Slowly we were being trained to fly the Sabre at its maximum level of performance. That meant you had to know your own capabilities as well as the aircrafts. If you exceeded either, you could buy the farm in a hurry. My tolerance for high G levels continued to get better, and my aerial combat techniques improved with every engagement. My gunnery scores were good, especially my air to groundwork, and as we neared the end of our training, I felt better about my chances to survive combat.

My First Promotion—Denied

About a month after I arrived at Nellis, I completed 18 months of active duty. This automatically placed me and several other Second Lieutenants in line for promotion. But no paperwork was submitted. So, I, and several others, went in to see the Squadron Commander, Major Skeen. He was a gruff, hard-nosed man who always seemed to be in a foul mood. No one liked to fly with him. He asked what the hell we wanted; we told him that we were eligible for promotion and that we would like to be submitted for it. But Skeen asked what made us think we were qualified for promotion. At first, we were stumped, but finally we reminded him that we had successfully completed flight school and had no unsatisfactory reports in our files. But Skeen told us that we were the worst group of student pilots he had ever seen, and, as far as he was concerned, we would never make it through combat, and he was not about to submit us for promotion. We were stunned. When we left, we were at a complete loss as to what to do next. But about a month later, I saw a notice that the Command Inspector General was coming to Nellis and that anyone who had a complaint could see him. On the day that the IG was on base about, ten of us showed up at Wing Headquarters. The Wing Adjutant was aghast when he saw us and asked us why we were there. We told him we wanted to see the IG, but we would not tell him the nature of our complaint. He immediately went into Col. Tice's office, and soon after Tice came out and asked us what this was all about. Again, we declined to tell him which unnerved him greatly. He invited us into his office. He was truly upset that all of us were going into see the IG. Finally, we told him that Major Skeen would not take action to promote us, and we were going to tell the IG that and see if he could help us. Tice flew into a rage and called the Adjutant back in, told him to takedown all of our information and get the paperwork flowing for our promotion. The last thing Tice wanted was for us to go into the IG with this complaint. So, it was done and on March 23, just before I shipped out to Korea, my promotion to First Lieutenant came through. But not receiving that promotion until two months past my eligibility had a lasting effect on my future promotions, all thanks to Major Skeen. I had learned a lesson about military bureaucracies. Don't depend on them and follow through on the things that are important to you. Assume nothing.

My combat training in the F-86 came to a satisfactory conclusion near the end of March 1953. I was now combat ready and prepared to go to war. I was confident in my skills, but I knew that I still had a lot to learn. I kept my ears and eyes open, and I especially listened carefully to veteran pilots who had come through combat successfully. If I could learn from their experiences, it would only make me better prepared for what was to come. At completion, I and three other fellow students jumped onto my Chevy and headed home for a few days leave before reporting to Camp

Stoneham, California for the Far East. We took turns driving and we did not stop except for gas and chow. When I arrived home in Great Neck, I was exhausted. I wanted to get as much time in with my family before leaving for Korea. I also realized that the coming year would be a pivotal period in my life and that anything could happen including my death in combat and possibly the death of my relatives. I was especially concerned about my grandmother who was now in her late 80's. She had raised me since the age of five. She was the central figure in the Cunneely Family, still taking care of her two spinster daughters and my father. She had suffered the loss of her husband and two sons. She sweated and prayed through WWII for her grandsons and now she would do the same for me. She had suffered two broken hips and was still doing the cooking for her family. I loved and admired her so much and will be forever grateful for what she had done for my brother and me. Saying goodbye to her was very difficult. As it turned out, I would not see her again.

I also had a few dates with Alice Marie. I left my car with my brother for his use, and I left NY by commercial air for the west coast and eventually Korea. The war there had turned into a stalemate with the two adversaries pretty well stuck in opposing positions around the 38th parallel north of Seoul. Still there was an air war going on and perhaps I could do my part to bring this thing to a climax. I flew in a TWA Constellation to San Francisco and then taxied over to Camp Stoneham for shipment. A few days there and my number came up. I flew out of Travis AFB on a C-97 contract flight to Hawaii, then to Midway, and finally into Tokyo. During our short stay in Hawaii, I sent an orchid to Alice Marie. All in all, it was an exhausting flight on a crowded plane, trying to sleep, eating box lunches that were terrible, and wondering if we were ever going to get there. In Tokyo we were housed in a former Japanese jailhouse where we scanned the Bulletin Boards twice a day waiting for an assignment and a shipping order. When we were cleared for the day, we set off for Tokyo where we enjoyed the sights and sounds of that city, still somewhat under American control. Prices were ridiculously cheap. Using Scrip, we ate and partied at the best watering holes in town. I also visited my high school and college chum Bill Collins and his wife Joan who were stationed at Tachikawa AFB just outside of Tokyo. But after about three days, my name appeared on the Assignment Board. I was assigned to the 80th Fighter Bomber Squadron, 8th Fighter Wing, (the Headhunters), at Suwon Air

Base (K-13). That night I boarded a C-124 and we rumbled over the Japanese Sea to Seoul. There, I caught a ride on a troop carrier truck for a long, bumpy ride to K-13. My combat tour in Korea was about to begin.

CHAPTER TWELVE: THE FORGOTTEN WAR

In War, there is no substitute for victory.

-Douglas MacArthur

When I arrived at Suwon, the Korean War was two years and nine months old. After terrible early defeats that drove our forces all the way to Pusan on the southern tip of the Korean peninsula, Gen. Douglas MacArthur initiated a very risky strategic maneuver. He had American forces invade Inchon, located west of Seoul, and began a pincer movement on the North Korean forces. Our ground troops, aided by close air support, drove the enemy back north to the northern border of Korea at the Yalu River. But MacArthur erred badly in believing that Chinese forces would not enter the fray. This error ended in disaster for our Marines and ground forces when the Chinese crossed the Yalu and drove our ground forces back. It was a most severe winter battle and our Marines and Army troops suffered terrible losses. It also caused MacArthur to lose his job when President Truman fired him. In April 1953, the war had stabilized, with our forces facing the enemy close to the 38th parallel. Our Air Force had almost complete control of the skies. With the introduction of the F-86 Sabre Jet, American fighter pilots had a 12 to 1 ratio of shooting down enemy MiG - 15s, piloted mostly by Russian pilots. Men like Jabara, McConnell, Kincheloe, Thyng and Davis became household names. But any ambitions I had to become an Ace were dashed when I was assigned to a fighter-bomber outfit where air-to-ground missions were primary.

The 8th Fighter Wing was stationed on the east side of the base. On the west side was the 51st Fighter Wing that flew the F-86 on sorties planned to intercept and engage the MiG-l5s flown mostly by Russian and Chinese pilots. These were the glory boys who made the headlines. One of the best, who was still there when I arrived, was Capt. Joseph McConnell, Jr. On May 18, McConnell shot down three MiGs in one furious air battle. I got to meet McConnell and I liked him right away. Although he asked to stay in Korea for 25 more missions, the Air Force ordered him home. He had 16 air-to-air victories when he left. But I was assigned to fighter-bombers whose primary missions were dedicated to bombing and strafing enemy positions. The 80th Fighter Bomber Squadron (The Headhunters), to which I was assigned, had a great history that went back to WWII. But to my chagrin, the 80th was still flying the F-80. Here I was ready to go in the F-86, but the squadron was just beginning its conversion to that aircraft. So, as anxious as I was to begin flying combat, I, and the others who just arrived operationally ready in the F-86, had to wait. It was big disappointment. On arrival in Korea, I had a total of 352 hours of flying time. I was a very junior birdman.

I was assigned to A Flight and the Flight Commander was Capt. Bill Schaumberg. Our quarters were a Quonset hut made of metal. It housed all members of the Flight. New heads got the smallest amount of floor space and a canvas cot to sleep on. I had no chest of drawers and so I lived out of my footlocker. The hut was heated by a small gas stove and, of course, new heads were in charge of keeping it going through the night. If you failed to keep it going and the hut's temperature plummeted, you were in big trouble. The latrine was outside and about twenty yards away. Going to the john at night was hell. Often pilots who had to pee badly just went outside the hut and let fly. This was a no-no, naturally, but one frequently violated. The latrine was a five-holer, and not the most elegant place in the world. You did not linger in there any longer than you had to. There was a shower room with overhead sprays. The trouble was that the showers operated only at certain hours and not more than about three days per week. If you missed those days and hours, you were out of luck. My fellow flight members were good guys. Some were nearing the end of their tour. Most pilots kept a visual check of the missions they had completed, either on a calendar near their bed or on their person. You could always tell the old heads from the new ones. When a pilot came close to his 100th and last mission, he became elated but also wary. The saddest thing was a pilot who was shot down when he was close to ending his tour of duty. When we were not flying, there was not much to do. New heads quickly were given additional duties in the squadron, and before I knew it, I was an assistant adjutant and personnel officer. You also were given such duties as runway control officer where you manned a small mobile unit near the end of the landing runway. You monitored each landing to ensure that each aircraft had its gear extended, was free of armament, had adequate clearance from other aircraft, and was landing in a safe manner. If things were not right, you had a radio with which you could call the aircraft and give advice or even send him around for another approach. You also had a flare gun to signal the pilot to go around if he did not acknowledge your instructions.

The Commander of the 8th Fighter Wing when I arrived was a black Colonel. His name was Benjamin O. Davis, Jr. and he later became a three-star general. Our squadron was still flying F-80s, while the other two squadrons in our Group, the 35th and 36th were in transition to the F-86. Us "new-heads" thought mistakenly that as we were already combat ready in the F-86, we would begin flying missions as soon as we got our aircraft. We were wrong. When the new aircraft arrived, the old heads began their training in the F-86. We had to wait until they got through their checkouts. That left very little for us to do but wait and perform ground duties. We were also sent to the 5th Air Force Survival School. This was a five-day TDY where we were taught escape and evasion techniques in Korea and how to survive off the land until we were rescued. We were also taught how to get around the system of "brainwashing" that the enemy was employing on captured American airmen. Part of the training included interrogation techniques and an escape and evasion exercise. Once in a while we got a T-33 flight to keep up to date. It was a frustrating period for us and, as it turned out, prevented us from completing our missions in Korea before the war ended. In 1952, Dwight Eisenhower was elected President and took office in January 1953. lke had been the Supreme Allied Commander in Europe in WWII. Before he took office, Ike visited Korea and let it be known that he was going to bring that war to an end soon. The North Koreans feared that Ike would employ nuclear weapons and so began to get serious about the peace talks. By April 1953, when I arrived in Korea, rumors were flying that the war was not going to last very much longer.

I can still remember the sights, sounds, and smells of Korea. The country was in a mess. The war had reeked devastation and ruin on most of the country. The North Korean army that had invaded and penetrated all the way to the southern coast and then had to retreat all the way back to the north after the Inchon landings, left an awful mess behind them.

While the south no longer was subject to invasion forces, the villages were poor and there were thousands of women and children who had lost fathers. Prostitution was widespread and this resulted in many American soldiers becoming infected with sexual diseases that prevented their rotation to the USA at tour's end. The odor of Kimchi, Korea's national food, was suffocating, even on the base where Korean workers cooked their own foods. In our hooch, we had a young Korean boy, approximately 14 years old, who did our housework and laundry. He was a great little boy. His name was Park J. Sil. He was dedicated to A Flight and was emotionally moved when one of our pilots was killed in action or rotated to the States. I frequently think of him and wonder how life treated him after the war. We did not leave the base often except on flights. But flying over the south I saw much war devastation and Seoul was mostly a scene of destroyed buildings.

Finally Combat

Up to this point, my knowledge of combat in Korea consisted only of what I heard and saw. I attended pre-mission and post-mission briefings and looked at the gun film taken during bombing and strafing runs by other pilots. It was clear that North Korea no longer had many strategic targets left to bomb. Once in a while, a flight would observe some ground movement of trucks, personnel, or even a train. When that occurred, flights that were airborne over the north would race to the area of the target and try to get in some good licks. Many of the missions were ground support when a flight was sent to the aid of an army or marine unit that was taking fire from the enemy. On these missions you worked with a FAC (fighter aircraft controller) who was either on the ground or airborne in a light aircraft. He would mark the target with phosphorescent flares and then direct the bomb runs of each flight. These were satisfying missions because you were helping the poor grunts on the ground, who were in dangerous positions, under fire, and who needed some help from above. In addition to bombing, we often strafed with our 50 caliber guns situated in the nose of the Sabre. On some rare missions, we carried napalm tanks that we skipped bombed into the enemy lines. These were very mean weapons that literally burned the enemy troops to death. If they were not hit directly, they usually died from asphyxiation as when the napalm ignited, it just sucked up all the oxygen in the area.

My First Mission

My first combat mission was on June 1, 1953. It lasted one hour and 15 minutes. My remembrances are that we struck a target composed of buildings on the Haeju Peninsula. These were supposed to have housed North Korean army troops who were occupying that area. I recall that my heart raced a little faster than usual when we crossed into enemy territory. I kept a close eye on my engine instruments and a close eye on the flight leader, Bill Schaumberg. Schaumberg found the target, made sure we all had it in sight and then began our attack. He made certain we all armed our bombs, and we then went in one at a time. I saw no enemy fire, or for that matter I saw no life at all on the ground near the buildings we bombed. Still, I was excited that at last I was doing some of the things I had trained for. I was also excited that just maybe there would be some anti-aircraft activity as we were hitting the target. But there was none, and we all came off our bomb runs intact. We then regrouped into tactical formation and headed home. I had completed my first combat mission. The one trouble was I missed the target. In fact, two of us did and the third new head just barely nicked one of the buildings we were aiming for. Not a great showing.

In the days ahead until the war ended, I flew 24 more missions and a total of over 30 hours of combat in North Korea. Most of my missions were routine and uneventful. There was a scarcity of targets left to be bombed in the north. We did hit a few significant targets near Pyongyang, and I clearly remember a Group mission I flew on where the target was heavily defended, and considerable enemy fire was encountered. In this mission, the number three man in the flight ahead of my own suffered a direct hit on his target run. I was just beginning my own dive onto the target when out of the corner of my eye an explosion occurred. It startled me and I believed that I was under fire and maybe had been hit. Then I realized what had happened. It happened so quickly that it took a few seconds for it to register. Several others of our aircraft returned with battle damage, i.e., holes in their wings or underbelly. That mission was deemed highly successful by our intelligence debriefers but losing an aircraft and a pilot took away any good feelings we might have had. Everyone shared the same feeling about it. Why did he get it and not me? There was never an answer to that question.

Many of my missions consisted of hitting a primary target and then rescuing, i.e., scouring the countryside for targets of opportunity and strafing them with our 50 caliber guns. We would divide the flight into two elements of two aircraft and each element would go off looking for anything that was moving. These missions were mostly unsuccessful, but occasionally we would find some trucks moving down a highway and we would take them with our guns. Once in a while we would spot some trucks and troops hiding under some trees, and then we would make passes at them until our ammo ran out or our fuel became too low to continue.

The missions I liked the most were the close support missions on the front line. We often sat on runway alert ready to take off in minutes. A call would come in from the TACC that some troops needed some air support because they were under intense fire, or they thought the enemy was about to mount an attack. We would scramble and then be directed to the target area. There, a ground controller or a FAC in a light aircraft would direct us to the target. They would spot the target with flares or phosphorescent bombs and then direct each aircraft as to what ordnance to choose, and where to place our bombs. He also warned of possible enemy fire in the area. These missions were fast and uncomplicated. They also were very rewarding as you could see the direct results of your work immediately. The troops on the ground were really appreciative of our efforts as usually our attacks caused the enemy to cease what they had been doing and retreat to safer areas. When we used napalm on these missions, the enemy usually disappeared in a hurry. I always felt good about help-ing the poor ground troops when we could.

Tom Curry Comes to Visit

Sometime in March, I returned from a mission that had not gone well. On top of that, I was very low on fuel, and when another fighter cut in front of me in the landing pattern, I had to execute a go-around and declare an emergency. I pulled up in a four G turn and reentered the base leg, dropped my landing gear, and made a winding final approach for a very fast touchdown. My heart was still beating rapidly as I taxied in, and I was swearing at the idiot who cut in front of me on my first approach. When I got back to the Ops Building, I was told I had a visitor. Who in the world would be visiting me in Suwon, Korea? It turned out to be Alice's brother, Tom, who had hitched a ride to K-13 from where he had been stationed near the front. He knew I was at K-13 because Alice wrote to me faithfully and had told him where I was stationed. Tom was a big sports fan, like me; we got along well and enjoyed each other's company. I respected him in many ways. I was glad to see him. I took him to our hooch, where we were able to take a shower and then off to the Officer's Club for a few beers and a steak. Tom really enjoyed the change and then I felt horrible that only a short while ago I was complaining about things that were relatively minor. I had it so much better than Tom and the other ground troops who lived in terrible conditions and ate a lot of K rations. Tom stayed a few days and then had to get back. I helped him get a ride to Seoul and we said goodbye. Tom helped my morale as much as a few days away from the front helped his. We wished each other good luck. It did not occur to me during his visit, that someday he would become my brother-in-law.

More Missions

In the ensuing days I logged more missions, but they came slowly as the war was obviously waning. Also, the guys who were getting close to completing their 100 missions began to hog the schedule leaving the new guys on the ground. I could understand this, but it was frustrating, nevertheless. Slowly I began to move up the line becoming assistant flight commander and squadron adjutant. These extra duties kept me busy when I was not flying. My saddest duties resulted when we lost a pilot. We had to pack up all his belongings and send them to his home. Letters had to be written for the Commander and I frequently drafted these too. One I remember was a pilot in my flight who had already completed his combat missions and was waiting orders back to the States. Bill Tallant was a good guy, very friendly and helpful to us new heads. He spent most of his time reading magazines. One day he was asked to ferry a sergeant to Japan on an emergency leave. They flew in a T-33. The sergeant had a B-4 bag and they crammed it into the space between the two cockpits above the instruments panel. Somewhere in South Korea their plane crashed killing both Bill and the sergeant. The plane appeared to be out of control when it hit the ground. The accident investigation concluded that the luggage bag came off the console and wedged between the control stick and the side of the rear cockpit. Bill was unable to control the aircraft as it went into a death spiral. Neither the pilot nor his passenger attempted a bailout. I was sickened over this accident because the pilot had just finished his 100 missions and was doing the sergeant a favor by flying him to Japan where he would embark on a flight back to the States. Whatever the reason was for his emergency leave, the tragedy was compounded and also took the life of my friend. I tried to learn something from every aircraft mishap I became acquainted with. You could never know too much about flying and how seemingly minor decisions could backfire into something unexpected and deadly.

Missing in Action

Occasionally when the weather up north was too bad for finding a target and bombing it, we flew radar-controlled bombing missions. I always thought these were silly missions in that they could not be very precise, and we were unable to determine whether the bombs hit the target or not. However, Intelligence had a list of targets that we would strike on these missions, and at least, we let the enemy know that we were still around to harass them. I was involved on one of these missions, which should have been a routine flight, but ended up quite differently.

I was the number four man in the last flight of a flight of twelve aircraft. We took off about mid-morning, joined up and headed north. The weather was overcast with a ceiling of 10,000 feet with rain. We entered the overcast before we invaded North Korea, flying in close formation all the way. We were being controlled by a ground radar site near the 38th parallel, and we had no idea of exactly where we were other than we were over enemy country. I was flying tight formation on my element leader when I realized that I could no longer hear the controller's radio calls. My radio had gone dead, and I could neither receive nor transmit. I hoped that my element leader might look over and I could hand signal to him that my radio was dead. But that did not happen, and we droned on to wherever we were going. I decided to stay with the formation until we finished the mission and returned home. I did not have any idea of where I was or how to get home anyway. The overcast was pretty thick, and I concentrated on my formation flying and not losing sight of my wingman. After a long period of this, I saw my leader bend over and I assumed he was arming his bombs for a drop. I did the same and put my thumb on the toggle switch. A few minutes later I saw his bombs peel off and I dropped mine.

Then the flight began a long turn into me and, being the last man in the last flight, I retarded the throttle to stay inside the turn being in close formation. However, when the formation returned to level flight I began to drop back. Pushing the throttle forward did not help, as the fuel control did not respond. Something now had gone bad with my engine fuel controls and I was falling back. RPM had dropped to 85 percent and would go no higher. My flight soon disappeared into the overcast and I was now alone, somewhere over North Korea, without a radio and with serious engine problems.

At 85 percent RPM I was able to maintain altitude (@20,000 feet), but I was afraid to retard the throttle any further as, once brought back, the fuel control would not recover, and I would be losing more thrust. I took up a southerly heading and flew as long as I thought I could to ensure I had returned over friendly territory. I was uncertain as to where I was, how low the cloud cover, and whether I would be seen on our radar as a possible enemy intruder. After a long period, I decided to start descending through the clouds. I did not retard the throttle, as I feared total engine failure might result. I broke out of the clouds about 5,000 feet above the ground and began looking for some familiar territory that would tell me where I was. After a nervous period, I saw the river that ran through Seoul. I followed the river to Seoul and then turned south toward Suwon. By the time I reached K-13, my flight had long since returned and landed. I came across the field and flew down initial approach wagging my wings, a sign that I had no radio. I then reentered the pattern and landed, with the tower controllers giving me a green light to land. Fortunately, my decision not to retard the throttle until landing proved wise. On the final approach, when I brought the throttle back, the engine unwound to idle. I would have had no power available should I have had to execute a go around. I barely made it off the runway and ended up on a taxiway where I awaited a tow truck to bring me back to the revetment. When I returned to Ops, my wingmen wondered what had happened. They did not realize I had left them until they were well on their way to home base. They assumed that I had either been shot down or crashed due to aircraft malfunction. I had been reported to FEAF as Missing in Action. Fortunately, that was corrected before it had been transmitted back to the

States. Reflecting back on that mission, I could only conclude that you never know what may happen in the air and in combat. Just be prepared for anything because anything frequently happens.

Did You Shoot Down Any MiGS?

Whenever I tell people that I flew F-86s in Korea, I am asked did I shoot down any MiGs. Unfortunately, I can only tell truth: no, I did not. But I did have one encounter with them, although it was not a very successful one. One day, we flew a bombing mission into northwestern North Korea. We were armed with two 2,000-pound bombs and had drop tanks (fuel) for endurance. But, due to conflicting operation schedules, we were without any top cover by fighter interceptors. We were heading north off the west coast of North Korea when we were jumped by a flight of four MiG-15s. They came down on us from above and out of the sun. Mayhem followed. As soon as they were called out, the flight leader ordered us to get rid of all armament and drop tanks. So, each of us, using the red button, jettisoned the drop tanks. Then we let go of the bombs, which floated down into the China Sea. We separated into tactical formation and turned into the MiGs that were diving on us guns blazing. I dropped in behind and slightly below my element leader, as it was my job to guard his six o'clock and call out any enemy attackers. But where were the MiGs? They flew through our formation and then faded away. Climbing back into the sun, they then headed north and presumably back home across the Yalu River to China where we were forbidden to chase them. We were a frustrated group. The enemy used a smart tactic, causing us to completely abort our primary mission, wasting our bombs and fuel tanks, and leaving us no other recourse than to head home. I have to say that for a few minutes my heart really began pumping at an accelerated rate, and the thoughts of possibly getting a shot at a MiG got all of us fighter bomber guys really charged up. It was not to be that day or any other day for me.

On some other missions I flew we provided top cover for some F-84 fighter-bombers but encountered no enemy aircraft. On a few sweeps, we flew up to the Yalu River looking for possible encounters with MiGs. As we flew just south of the river, we could see the MiG bases on the other side. On the ramp were MiG aircraft parked, just like sitting ducks. But we could not attack them. Naturally, this gave the enemy a great tactical advantage. Because of these rules, they could take off, form up and climb to high altitude in Chinese airspace without fear of attack. Then, when they were at high altitude, (often 50,000 feet for the light weight MiGs), they would sweep across the river and attack USAF aircraft. In that airspace, ensued great aerial battles between USAF pilots and MiG pilots who were mostly Russians. These battles took place in an area in northern Korea called MiG Alley. The 12 to 1 victory ratio our pilots attained was a tribute to the aggressive bravery of the American pilots and the training they received at Nellis AFB, Nevada. You had to be proud of the accomplishments of the USAF and US Navy pilots in taking total control over enemy airspace, which made life a lot easier for the ground troops.

Combat

Going into combat a year and a half after leaving home and college had a most serious effect on me. Although I had witnessed death at home during my early years, living every day with the possibility of dying suddenly and perhaps painfully brought many things into perspective. Also, having some of my friends not come back from missions and sometimes seeing them die from enemy fire or collision with the ground in a huge fireball, made me realize that my humanity was very fragile and that I had only little control over what would happen to me. Many of those I saw die were truly good guys, competent and skilled fighter pilots, and as dedicated to living through their combat tour as much as I. Yet, circumstances dictated otherwise. The mystery of it all was both intriguing and depressing. Why them and not me? Or would tomorrow's mission be my last. Most disheartening was my duty to pack the personal effects of a dead pilot and send them home to his family. In my additional duty as Squadron Adjutant, I often drafted letters from the Commander to the parents of the deceased. These were very difficult to word, and I never felt satisfied with the results of my editing. I could only picture the face of the parents or the wife when they received this most unbearable missive. It occurred to me how my father and grandmother would have reacted should they have received that dreaded news. During World War II, our family was always frightened about the possibility of receiving tragic news about my brother and my cousins who were in the combat areas. Thank God that never came, but there were thousands of Gold Stars in the windows in our town signifying a soldier or sailor from that home had died in combat. It certainly brought me to doing a lot of thinking about life and how fragile it was. I also thought much about death too and even though I was in my early twenties, death was close at hand for me and my fellow wingmen every time we took off. I was truly thankful for my religious upbringing and for the marvelous education I received from the Sisters at Blessed Sacrament, the brothers at Chaminade, and the Jesuits at Fordham. Especially helpful were the philosophy courses I had from the Jesuits, where we contemplated the existence of God, and the deepest questions we must ask ourselves: Why were we given life? What are we to do with our lives? What is the relationship between God, and good and evil? And is there a life after death, as Jesus promised. I tried to attend Mass often and I was helped tremendously by a great Irish Catholic chaplain who was always there for us. Father Duncan would attend our pre mission briefings, even at 0400 hours, and then have breakfast with us before we went out to our aircraft. He always offered Confession (which I took him up on a few times). Then as we taxied out, he stood on the revetments and blessed the flights as they left. That we made me feel good and ready for whatever God had in store. (Note: A new Wing Commander came in and was furious that a Chaplain would do such a thing and got him transferred out. Years later, I told that commander, who was now a two-star general, how wrong he was. His name was Elder.)

My Grandmother Passes On

In late April 1953, I received a letter from Alice Marie. It contained the saddest news I would ever get. My dearest Grandma, who took my brother and me in when our mother left our home and raised us until we both left home for war, died on April 18. By the time I got the news, Grandma was already buried next to her husband in Calvary Cemetery in Long Island City. I recall the dire thought I had when I last said goodbye to her and kissed her. This might be the last time I would see her. And so, that is what had happened. I was devastated. Death was no stranger to me at this point having seen so many of my comrades die. I had also witnessed the loss of some boyhood chums and a few uncles. But my Grandmother was the most precious human gift God had given me. That I would not ever see her again tore at me and for one of the few times in my life to date, I sobbed. I prayed for her soul and asked the Chaplain to say a Mass for her, which he did. I had lost the most important female in my life, and I sensed I would never see her likes again.

The War Ends

Rumors increased that the war would be ending soon. Peace talks had been ongoing and neither side had taken any offensive actions on the battlefield. Strategic targets were scarce and most of our missions were close support or MiG sweeps. Emphasis on safety and lowering casualties increased as no one wanted to die in a war that was about to end. Ground forces strove to consolidate their gains and fortify their positions before the ceasefire took effect. As I flew more missions my confidence grew, and I found my dive bombing and strafing became more accurate. I was always concerned on close support missions where our troops were close to our target, that our aim was dead on. (Often the enemy would try to confuse us by setting off their own phosphorous flares similar to those used by our FACs to mark the target.)

On July 23, 1953, it was the last day of hostilities in Korea. The 8th Fighter Bomber Wing was ready to go all out to inflict the most damage we could on the enemy. On that day I flew three missions. All were close support as the ground forces fought bitterly to secure as much ground as they could from each other. It was the most hectic day of combat flying I ever witnessed. F-86s were taking off and landing from dawn to sunset. There was no time to eat, and little to debrief the mission from which you just returned. The skies above North Korea were crowded with fighter aircraft headed in all directions, at all altitudes, and the radio traffic was jammed with all manner of combat communications including MAYDAY calls from pilots who had been hit by anti-aircraft fire or MiGs. One desperate pilot jammed the emergency channel with a continuous plea for help that became so distractive, he was finally told by some old hand to shut up "and die like a man." When the day ended and the war concluded, celebrations on base began but were rather muted. Those of us still alive felt good to be so, but it was a war the USA did not win. The enemy still occupied the ground we took, and things in the Korean peninsula would not be the same for those who lived in the north again. I concluded my combat tour with 25 missions and 30 some odd combat hours.

Post War

The end of the war brought huge changes. The military now came under the control of the United Nations who oversaw everything including our armament. Each side was required to have their weapons counted and not permitted to bring into the country any new ones. UN blue hats were seen everywhere around the base, doing their tasks and generally making a pest of themselves. Aircrews who had accrued 50 missions or more were put on the list for rotation back to the states as well as enlisted who had been in country a year. Replacements in bound had their orders cancelled. Shortly after the ceasefire the 80th FBS was rotated to Tsuiki AB in Japan. Tsuiki was our depot base where we flew our Sabres for their 50 hour and other major inspections. It was located on the southernmost island of Japan, Kyushu, not far from the city of Fukuoka. It was a small base that had a runway composed of steel pierced planking (PCP). However, it was a lot more comfortable than K-13 in that it had decent living quarters, a club with good food and entertainment. We stayed there a few weeks before rotating back to Suwon to pick up alert duty and training. It became soon clear that we were now facing severe personnel shortages especially in maintenance and armament. As winter approached, aircrews were now required to refuel their aircraft and assist in the arming. It was during these months that my appreciation for the work of the ground crews expanded and never wavered throughout my flying career. I also picked up a host of additional duties. Gradually I became A Flight Commander, Squadron Adjutant, and Assistant Operations Officer. I also became the squadron Instrument Instructor Pilot in both the T-33 and the F-86.

Christmas 1953

Christmas 1953 was difficult to get through. It was a cold snowy, winter and there was not much going on at K-13. There were few days we could fly, although we sat alert daily. When we did fly, the flights were routine and rarely were we able to get much training in. I recall one time I took my flight on a cross-country trip around South Korea. The weather was not great, and we flew above an overcast all the way around a round robin course of approximately 400 miles. Returning to K-13, our flight made instrument approaches (GCAs) to the field and landed without incident. But the Wing Commander had a conniption. He was frightened that some of his aircraft were actually flying in inclement weather, and he wanted to know who had authorized this flight. Tom McElmurry said he had authorized it and saw no reason not to have done so. As we were already on the ground safely, the Wing CO could not say much more about it.

On Christmas Eve, a USO Show came to K-13 and entertained the troops. The star was Dick Contino, an accordionist, who had been a great radio favorite, but whose fame was in decline. (He had fled induction and had served six months in jail for desertion. However, he put on a great show and was very generous. After the show, he came over to A Flight's hooch and spent the night with us, drinking and singing. It wasn't the worst way to spend a Christmas away from those you love. I was with the guys I fly with and live with and depend on. I was doing the thing I volunteered to do and serving my country. Tomorrow I would go to Mass and receive the Body and Blood of Our Savior. I survived the war and, God willing, I will survive Korea and see my loved ones again soon. Merry Christmas!

Japan

Our CO, George Halliwell rotated, and we got some new personnel. Tom McElmurry was our ops officer and he was a great role model for me. During a slack period, he and I took a week off and went to Tokyo on an R&R. We had a great time sightseeing. We stayed at the Tokyo Electric building, one of the buildings that the Americans occupied since the end of WWII. There we had everything we needed; a great restaurant, entertainment, a central location from which to see Tokyo. We even took a few days off and took a train to a Japanese ski resort, where the two of us had a ball floundering around the slopes trying to ski to the delight of the Japanese onlookers. All in all, it was a great chance to get away from the dreariness of Korea and it lifted our spirits.

My War Ends

The end of the war brought many different reactions. Some pilots were upset that they were unable to fly more missions against the communist forces of the north. The interceptor pilots, all of whom aspired to become Aces, were truly disappointed. With their mission, casualties were few and air victory was a path to promotion and fame. But to those of us in the business of low-level strafing and bombing against an enemy that was firing back at us, there was relief that the conflict was over and our chances of surviving and going home greatly enhanced. Looking back at my combat missions, as few as they were, I realize how much impact they had on me. Each was a command to drop death and destruction on the enemy, who in turn was called on to execute death and destruction on me and my plane. Each mission is a duel of death. My religious faith was most helpful. While many of my missions were uneventful, some were extremely hazardous, and I saw wingmen hit by enemy fire and die. I saw some fly into the ground and explode. I never flew a mission that I did not pray beforehand, that God watch over me and instill in me the courage to do my duty as I was trained to do.

Disappointing News

Col. Halliwell took command of an F-86 squadron at Stewart AFB, NY. Stewart was about 60 miles north of NYC and I asked Halliwell if he could get me assigned to his squadron. He told me he would work on that and he assured me he could fix it for me. I was really excited about getting to Stewart, close to home, and working again for Halliwell and flying the Sabres. But in March, I received orders sending me to McChord AFB, in Tacoma, Washington. That was about as far from home as the Air Force could send me. What had gone wrong? I wrote Halliwell and he replied that he would get my orders changed. So, I remained hopeful. But, alas, it never came to pass.

I Had a Guardian Angel on My Wing

Little did I know that when I returned from my tour in Korea, I would not fly another mission during my military service where I would fire against enemy combatants or be fired at. Nevertheless, I would go on to fly over 4,000 more hours in some of the world's fastest and most advanced fighter aircraft ever built. Still, I had come through my first test against an enemy determined to take over our ally, and we repelled them successfully if not totally defeating them as we could have. In doing so I certainly learned a lot about myself and came away much relieved that I had comported myself well and had survived. I also knew that my survival could not be attributed entirely to my overwhelming piloting skills. I also had the support and prayers of many others which contributed to my welfare. Not least were the prayers of my future spouse Alice Marie Curry. Shortly after arriving in Korea, I received a letter from Alice with an enclosure that indicated a Mass was being offered for me by *The Franciscan Fathers* at the Church of St. Francis of Assisi in New York City. Also was the Lord's Blessing and a note from Alice. I retained that note and a copy of it is seen at the bottom of this page. I am certain that I am here today—despite the thousands of opportunities during my flying days to not be—because of my wife's prayers.

The Franciscan Fathers
have been requested to offer
The Holy Sacrifice
of the Mass
for the
spiritual and temporal welfare
of
Eugene Cunneely.

Church of St. Francis of Assisi
135 West Thirty-first Street
New York 1, N. Y.

Dear Gene,

May the Lord bless thee
and keep thee;
May He show His face to thee
and have mercy upon thee;
May He turn His countenance
to thee, and give thee peace.
May the Lord ✠ bless thee:
Amen.

THE BLESSING OF ST. FRANCIS.

& Alyce

P.S. If you should see a vision of your Guardian Angel on the wing of your Sabrejet you'll know why.

My letter from K-13, Suwon, Korea to Alice Marie dated 28 April 1953 after receiving news of the death of my grandmother, Mary Cunneely:

Alice,

Your letters awaited my return from K-46 Monday. There were also two letters from Dad, the later one postmarked the 13th of April. I read the 4 letters in their chronological series. The shock delivered by your last letter is still with me. It will be long with me!

Though Dad wrote that Grandma was still weak and swollen in the limbs, there was no hint of urgency. Apparently, he himself became too stunned or overwrought to notify me properly. Of course, there was nothing for me to do - if her coma came about suddenly or if she was unable to communicate, my presence would have been to no avail. It is torturous to think about how distant I was when she may have needed me most. But it might have been worse to watch her go with no means of aiding or warding off the inevitable.

Mary Cunneely was the greatest person I have ever known or expect to know. Her inner strength, her wisdom, her faith, the beauty of her soul, her indomitable courage never wavered despite the numerous afflictions life gave her. As her life span lengthened, so also did torments of mind and body multiply. In spite of the true greatness of her character and the warm goodness of her very being, she was one of life's tragedies-"a flower born to blush unseen" -but unforgettable to the few who knew her well.

She was the only mother I knew, fate's phenomenal substitute. No real mother ever gave more to her sons than she- was more selfless, understanding, and tender. This you know. Though I knew that I would never see her again on departure from home, as you say, when the news arrives, so also comes dismay. Last night, I wept, I hadn't done so since I was 10.

I know tonight she is with the Lord and the angels-surely so, for that's where one like her could only arrive. But she is also here in my heart.

Thanks for your letter and I will write again soon when things fall back into place.

Till then.

Love and gratitude,

Gene

CHAPTER THIRTEEN:
BACK HOME AND TO MY NEW ASSIGNMENT

Home is the place where, when you go there,
they have to take you in.

-Robert Frost

Heading back to the states from Korea was such a great experience. A friend of mine and I took a "bump" at Hickam and spent a few days, free of charge in Honolulu i.e., not charged against my leave. It was great but it rained much of the time. In San Francisco, I checked into the St. Francis Hotel for a few nights of luxury living before proceeding on to New York. It was so great being back in the USA after 12 months in the Far East. Arriving in NY was super except for the bad news that my Aunt Mae had died while I was in transit. Mae was the mother of my two cousins Lou and Jim and was very close to our family. Of course, our home was different as my grandmother was no longer alive. Her absence was deeply noticeable, and I visited her grave. During my leave, I saw Alice Marie several times and we had some good dates. Alice was a faithful correspondent to me all the time I was in Korea, often sending me goodies, books and, I know, saying prayers for my safe return. I was pleased to know that Alice was still unattached. She was still working at Harpers, and I found her to be more attractive and pleasing to be with than ever. I certainly did not have any other women in my life, and due to the circumstances, I had been through in the past two years, I had not given much thought or had much time to think about establishing a steady relationship with anyone. And now I was about to leave again on another assignment that looked like it would be far from home and Alice Marie. Things were just not falling in place for making any promises.

During my stay, Dad and I flew to Miami for a week's stay there at the Robert Richter Hotel right on the beach. It was good to spend some time with Dad. His mother's death had affected him deeply and I knew he had worried about me. We had a good time together. He did not like the thoughts of my leaving again and certainly not to Tacoma. He spoke to me several times about resigning from the Air Force and coming back to civilian life. He thought I should consider law school. I had to remind him that my commitment to the Air Force still had over two more years to go because of my completion of pilot training. My brother Joe had now become a New York City police officer and his wife Patricia was working for Prudential Life. They were living in Jackson Heights and doing well. I believe Dad wanted me to settle down in NY like them.

On our flight down to Miami on National Airlines, I wore my Air Force khaki uniform. A few hours into the flight, the stewardess came to me and told me that the Captain invited me up to the flight cockpit. When I entered, I found the pilot and co-pilot enjoying a game of cards and the aircraft was on autopilot. It was a beautiful clear day. I sat in the co-pilot's seat and the three of us chatted. They were interested in my Korean experiences and what it was like flying jets. (We were in a prop C-54). After a while, the Captain asked if I wanted to fly the bird. I said sure, and I did. For about an hour I flew National Airlines to Miami with a whole bunch of passengers in the back including Dad. They never knew. When I returned to my seat and told Dad what I had been up to, he couldn't believe it and just couldn't stop laughing. One thing I learned from that experience stuck with me: Becoming an airline pilot was not something I would ever consider doing. Just flying a big old thing from A to B and then from B back to A was not for me. Like driving a bus.

Dad and I had a good time together in Florida and I again appreciated what a great father he had been to my brother and me. When our mother deserted us, he made it clear that his sons would stay with him. Of course, he needed his mother to provide substantial help in the maternal department and she did that beautifully. We spoke about her at length and how we missed her and what a marvelous human being she was. Grandma had dedicated her life, and a long life it was, to her family and to her faith. And after raising her own children to adulthood, she raised her two grandchildren, my brother and me. This she did without complaint or hesitation. She was the living model of what a mother should be in every sense.

Another thing I came to realize that though I was what they called a "depression" baby, born in the year of the great stock market crash, and lived my childhood during an era of the worst years of unemployment, poverty, and national hard times, thanks to my dad and grandma, this never touched me. I never knew a day that I did not have three nourishing meals, a roof over my head, a soft bed to sleep in, and a peace filled loving home. We were by no means rich, but we did all right and better than many. We lived in apartments with only one bathroom for seven people. My grandma and aunt slept in the dining room in the same bed. Dad, my brother Joe, and I shared the same bedroom, and sometimes Uncle Frank joined us. But my dad always had a job, and so did my Aunt Anna. We always had a car and enough money for a ballgame, a movie, a birthday present and a Christmas tree and one present. God had blessed us.

My Luck Runs Out

While I was home, I was in touch with Halliwell regularly. He had greased my reassignment to Stewart through Eastern Air Defense Command and through Air Defense Command, but somebody at McChord was holding it up. He had a pilot in his squadron who had volunteered to take my place at McChord, but someone at McChord would not agree with the exchange. It was frustrating and time was running out. I had my orders to report to the 465th FIS at McChord and McChord was a good five days driving time from New York to where I was. I truly did not want to go to Tacoma, Washington and fly F-86Ds when I was so close to joining my old commander at Stewart Field, a short way up the road from home and fly the F-86Fs, a fighter I really loved and had experience in. But despite many phone calls and much pressure from Col. Halliwell, pulling all the strings he could, his efforts failed. It was a most disappointing setback for me when Halliwell called and told me that we had lost the fight to get my orders changed. Reality set in on me that I would soon be leaving home again. My time with my family had run by swiftly and thoughts of leaving were hurtful. I would also miss seeing Alice Marie.

Dating her again was especially exciting. Not having seen her in well over a year had left me in doubt about how things were between us. And now I had to leave again. On our last date I promised to keep in touch and that I would try to see her again soon. How I would do that, I was not sure at all. The next morning, I packed my meager belongings into my green Chevrolet and headed west. In 1954, there were no interstate freeways and most of the roads were two-lane. It would take me five lonely days to drive from New York to Tacoma. It was a long sad drive away from home again to an unknown place.

Air Defense Command

My drive across the country was arduous but rewarding. I was again able to visit the various geographies that make up our wondrous nation. I rolled into McChord AFB on a wet, foggy night, checked into the BOQ and hit the sack. I awoke in the morning to the sound of jet engines. The BOQ was close to the taxiway, and I could hear what sounded as jet aircraft taxiing. The fog was pretty thick and low, and I surmised that these aircraft were headed for the engine runup pad. I was wrong. I soon heard these aircraft take to the air. I then realized that I was now assigned to an all-weather squadron that sat alert 24 hours a day and flew in all kinds of weather. While I fancied myself a fairly competent instrument pilot, I had never flown in the kind of low ceiling I saw outside my BOQ window that morning. The thought occurred to me: I still had a lot to learn about flying.

At this point in my flying career, I was still a junior birdman. I had a total of 366 hours, but most of them were in the F-80 and the F-86, the latest and fastest USAF fighter aircraft. Since leaving pilot school and getting my wings, I had served in various administrative positions and command positions, including squadron adjutant, flight commander, and assistant squadron operations officer. I also had served as the squadron Instrument Training Officer in the 80th FBS. These positions provided me with additional insights as to what makes a unit tick and how everything has to fit together for an organization to meet all of its various functions and responsibilities within and without. I now felt ready to move on and bring my experience to my new organization and help it become a better one in every way possible that I could. I still had two years to go to fulfill in my commitment to the Air Force as a result of my attending flight school. I would then have to make a decision about my future life, principally whether I should make the AF a lifetime career or return to civilian life. I was close to 25 years of age. There were many options open to me, but it was up to me to decide the direction I wanted to aim at. How things would turn out in my new assignment with the 456th FIS would provide a key.

The 456th Fighter-Interceptor Squadron was housed in the back of one McChord's large hangars. The unit had 25 F-86Ds and three T-33s. The F-86D was the all-weather version of the aircraft I flew in Korea. The D had a black nose cone in its nose that housed a radar system employed to detect enemy bombers at long range. The pilot employed the E-4 fire control system to locate and lock on to the target. He then, using the fire control system, brought the fighter into a position where it would launch its 2.75 FFAR rockets at the proper range, any one of which would bring down the bomber. The F-86D also had an afterburner for greater acceleration on takeoff, climb out and closure on a fast-moving target. McChord had two squadrons on base, the other being the 317th FIS. The two squadrons were competitive and alternated alert duty on a monthly basis.

My experience with the 465th FIS was very unhappy. Thankfully, it was short. After coming from a combat squadron that was really gung-ho the 465th was bureaucratic and stodgy. I liked the Ops Officer, Major Charley Garino very much. He was a great guy, very friendly and helpful, but I was still unable to get on the flight schedule. Although I had over 250 hours in the F-86F, the squadron refused to let me fly the D. In fact, they would not let me fly the T-33 either unless there was another squadron pilot with me even though I had more flying time than the other pilot. I was very frustrated with this treatment and began to complain. Even worse, my efforts to affect a transfer to Stewart AFB was finally turned down. When I met the Colonel who refused to agree to my transfer, an old ground-pounder type, he told me he refused it because he felt that every AF type should accept the assignment given him regardless of whether he likes it, and that was that. My morale at this point was at the lowest it had been since I had entered the service. I was 3,000 miles away from home and family, in a squadron that seemed dysfunctional. The squadron did not seem interested in me or putting to use my experience. They refused to let me fly the F-86D and do my job. Going to work each day was a downer and I was unable to do anything about it. I was sitting around on the ground and wasting my time and talent. Finally, I saw a chance to make at least a temporary change. I put in for F-86D school at Perrin AFB, Texas. Major Garino agreed to it and off I went again. At the end of May 1954, after two months in the 465th, and having flown only 30 hours in the T-33 in those two months, I packed the car and drove off to Texas. It was not what I wanted, but it was a chance to begin flying fighters again.

It took me three days to drive from Washington to Texas in my green Chevy. I enjoyed the drive and a chance to see more of our great country. This time I drove through Salt Lake City, over the Rockies and down through Oklahoma. Texas was dry, flat and hot. Perrin AFB was a training base for F-86D interceptor pilots. With my experience in the F-86F, I had no trouble checking out in the D. Learning the fire control system was more of a challenge and it required many hours of simulator time and intercept missions over Texas. I had hoped to have seen the last of Texas having spent the summer of 1952 at Bryan AFB, but here I was again back in the Lone Star State. On weekends I was off to Dallas where the sights and sounds were more entertaining. Usually, I departed after my last mission on Friday and returned late Sunday night. It was at Perrin that I decided to upgrade my Chevy and traded it in for a yellow and gold Pontiac Catalina Coupe. I really loved that car and felt it better suited the lifestyle of a fighter pilot.

The F-86D

Compared to the F-86F, the D was a dog. The large unsightly nose cone in the front slowed its performance in climbs and turns. Offsetting that was the afterburner that got you off the ground in a hurry and helped you catch a fast-moving target in the air. But once you hit the afterburner, your fuel gauge dropped in a hurry. Thus, the D had limited range. Other than these disadvantages, the D could do what it was designed to do: find, intercept, and destroy incoming bombers in any kind of weather, day or night. I realized right away that I needed to sharpen up my instrument flying skills quickly to stay alive.

Those who designed the specs for the D wanted the capability for the pilot to move the throttle from idle right into afterburner on startup to permit a quick engine start and takeoff meeting ADC's five-minute alert standard. But aircrews quickly lost faith in this procedure and fell back on starting the engine using a more cautious method whereby you monitor the exhaust gas temperature (EGT) as the RPM wound up. The instructors at Perrin, who were really pushing the F-86D as a great machine, decided to demonstrate the fast start capability to the students. So, on a hot, sunny,

Texas afternoon, they gathered us all on the ramp and proceeded to show us how you could run the throttle from idle directly up into afterburner and let the automatic engine starting system take care of the acceleration. But as the engine began winding up it ran out of control and belched a huge ball of fire out the tailpipe with a bang that you could hear in Big D. We all scattered in every direction and the fire engines appeared on scene squirting foam all over the ramp. This was not a successful demonstration of the F-86D's fast start capability and from then on, we were ordered to start the aircraft using the emergency fuel system only.

Life at Perrin was like being at Flight School again. I had an instructor who determined my schedule. I attended classes and spent hours in the simulator mostly learning how to employ the E-4 Fire Control System. We lived in WW II buildings that were not air conditioned and I ate meals in the mess hall or the O club. I had no social life, except for my weekends in Dallas. There I did manage a few dates, mostly with airline stewardesses, but all of us were coming and going on differing schedules. Of course, I continued corresponding with Alice Marie, and she came down to Dallas and spent a weekend there that I enjoyed immensely. It was so good seeing her again.

I Get into Trouble

It was a hot July night, and I was scheduled for two high altitude intercept missions. The targets were usually T-33s or other F-86Ds, flown by instructors. It was a clear night, as it usually was in Texas in the summer following afternoon thunderstorms. As I broke ground and raised the landing gear, a red warning light illuminated just to the left of the radar scope. My heartbeat doubled as I rechecked to see what it was. As I feared it was the engine fire warning light. This, as you might guess, is one warning light that you fear most. But a quick check of my engine gauges revealed that nothing was wrong with the aircraft. Though I quickly throttled back, the engine RPM responded as expected. More importantly, the EGT was normal. Also, there were no other indications of any other problems with the aircraft. I quickly determined that the problem was a faulty fire warning light. This was not the first time that I had experienced a bad warning indicator light, so I decided to ignore it and continue the mission. I stuffed my scarf over the lamp to dim it, as a fire warning light is not one a pilot can dim, and the light was blinding me to the point I could not see my other instruments or the radar. On landing, I wrote in the aircraft logbook that the fire warning light came on and would not go out. I then proceeded to Ops for a briefing for my second mission. As I was about to go out to my aircraft for the second mission, all hell broke loose. In came the Chief Maintenance Officer, the Squadron Ops Officer and my instructor. They wanted to know why I flew a mission with a warning light on and I told them. My explanation did not satisfy them, and they reminded me that the Tech Order stated that when an engine fire warning light came on, the pilot should declare an emergency, and land as soon as possible. On that they were absolutely correct. But I did tell them that the pilot also, being the aircraft commander, could use his discretion and experience in some instances. In this instance, it was clear that no engine fire existed. My explanation did not satisfy them, and I received a terrible verbal chewing out. I decided I would adhere to the tech orders from here on out. Still, I was right, there was no engine on fire; if there had been, I would have been back on the ground post haste.

I Am Transferred

About a month before completion of my training, I received a call from McChord personnel. I was asked if I would agree to be transferred from the 465th FIS to the 317th FIS. The 317th was in need of some experienced pilots and my name came up because of my hours in the Sabre and that I was a First Lt. It took me less than a Manhattan minute to agree to the transfer. My dissatisfaction with the 465th organization and the way that it had treated me in the two months I was in it made my decision an easy one. Besides, the 317th's reputation was far and away better. I had met many of their pilots and I envied the esprit that their organization eschewed. I was delighted with this turn of events. Shortly after, I received a call from Major Garino. He was upset with my decision to leave the 465th and wanted to know why I agreed to it. I did not hold back in telling him why I decided to go. That did not make him any happier, but I felt better about giving him an honest answer, and I got the feeling he understood why I felt as I did. It was one of my better decisions and one I never regretted.

At the end of August, I completed my training and was declared Operationally Ready in the F-86D. I packed my car again and headed back to

Tacoma and my new assignment. The 317th CO was Major Hubert Williams. He was much more operations oriented, and I had a great deal of respect for him. He eventually had much influence on my decision to become a Regular officer in the AF and therefore make it my career. I was assigned as an assistant Flight Commander of a Flight; the Flight CO being Capt. Archie Young. From the very beginning I enjoyed the 317th and my fellow pilots. I began flying on the regular flight schedule and our flight sat air defense alert every fourth day when the 317th had the alert duty (every other month). Sitting alert was different here than it was in Korea. Pilots on five-minute status had to stay awake and be airborne within five minutes when the klaxon horn sounded. During the day this was not difficult but staying awake at two or three or four in the morning was much harder. In 1954, the Cold War was on, and air defense was big. We scrambled on any bogey that entered the ADIZ (air defense identification zone) that military ground controllers or FAA could not identify, that is, align the bogey with a flight plan of an incoming airliner or military aircraft. When scrambled, (GCI) ground control radars would direct us to intercept the unknown aircraft. If not identified before we got there, the flight leader would approach the bogey from the stern and call out the type aircraft and tail number. Usually, it was an airliner blown off course or behind schedule. When it was identified as friendly, we returned to base. The trouble was, we often were far away from base when the ID was completed and with weather as bad as it was in the Pacific Northwest, returning with sufficient fuel to perform low ceiling approaches made for some exciting stories. It was vital that all of us became as sharp as we could be in our instrument flying and in making GCAs (ground-controlled approaches). We practiced them on every mission, and it was common at the end of every training sortie to finish with two or three practice GCA runs. Staying on the center line and on glide path right down to the runway was the goal because many of our scrambles ended in returning to a field that was socked in, i.e., below minimums and in some cases, zero, zero, i.e., ceiling zero, visibility, zero. The F-86D, having such short legs, frequently left you with insufficient fuel to reach an alternate airport. Thus, you had no choice, you had to land at McChord or bail out. Flying the F-86D at McChord AFB made me a complete pilot in that I soon was able to position my aircraft exactly where I wanted it, at the exact airspeed and altitude I wanted it at all times. I became confident, but not overly so, that I could fly in any kind of weather, day or night, and bring my aircraft safely home every time. It was a skill that I had to put to use many times during my career.

Air Defense

Every fourth day our Flight took over alert for a 24-hour period. The alert hangar was situated at the south end of the field. It had eight stalls for the aircraft and in the center were the living quarters for the pilots and crew chiefs. When you relieved the previous flight, each pilot set up his aircraft by doing a walk around inspection, then placing his parachute and helmet in the cockpit. After that, no one was permitted to go near that aircraft as it was loaded with armament and ready for combat. Two aircraft were on five-minute status; they had to become airborne within five minutes whenever the klaxon horn was sounded. Aircrews on five-minute status were not permitted to sleep. Those pilots on the midnight to four AM shift were hard pressed to stay awake. The sound of that klaxon was a dreadful one that brought a surge of heart pounding and nerve shattering energy that almost left you incapacitated. You fled to the staircase, (some alert hangars had a fireman's pole) and descended in a hell-bent fashion, three or four steps at a time, and raced to your aircraft. As soon as you climbed into your cockpit, the crew chief, who was always there before you, helped you strap in while you started the engine.

ADC aircraft scrambled were given the highest priority for takeoff and departure clearance. The theory was you were never sure what the unidentified aircraft could be, a friendly aircraft off course or an incoming enemy bomber intent on dropping its load and the commencement of World War III. Many times, the target was identified before you arrive at the point of interception, in which case you often stayed airborne and did some training. More often, when you completed the intercept, it was some airliner off course or behind in its scheduled arrival time at the ADIZ. Sometimes it was a military plane with the same problem.

When intercepting, you picked up the 'bogey' on your radar and closed in on it, usually from the six o'clock position, until you had visual contact. Then the leader would identify the plane by type and tail number. When ground control ascertained that the aircraft was friendly, you would break off and head home. Airborne intercepts were usually routine when the aircraft was in the clear, but when the aircraft was in the clouds, the entire approach had to be made using airborne radar. This had to be done slowly and carefully with some vertical separation to avoid collision. Nighttime intercepts were also tricky. Many an airline pilot was startled to see an F-86D pass by his right wing at 35,000 feet in the clouds and wondered what the hell was that all about.

These were the years of the Cold War between the Soviet Union and the western nations led by the USA. Both sides had nuclear capabilities that could be delivered by long-range bomber forces and or ballistic missiles. To counter both sides had air defenses consisting of long and short-range radar that could detect, and track missile launches and bomber penetrations. In the 50's, the US had over 50 air defense squadrons on alert from Alaska, across the northern mainland and into Greenland and Iceland. Canada also had air defense units on alert to meet this threat. But there existed no capability to shoot down incoming missiles aimed against American military and civilian targets. All we could do was provide warning to our population from North American Air Defense (NORAD) Command that we were under attack. The Cold War continued for decades despite an endless series of agreements calling for nuclear disarmament and variations thereof, most of which were fruitless. Only when the USA, under President Reagan began to go forward with its anti-missile system, dubiously named 'Star Wars', did the Soviet Union give up its ambitions to compete with the West and the Cold War ended.

But in the 1950's, air defense and the need to maintain a capability to take a first strike by the Soviet Union and retaliate in an overwhelming fashion was a primary military mission that was costly in terms of manpower and resources. Each night I sat alert, I would look out from the alert hangar window and wonder if this would be the night, would this be the beginning of what many people believed would be the worst and most deadly war ever fought. And would I be a part of it? Would I be up to it, and would I do my very best to defend my country and those I love from a surprise attack on our way of life? Certainly, my chance of survival was slim should such a catastrophic occurrence take place, but that is what I had signed up to do, what I had been trained to do, and what I believed in. The Strategic Air Command's bomber force and intercontinental missiles would then expect retaliation. The results would be worldwide mass destruction and deaths into the millions. It was not a pleasant thing to think about nor live with. But those were the times we lived in.

The 317th FIS

I was really happy with the switch to the 317th. It was by far, a much better organization than the 465th. From the CO on down, this outfit was operationally sound. Morale was high and we generated more sorties and more successful sorties than our sister squadron on every exercise. Shortly after my return, our squadron was deployed to Yuma, Arizona for gunnery. We deployed for 30 days during which we flew missions over the desert south of Yuma and fired rockets at a target towed by B-26s. For safety, each fighter was chased by a second who made certain that the firing aircraft was locked on to the target and not on the tow ship. The pilot could not squeeze the trigger until he was "cleared to fire" by the chase aircraft. The deployment was a great time for the squadron because we flew exciting missions then headed to the golf course or the officer's club bar or downtown for Mexican food etc. The 317th set a new ADC record for hits on target and we returned home in triumph.

Up to this time I was living in the BOQ next to the McChord Officer's Club. It was a convenient place to reside, but it had some disadvantages. Ninety percent of the fighter pilots on base were single and when they were not on alert or night flying, they ate and drank at the Club. Also on base was a MATS C-124 Group that flew the huge "Old Shakeys" back and forth across the Pacific. These missions usually took about 10 days to two weeks to complete. Unfortunately, some of the MATS wives began to drop by the Club while their husbands were away and fraternize with the fighter jocks. As you can guess, some bad scenes took place. The MATS commander called the wives together to read them the riot act and told them to stay out of the Club unless they were with their hubbies. I decided to find living quarters off the base and so I moved into a two-bedroom apartment in Lakewood with a Captain who was with the weather squadron. I found this more suitable as it provided more privacy and peaceful nights for sleeping.

I Become an IP

Something happened now that really had an impact on my flying career. Gen. Puryear, Commander of the 25th Air Division at McChord wanted all of his rated staff officers to be jet qualified. These officers, mostly Majors and Lt. Cols, had flown prop planes, in some cases fighters, but had never been checked out in jet aircraft. Some were still flying the Gooney Bird

(C-47) to get their flying time in. Puryear thought that they would be better qualified in their staff positions if they became familiar with jet flying. The base had two T-33s and two F-80s that were primarily used for non-squadron personnel to get their flying hours in. What they needed was an instructor pilot to check these staff officers out and my name came up. At that time, I had close to 600 hours total time, all of it in jet fighter aircraft, and much of it in T-33 and F-80 aircraft. I was also a certified instructor pilot in the T-33. So, I was given the job, in addition to my flying duties with the 317th. What I did was schedule these four aircraft and begin the process of checking out these old heads in the T-33. Many of these pilots were not happy about having to checkout in jets at this stage of their career. They let me know it by frequently not showing up for their scheduled flights. While some transitioned pretty well, others did not and gave me some very scary moments. Though I was a First Lt., I was the IP, and the aircraft commander, even if the pilot in the front seat was several grades higher. Some did not appreciate it when I had to take the stick away from them or give them a stern debriefing about their shortcomings. Some complained to their bosses about me, but I was always backed up. At least a few of them admitted that I had kept them alive and that they would have killed themselves had they been alone. But in the long run, it was I who benefited the most from this experience. Being an instructor is a truly rewarding learning experience. You have to stay ahead of the student and expect that he will do something totally unexpected and stupid. They often do. Being prepared was what saved them and me and the airplane. The result was that I became a better all-around pilot. My confidence was very high because I had to handle so many life- threatening situations and bring the aircraft back to base unharmed. If a pilot lacks confidence that he can put his plane in the exact position, attitude, and altitude that it must be in, then he will get into trouble soon. I also learned to detect weaknesses and strengths in my student pilots, and I used that knowledge to help them work on their weaknesses and overcome their lack of self-confidence so that they could fly alone safely. All in all, this was a good experience for me, but it got me into trouble. This job required me to report to base ops every day to brief my student pilots and get them off on time and safely. Juggling my instructor duties at Base Ops with my squadron duties involved many conflicts and before long the 317th squadron ops officer began to complain about it. His complaints to the Division went unheeded, so then he came after me.

I tried to explain that the Division was pressuring me to meet their schedule of moving these officers through the checkout schedule as quickly as possible, and I was already flying weekends to keep up. But the ops officer, Major Porter, was really steaming about it. I finally had to tell the Division that they should find another instructor to take over my duties, which they eventually did. But in the months of October through December 1954, I was one busy guy. And I really racked the flying hours, frequently flying cross-country flights on the weekends to New York, Las Vegas, Dallas, and San Francisco. The F-80 was especially good for long distance flights as it carried extra fuel, having no rear cockpit. I used to file a flight plan to Grand Forks AFB, South Dakota. About 200 miles out of Grand Forks, I would check fuel status and weather ahead. If all looked good, I would refile in the air and extend my flight to Sioux City, Iowa. Putting that extra distance on my first leg allowed me to fly the second leg into Mitchel Field, on Long Island. I would then have a nice weekend at home, visit with my family and see my first girlfriend Alice Marie.

November 9, 1954

Col. George Thomas was scheduled for a night flight along with Major Green in one of the T-33s. But late in the afternoon, Green called and canceled. Sgt. Smith was unable to get a replacement and so it appeared that Thomas might have to fly solo. Thomas had been a slow student for me. He handled the aircraft in an acceptable manner but did seem slow in some of the maneuvers especially in the landing pattern. He had completed a night checkout with me, and he was cleared to fly solo on this mission. I waited in ops until he showed up and I gave him a short rundown on the mission. He seemed disappointed that Green would not be with him but decided he would fly without him. I began to worry about Thomas going alone. It was a typical McChord night, overcast with light rain, although visibility was pretty good. I volunteered to fly with him. Thomas said I did not have to do that, that he would be okay and began to suit up for the flight. I was about to leave ops when something inside said to me, don't let him go alone. I told Thomas I would go with him.

With Thomas in the front seat, we took off and climbed up through the clouds, breaking out on top at 25,000 feet. It was a dark night, no moon, but there was a horizon, and we began to bore some circles. We checked in with GCI and volunteered to be a target for any interceptors that became airborne, but there was no action for us. So, we tooled around for an hour and a half, listened to the radio, and then prepared for our approach and landing. I was going to have Thomas do some night touch and go landings, as we had the fuel, and it would be excellent practice for him. We made a teardrop letdown off the McChord beacon, broke out at about 2,000 feet and headed into the field for landing. Thomas decided to make an overhead pattern approach. His entry onto the initial and downwind legs was fine and I felt very good that my student was doing okay and that my concerns had been misguided. Then as he turned from base leg to final, I noticed that he aligned the aircraft with the left side of the runway. Peeking out from the back cockpit, I could see the runway lights on the left edge of the runway disappear under the nose of the aircraft. We were still a few hundred feet in the air but descending swiftly. Finally, I told Thomas to "center the aircraft, we are too far left." Thomas did not acknowledge my call, nor did he take action to put the T-33 on the centerline. This time I shouted into the intercom, "you're landing off the runway." Still nothing. By this time, we were close to touchdown. I had no choice. I shoved the throttle forward and grabbed the stick.

Thomas gave no resistance to my actions, and I completed the go-around. We could not have been more than a few feet from hitting the ground and disaster. When my heart finally slowed down, I asked Thomas what in hell was he doing; he answered that he saw no reason why I took control of the aircraft. He denied that he was lined up with the left edge of the runway and felt that everything was under control. He was clearly upset with my intervention. We came around again, and this time I landed the aircraft. We had quite a debriefing and I made out a report and sent it to 25 Air Division. I recommended another instructor give this pilot a flight examination to determine his safety. They did and the Colonel was sent back to C-47s. Whenever I think of that flight, I come away with mixed emotions. If I had not flown with Thomas, would he have crashed, as he surely was about to do when I took the controls from him? Or would he have caught himself in time? Then, wasn't it better that this Colonel, in the twilight of his career, not get himself killed flying jets in which he would never be asked to fight again? While Col. Thomas did not like me and did not like what I had done, I know I saved him that night from something very bad. It was a very close call.

PART III: ALICE MARIE CURRY BECOMES MY WIFE— THE CUNNEELY FAMILY GROWS

CHAPTER FOURTEEN: I GO TO IWIS, I COME BACK MARRIED

Back to full time flying duties in the 317th, I really enjoyed what I was doing. I was now an assistant flight commander, was living off base in Lakewood, and enjoying the Pacific Northwest. I truly loved flying over this great landscape that included the Cascade and Olympic Mountain Ranges, the blue Pacific Ocean with its wide beaches, the unending evergreen forests and the snow covered glacial Canadian Rockies. I began to think less and less about my earlier days in New York and I totally laid aside my bitterness at missing out on the assignment to Stewart AFB, NY. I made many friends with my fellow pilots, and I began to play golf fairly regularly on the many courses in the Tacoma area that were open and green all year. I was getting lots of flying time and in both the F-86D and the T-Bird and at least one cross-country a month to a favorite destination. After the turmoil of the previous few years, life was beginning to be good. Also Col. Williams, our CO, was pressing me to apply for a Regular Commission, i.e., become a career officer. I was undecided about that but giving it some thought.

In June 1955, I was selected to attend the Interceptor Weapons Instructor School at Tyndall AFB, Panama City, Florida. This was quite an honor as IWIS grads were the cream of the crop. Their job, when returning to the squadron, was to impart their knowledge to the rest of the aircrews and oversee their abilities to utilize the aircraft's complicated armament systems. Being single and a fairly experienced new head, I was a logical choice for this training. I was pleased to go. And so, I was off again for a four-month TDY to Florida.

On the Road Again

At the end of June 1955, I headed southeast in my Pontiac Catalina Coupe for Florida, a journey of close to 3,500 miles. I headed across the state of Washington and over the Columbia River, by barge no less, and down through Idaho and into Salt Lake country. Crossing the Rockies, I then headed south to Texas to visit some old friends for a few days, then on through Louisiana and into the Florida Panhandle. Tyndall was an old base, and the housing was poor, no air-conditioning yet, just window fans. I roomed with two other guys who would be in my class at IWIS. It was already very hot and humid. The classes were under whelming and some of the older heads in the class, I include myself, thought that the missions were not well organized and planned. The instructors were good heads, but those in charge were truly scared of having an accident. Thus, they cancelled many missions for weather worries when there did not seem a real need to do so. Some of us who had been flying real air defense missions in the field in all kinds of weather, especially in places like McChord where the weather was always a factor, did not see what the problem was. Consequently, we began to fall behind in the schedule. Nevertheless, I enjoyed the training and did not miss sitting alert all that much. It was a five-day week schedule with a half-day of classroom. We had weekends off and sometimes I finagled a cross-country trip. One weekend, I was able to get a T-33 and fly to New York where I had a date with Alice Marie, whom I had not seen in some time. Although we corresponded frequently, the distance between us was quite far and it was not easy getting all the way across the country that often. Seeing her again awakened me to the fact that my social life was not all that great. I had been a busy guy the past few years with lots of travel, changing assignments, and building up my flying time. I also was reaching a point when I had to make a decision about my future. My commitment to the Air Force would be coming to an end in October 1956. Being a Reserve Officer, the AF could deactivate me any time it wanted. The Korean War having ended, Congress was carefully scrutinizing the military budgets. It was time for me to decide what I wanted to do with the rest of my life. And soon.

The IWIS flying program was not as exciting as I had hoped, but it did have its moments.

July 27, 1955

We were scheduled for a night intercept mission over the Gulf against low level targets, T-33s with chaff. The weather was forecast to be clear, visibility good. Our flight of four took off and climbed on a southerly direction over the Gulf. We were then picked up by ground radar and directed to the target area where we descended and began our intercepts against the incoming targets. As we descended, we noticed that not far off the coast of Panama City was fog bank lying over the water and extending as far out as we could see. We continued with our intercepts for about 30 minutes when GCI informed us that the fog was now approaching the base and that all aircraft were directed to return to base as soon as possible. This required us to break off our intercepts and climb to 20,000 feet and head for Tyndall which was about 100 miles to the north. But we were now informed that visibility at the base had deteriorated rapidly. Tyndall was now calling a ceiling of 500 feet and visibility of one-half mile. This was not below minimums for a GCA approach; however, the GCA was not up and running. It sees that everyone was caught by surprise at the faulty weather forecast. By the time we reached Tyndall, the base was socked in and there was a bunch of aircraft ahead of us trying to get down.

Several were in deep trouble and had made missed approaches. One T-33 in particular, crewed by inexperienced pilots began to show signs of panic as they failed to find the runway and were running out of fuel. While this was going on our flight was in a holding pattern above the base awaiting clearance to letdown and land. But time began to run out on us too as fuel became a problem. GCA still had not come on. To land, pilots had to make a letdown from the Tyndall beacon, hit the outer marker at 2,000 feet and letdown on the runway heading until they broke out under the cloud deck. If the pilot did not see the runway at 200 feet, which was the legal minimum altitude, they had to make a go around. But some pilots did not press to 200 feet and started their go around sooner. Now the traffic pattern was filling up with aircraft trying to get down and running low on gas, including our flight of four. More and more aircraft were declaring minimum fuel or emergency fuel, and some started to headed for alternate airports further inland away from the fog.

Our flight now had serious low fuel problems and had to get down soon. As flight leader, I relayed this to Tyndall approach control. We were holding on the Tyndall beacon at 20,000 feet. Looking down, you could see the lights of Panama City through the low cloud cover, but there was very little horizontal visibility. We were instructed to begin our approach in trail with one-interval between aircraft. On the basis of fuel remaining, I was number two to begin letdown. When beginning my straight in approach to the runway from the outer marker at 2,000 feet, the number one aircraft made a go-around. I began my letdown into the soup at 300 feet per minute and holding steady on the runway heading. Entering the fog, everything outside went black. I held my course and pressed on. I was down to less than 500 pounds of fuel, not enough to make a go-around and land anywhere else. Besides, there was a bunch of planes behind me trying desperately to land on this runway or punch out. I kept cross checking for the runway lights. No sign of them at 1,000 feet, nor at 500 feet. Everything outside was still black. I held the course steady and pressed ahead. I was determined to keep descending until I saw the lights: 400 feet, 300 feet, still nothing. Now minimums, 200 feet. Was that a light off to the right? I kept pressing, I had to or else go back up to God knows what. Yes, it was a light, a runway light about 50 feet to my right. I was too much off to the left side to swerve and land and only about 100 feet in the air at 150 knots. I told the tower I had the runway, and I would be making a 360 for a landing. The tower operator gave me a Roger then told me of another aircraft approaching about a mile out. I pushed the throttle forward to the stop, retracted the landing gear, and banked into a 90-degree left turn only about 50 to 70 feet above the ramp, no doubt scaring the wits out of the tower personnel and anyone else on the ramp. The nose of the aircraft kept rising into the clouds and I lost sight of the field, the runway and everything. I then had to push the nose back down until I could again gain sight of the runway, all the time in this wild turning maneuver that I hoped would eventually take me back to a spot off the end of the runway where I could line the aircraft up and land. When I finally finished turning, I pushed the nose over and broke out at about 100 feet altitude, but also about 200 feet to the right of the runway. I dropped the speed brakes, the landing gear, put on the landing lights, and with a 60-degree banking turn, lined up just before touchdown. It was the wildest landing I had ever made, bar none, but one that I had to make or suffer the worst consequences. Looking up I saw an F-86 on the runway in front of me so I hit the brakes hard so as not to overtake him. The airwaves were crackling with pilots calling for emergency landing instructions and some breaking off and heading for alternate airfields. But I was on good old terra firma after a wild ride in and out of a fog bank, a 360-degree banking turn at 100 feet altitude, all the way flirting with a low altitude stall out, and a frantic touchdown on the side of the runway, leveling the wings just an instant before touching down. My heart was still pounding rapidly until I was off the runway and in the chocks. Tyndall lost a T- 33 that night and two pilots who apparently flew into the ground on the west side of Panama City. It was a needless waste of life. The decision to launch aircraft into the night sky, with a fog bank lying so close offshore, without all approach facilities up and running should have resulted in a court martial. It did not. It was a flight I will not forget. But it was a sortie that I learned from. A smart pilot learns from every flight he flies and every mistake he makes or perceives. He stashes it away in his memory bank and sometime later, he will recall it and put what he learned to use. It will save his life.

Alice Marie Curry

As my time at IWIS passed the midway point, I began to think seriously about my future plans. I had signed up to take the test for a Regular Commission, at the strong urging of Major Williams, my CO, and in early August I was called up to Wing Headquarters at Tyndall to take the test. It was one of those three-hour multiple-choice exams that covered many subjects, and I took the test half-heartedly with a sergeant sitting there as proctor. But I realized I had to make a decision if I passed the test, and that decision would decide what the rest of my life would look like. My dad still hoped that I would return to NY and enter law school. My dreams of teaching and coaching basketball were now a little outdated. I was now beginning to yearn to at least settle down from all the traveling I had been subjected to. My social life was almost nil other than a few scattered dates here and there. I had been on the move so much and spending long days and many weekends flying, that I had few permanent close friends outside of the unit. My personal possessions fit into the trunk of my car. And my bank account was not overwhelming either. The airlines were a possibility as they were hiring plenty of pilots, especially those who had jet experience. Having been in one of the early jet classes and with over 700 hours of jet fighter time, the airlines offered a good opportunity. But airline flying held very little appeal to me. I loved flying but flying big airplanes from point A to point B, seemed awfully dull to me. I loved the fighters, and if I stayed flying, it would be fighter aircraft that I would fly first and only.

Throughout my years in the Air Force, since leaving NY, I had continued corresponding with Alice Marie. Whenever I got back to NY on leave or on a weekend trip, we dated. Fortunately, (for me) she was still single and was working as a secretary at Harpers Publishers. I always enjoyed our dates. Alice, in addition to being a very good-looking woman, was just easy to get along with. We had similar interests in music, books, theater, sports, and values. We were both Catholic. I began to realize that while I was making up my mind about my future career, I should also be thinking about marriage. And when thinking about marriage, I began to realize that Alice

Marie Curry would be a first choice and a great choice. That is, of course, if she would say yes.

I finally made my move. In mid-August, I called Alice and asked her if she would come down to Panama City in September and marry me. She was clearly taken by surprise and said that she could not do that. I told her to think about it. After hanging up, I was quite downhearted. I was unsure whether her answer meant that she would not marry me, or that she would not come down to Panama City and marry me. I felt that I had botched the proposal. However, about ten days later, we talked again. Alice had talked to her mother, and she told me that she would do it. I was very happy. I immediately went over to the Base Exchange and bought an engagement and wedding ring combination and mailed it to my fiancée. I had made another big decision in my life. But I felt good about it. I realized how much I loved Alice, and I knew what a great wife she would be for me. Whatever life had in store for us, having her by my side would be a great help in overcoming problems and bringing peace and stability into what had up to now had been a rocky and turbulent existence. I had done the right thing at age 26.

Completing IWIS

After becoming engaged, I realized that I still had a lot of things to get done. I had over a month to go in the IWIS and I had no idea or firm plans on what Alice, and I would do after we married, and I returned to McChord. We had nowhere to live, and I certainly was not making a lot of money and had little savings. I also had to make arrangements for our wedding and what it took to get married in Bay County, Florida.

I had another frightening episode in late August. We were scheduled for a mid-morning flight over the Gulf to intercept and fire rockets at a target towed by a B-25. The forecast was for thunderstorms in the area south of the base, but clear in the area of our intercept work. We took off in trail formation, that is, one aircraft following another with a spacing of 60 seconds. I was the lead aircraft of a flight of four. As I broke ground to the east I turned south and entered the clouds at approximately 500 feet just as I passed over the beach area. Immediately I encountered severe turbulence, heavy rain mixed with hail, and lightning. I notified radar control of this situation to warning the aircraft following, all the time trying to keep my bird under control and hoping to climb out of this mess as soon as possible.

Then, my aircraft was struck with a lightning bolt that knocked out my instruments, my radar and God knows what else. The sound deafened me, and my heart just about stopped beating. I don't believe anything I ever experienced before in my life scared me as much as that lightning strike that lit up the entire cockpit and sounded as though some giant had struck my aircraft with a giant mallet. I held on for dear life and keeping the wings level and maintaining a climbing attitude. I called GCI and reported in the calmest voice I could muster that I had an emergency situation. I was declaring MAYDAY, and that I would be requiring an immediate vector back to base, but not through the area I had just come through. After what seemed an awfully long time I broke out into the open where I was able to assess the damage. It looked like a good part of the radome was gone. My electrical instruments were out, but my emergency UHF radio was operating. I placed my IFF on emergency and ground radar picked it up and began giving me directions back to Tyndall. By this time, my wingmen began checking in. Two of them had also suffered lightning strikes and had to return to base with severe damage. Tyndall now had three emergencies to handle. Fortunately, none of us were in a dire situation that required an immediate landing or bailout. But each of our aircraft had suffered severe damage and all of us had been frightened, as we had never been frightened before. To this day, I can recall that instant when the cockpit lit up enough to blind me and my aircraft was pounded to the point, I thought it was going to come apart. My nervous system was tested that day and passed, thank God. All of us got down okay. I could not believe what that lightning strike did to the front of my F-86D. You could look right through where the radome was and see the electronic boxes inside, many scorched black by Mother Nature's sting. After debriefing, we all went to the O club and had a few stiff drinks. It was only noon. I decided not to tell Alice about that mission.

SEPTEMBER 17, 1955 "I WILL"

As the days wound down to the end of IWIS, I began to concentrate on my forthcoming marriage. There was more to be done than I realized. I first notified my family that I would be getting married at Tyndall. I did this by calling Dad. While he was pleased to learn that I had chosen Alice Marie, he was not happy that we were to be married away from home. Next, I made arrangements with the Base Chaplain for use of the chapel and for his services. This was more complicated than I had anticipated. Father Heard needed Alice's baptismal and confirmation certificates and a letter of approval from her pastor in New York. He also needed mine which fortunately I had. We also had a problem with use of the chapel as September 17th was a Jewish holy day and we would have to be out of the chapel by nine o'clock in the morning. He wanted to see us on the night before the wedding at six o'clock sharp.

Although our families wished it to be otherwise, they both supported our plans. My dad drove Alice to the airport as she flew off to Florida on September 13th. Alice had to give two weeks' notice to her employer, Harper Bros. Publishers, tie up her affairs, say goodbye to many friends, buy a wedding dress, and say farewell to her mother and family as she embarked on this very unusual and risky venture. When her small airplane arrived safely at the Panama City airport, I was there and very happy to see her. We were together at last and would be together for many years to come.

Alice stayed in the Guest Quarters, which sounded better than they actually were. I still had a few missions to complete, and we had to appear at the Bay County Courthouse for a blood test to be taken three days before the wedding. On the Friday evening before our wedding, Alice and I went to the Club for dinner when she asked me who would be the best man and bridesmaid, I was flabbergasted. With all the details I had been working on for the wedding, I had completely overlooked the need for witnesses. In desperation, I called one of the IWIS instructors, Capt. Bill Cunningham, whom I knew was a Catholic. Luckily, I caught Bill at home. I asked Bill for a big favor. He asked me what it was. I said, "What are you and your wife doing at eight o'clock tomorrow morning?"

Bill wanted to know why I wanted to know that.

"Because," I said, "I am getting married, and I need a best man and a bridesmaid, badly."

"Are you drunk, Cunneely?" Bill came back.

"Bill, I'm serious and I need help," I pleaded. Thank God, Bill and his wife came through.

When we met with Chaplain Heard we gave him the necessary papers, went over the next morning's proceedings and then he asked to speak to each of us privately. When we each came out, we were both shaken. To me he spoke about birth control and then heard my confession. I guess he did the same with Alice. I didn't understand much of what he was talking about, so I just pressed on. At 0745 the next morning, I picked up Alice

Marie at her quarters. When she came out, she was wearing a lovely white dress with a little veil and white shoes. She was beautiful and I was feeling very proud and happy. I wore my suntans and we drove to the chapel where we met the Cunningham's. As we entered the church, there was one lady sitting in the rear praying. When Father entered, I walked.

CHAPTER FIFTEEN: I MAKE ANOTHER MAJOR DECISION

One of our first jobs was to find a place to live. Neither Alice nor I had any furniture to speak of, so we needed a furnished apartment. My salary was $400 a month (including flight pay), so our choices were limited. After a few days of searching, we found two places we liked. One was a nice place in Steilacom, about a 20-minute drive to the base that was $75. The other was a small one-bedroom house on Lake Steilacom, with a little beach front, a canoe, about 15 minutes from the field. However, this was more expensive, $100 a month. We were torn between the two, but finally did the impractical thing and took the one on the lake. It was our first home and we lived there two years and truly enjoyed it. We had great neighbors, who became life-long friends. Our memories of our days there still linger fondly and always will.

I settled into squadron work quickly. The 317th was a great unit, well lead with high morale and professionalism from top to bottom. I was proud to be part of it. In addition to being an excellent flying unit, we had wonderful spirit and strong competitive drive. Our unit always broke Air Defense Command records for most hits scored on our annual squadron gunnery deployments to Yuma. As the squadron weapons instructor officer, this was my area of responsibility. I spent hundreds of hours in the debriefing tent assessing gunnery film of our younger pilots to help them bring their techniques up to speed. I also took pride in my own firing scores as I was expected to show the way to the rest of the squadron in firing accuracy. I was expected to be a leading pilot in that area, and I always was.

At about this time I received word that I was offered a Regular Commission, having passed the test I took while I was at Tyndall. This would mean that I would never be forced out, i.e., RIF'd, by the AF whenever they decided they had too many officers. Now that I was married, I wanted a permanent job and a chance to have a successful career without the worry of being sent out into the civilian world with short notice. Unlike many of my peers, I had decided that flying airliners was not for me. I did not like big, multiengine planes, and flying from A to B and back again would be an awful boring job regardless of the money. I loved flying fighter aircraft and I wanted to continue doing that as long as I could. My ultimate goal: becoming a commander of my own fighter squadron. After talking it over with Alice Marie, I accepted a Regular Commission. I became a "lifer". Now more than ever I went about the business of being the best officer and best pilot I could be.

Looking back, these days in the 317th FIS, with my new wife in our little home on the lake, were some of the best days of my life. Each day, I realized what a great move I had made in marrying Alice Marie. We had known each other for many years but had spent many of them miles and miles apart, seeing each other infrequently, but always staying in touch. My dates with other women were never satisfactory and I never gave serious thought to marrying anyone else. When I decided to marry, I proposed to Alice Marie. Thank God she was still single and took a huge gamble in leaving her career and home and family to become an Air Force wife with all that that entailed. Each day, I looked forward to coming home to my bride and sharing our lives together. My love for her grew daily. Also, my dedication to my job also began to grow. I dedicated myself to my duties as an instructor in the weapons systems of the F-86D and in sitting air defense alert on a periodic basis with one of the flights. I also continued to perform duties as an instrument instructor pilot in the T-33. In short, I was busy at work and very happy at home. The squadron had an active social program that involved all of the wives and Alice soon got to know most of them. There were always parties and get togethers of some sort to look forward to. The pilots, of course, had their beer calls and the 317th was a big golf squadron. Almost every week, and on some days off, we played golf on any one of the several excellent courses in the Tacoma area. We played in all kinds of weather and the competition was keen and sometimes ruthless. All in all, it was a great time for us.

Air Defense Alert

During my years in ADC, I sat alert hundreds of times. From the end of WWII, the USA realized that the threat to our country and to world peace had not disappeared. With the Nazis and Japanese no longer on the scene, we now were concerned with the Soviet Union which had used the war ending treaties to expand its influence throughout eastern Europe and Asia.

It had also acquired nuclear capability along with long range bombers and then missiles to deliver atomic bombs to anywhere in the world. The Cold War had begun and with it the fear of nuclear war with the Soviet Union became the nation's major fear that dominated our foreign policies and our defense strategies. Strategic Air Command dominated the military scene with its long-range bomber force and then ICBM silos. Alongside was Air Defense Command with its fighter-interceptor squadrons, air defense identification zones, ACW sites, and distant early warning systems. In the 1950's, ADC had over 50 FIS's across the USA, plus those in Alaska, Iceland, Greenland, Canada, Labrador, and Newfoundland. The cost to the nation was huge in dollars and manpower. Pilots and radar controllers were on 24-hour alert seven days a week. Whenever I sat alert, I took the duty seriously because I never knew whether I would be scrambled and why I would be scrambled.

On alert day, I reported to the alert hangar about seven in the morning to relieve the flight coming off duty. After a weather and intelligence briefing, you went to your assigned aircraft, gave it a preflight, then set up the cockpit with your helmet, parachute, flashlight, etc. The alert hangar was situated at the end of the active runway. It was a restricted area. There were normally six to eight stalls with front and rear doors to keep out the weather. In the center space was the living areas for the crew chiefs (downstairs) and for the pilots (upstairs). The living areas contained a sitting area for reading and TV watching, and a kitchen with eating space. In the rear were the sleeping quarters. Many of the alert hangars had a fireman's pole for the pilots to quickly descend to the aircraft on a scramble. A typical day on alert would include at least one training flight, but there were always two aircraft on 5-minute status. Pilots on 5-minute status had to be awake and were usually relieved every four hours. Staying awake through the early hours of the morning was always difficult for me. The aircraft were fully armed and ready for combat in the event that a scramble would involve interception of enemy aircraft. In the years I sat alert I was scrambled many times. Most times I intercepted civilian aircraft that were off course or late in meeting their flight arrival time at the ADIZ. Sometimes I intercepted military aircraft with the same problem. I never heard a scramble klaxon horn that my heart didn't pound. Racing to my aircraft climbing in and starting up, I usually hoped that the scramble would be canceled. Air defense scrambles receive priority over all other traffic, so it all happened quickly, and from the time you were sitting in the lounge to the time you were climbing up through the clouds did not give you any time to think about anything except what you were doing to stay alive.

Weather was always a factor when making intercept. If it was a daylight or VFR intercept, no problem. But many were at night, and often the bogey was in the clouds. Then you had to approach using radar. Once you had the unknown aircraft on your radar you approached it from its six o'clock. Your wingman stayed with you but stayed low and behind for safety. As the bogey got inside of one mile you throttled back to close at a very low rate and then put your windscreen in your cross check hoping to see its lights. Once you saw the aircraft's lights you closed using formation flying techniques, trying not to get vertigo. You then called in to ground control the aircraft type, number and other ID. Usually the aircraft was aligned with a flight plan that was on file, or had been mislaid, or the aircraft had just run into headwinds and was off course or behind schedule. Scrambles were always nerve-wracking flights because of the time constraints, the unknowns, and the pressures they placed on pilots, crew chiefs, ground control etc. In 1962, I scrambled in a fully loaded (nuclear armed) F-106. This was during the Cuban missile crisis. Happily, the unknown aircraft was a United Airlines airliner.

Sitting alert was a duty I enjoyed doing, and I volunteered for it even when I was a Colonel and a squadron commander and did not have to do so. I felt it a great thing to do, akin to defending your country even though we were not at war. Just being there in case, to utilize the training and the skills I had to prevent a potential enemy from hurting the citizens of my country, was an honor. Thank God, I never had to intercept an enemy approaching our shores determined to kill our families and way of life.

During my tour with TACEVAL, I sat alert with squadrons throughout the United States, as well as Labrador and Greenland. When I visited these units to fly with them and keep up my currency in the F-102, F-106 or F-86, I always volunteered to sit alert and give one of their pilots a night off from alert duty. Even when I was stationed at North Bay, whenever I flew the F-106 with the 29th FIS at Loring AFB, Maine, which was within Northern NORAD Region's area of command, I sat alert. The squadron pilots appreciated getting a night's off from alert duty and it was my way of paying the squadron back for giving up some flying time to a "headquarters weenie." Sitting Air Defense alert through the Cold War years was expensive to the nation. Many thought that air defense was wasteful and that the Soviets would never attack the USA utilizing manned bomber especially after the entrance of the age of nuclear missiles. Nevertheless, the Soviets did have a huge bomber force which they flaunted and consistently used to approach our boarders and threaten our sovereignty. But, Air Defense Command provided a major force of experienced fighter jocks that our country needed for Vietnam and the later wars in the Middle East.

One day, I sat alert with B Flight after completing my duties as Weapons Control Officer. I reported to the alert hangar at four o'clock. I relieved one of the B Flight pilots, setting up in his aircraft with my chute and personal equipment. I then checked into the alert hangar living quarters where I reported to the Flight Commander, received a weather briefing, and paired up with another pilot who would be my wingman. We would be taking over five-minute alert status at 2000 hours, so we had supper and then sat around reading. At 1950, we received a weather warning that visibility at the field was deteriorating as fog was forming. At 2030, soon after we assumed five-minute alert status, weather informed us that that the ceiling and visibility were getting worse and may reach minimums soon. I looked out the alert center windows located about 50 yards from the end of runway 34. I could see fog forming around the runway and the blue taxiway lights. Things were rapidly getting worse. We both hoped that it would be a quiet night with no scrambles. The only alternate field open was Larson AFB located in Moses Lake, which was on the eastern side of the Cascades. To make it to Larson from McChord, you had better have at least 500 to 700 pounds of fuel left from a missed approach at McChord. The entire west side of the mountains, from British Columbia to southern Oregon, was socked in.

Then at 2120, 25th Air Division called on the red phone. They had an unknown entering the system. They were trying to identify it through FAA and military flight control. If unable they would have to scramble us. I asked the Division Senior Controller if he was aware of the weather. He said he knew the situation. By now McChord Field was at minimums, 200-foot ceiling, one-half mile visibility. I again looked out the window. Through the darkness I could barely discern the white runway lights. I was having a sinking feeling in my stomach. And then, the KLAXON horn sounded. I had heard that horn many times before, but on this night, it had never sounded so terrifying. They can't be serious! I flew down the stairs and on reaching my aircraft, I found both hangar doors rising, the APU (starting generator) running, and the crew chief standing by to strap me in. After startup, I checked in with the tower on the radio, hoping against hope, for a cancellation. Tower gave immediate takeoff clearance for our flight, the field barometer setting, and instructions to contact Mercury Control after breaking ground. By this time, our flight had entered the runway and was lined up for takeoff. I peered down the runway but could only see a few feet ahead of the nose of the aircraft. I lined up on the centerline and ran the engine up to 100 percent RPM. Everything looked okay, so I came off the brakes, hit the afterburner, kept the nosewheel lined down the centerline and let her fly. It did not take a long runway roll before I broke ground and then I was on instruments using the attitude indicator to raise the nose and keep the wings level as I rose up through the fog. The fog eerily reflected my external lights and afterburner lighting up the clouds around me as the F-86 climbed rapidly upward. In no time I broke out on top of the fog layer, and I saw above me a clear sky filled with stars. Looking down I could see the lights of Tacoma and Seattle blurred through the fog covering the Puget Sound region. My wingman soon joined up with me and checking in with Mercury we were vectored on a heading west by northwest, signifying that we would be intercepting an incoming bogey over the Pacific.

But this mission was fated to be trouble from the start. My wingman soon called in that he was experiencing engine RPM surges and had to switch to the emergency fuel system. I told him to declare an emergency and return to base, which he did. That left me alone to continue the intercept. The further north I flew the more nervous I became. I knew that weather conditions at home base were only getting worse and I really could not afford to waste fuel on a mission that ended in my intercepting a friendly aircraft off course and cause me to bail out of a perfectly good fighter aircraft due to running out of fuel. But Mercury Control still did not have ID on the bogey and now I was over Vancouver Island and still heading away from home base. It was one dark night with no lights on the ground to be seen as the fog had now completely covered the northwest including British Columbia. The bogey was now within 100 miles as I turned into my intercept heading. At 25 miles I had a radar contact and began closing. At 10 miles I had a visual contact. The aircraft was well lit and was clearly not an enemy bomber with evil intent. Then Mercury informed me that they had identified the bogey as a SAC B-47. I continued to close and passed the bomber on its port side and confirmed the aircraft type. I asked for immediate pigeons to home base. It would be close as I was over 200 miles from McChord with less than 500 pounds of fuel left. Weather at TCM was now below minimums, the nearest base open was Larson that was beyond my range. I throttled back as much as I could to preserve fuel and obtain the best fuel/range advantage. With no other field open I had no better choice than to try and make it in to McChord. I declared an emergency, which did me little good as there were no other aircraft airborne for me to get priority over. Seattle Center and Approach Control brought me on a straight in vector to the field and then gave me a long idle throttle letdown to the outer marker for a GCA approach to the runway. Many thoughts were dancing through my head. If I missed this approach, I would have to eject, but where? Landing to the north, I would have to turn the aircraft to the south and east before bailing out in hopes that the aircraft would not hit a populated area. Preferably, I would try to eject somewhere south of the base on a southerly heading. But that all depended on how much fuel was still remaining.

GCA picked me up at the outer marker, five miles out from runway 34 at 2000 feet. The voice of the final controller was calm and experienced. He had my life in his hands. He gave me the weather as 100 feet ceiling, one-fourth mile visibility, fog.

GCA: Red One, you are approaching glide path, your gear should be down.

Red One: Roger, gear down and locked.

GCA: Roger, Red One, you are cleared to land. Begin your glide path, heading 340.

GCA: Red One, you are glide path, on centerline, four miles from touch down.

GCA: Red One, on glide path, slightly left of centerline, turn starboard to 342 degrees.

GCA: Red One: on glide path, on centerline, turn port to 340 degrees, three miles from touchdown.

GCA: Red One, you are above the glide path, increase your rate of descent, on centerline, two and one-half miles from touchdown.

GCA: Red One, coming back to glide path, on centerline, one and one-half miles from touchdown, cleared to land.

GCA: Red One, on glide path, on centerline, one mile from touchdown. Approaching minimums.

(Back in the cockpit, I was concentrating on my instrument cross check as I had never done before. I was going to have to land this aircraft, coming down a glide slope at a 150 mph on a runway I would not be able to see before the wheels hit the ground. What if I missed the runway, or hit it too hard, or lost my composure, hit the throttle, and went around? My fate was unknown but probably bad. I tried to steady my nerves, stay cool and listen carefully to the GCA controller. I was determined not to look for the runway lights in my cross check until I hit the one-mile notice. Now I was there.)

GCA: Red One, on glide path, on centerline, you are now below minimums and over the end of the runway contact tower on channel one.

GCA had now lost me as I was under their radar. I still could not see the runway, but I was determined not to go around yet. I pressed on, keeping my wings level on the artificial horizon. I was now looking through the windscreen for some lights. But there were none. I knew that I had to go around before this bird smashed into the ground. My left hand went to the throttle and just as I began to go forward with it, I saw a light out to the left side, then another, and another. Then there was the runway coming up at me in a huge rush. I brought the throttle to idle and brought the nose of the aircraft up as abruptly as I could. I was committed now and, to my bewilderment, the F-86D landed in the fog. I could not see anything ahead, so I lowered the nose wheel to the runway even though I was going too fast for that. But now, at least, I could see the runway lights and the centerline markings so I could finish my rollout. I can't describe the feelings that went through me, of joy, relief, awe, mystery. I was so thrilled to be on the ground in one piece that I forgot where I was. The tower operator finally woke me up.

Tower: Red One, state your position.

Red One: Tower, I'm on the runway, but not sure of my distance as I can't see the taxiways. Also, please notify Alert Hangar that I may not have sufficient fuel to taxi back there and may need a tow.

Tower: Roger, Red One, give us a position report when you can.

The story ended with my finally finding a taxiway and getting off the active runway, which was not very active that night anyhow. My aircraft showed less than 100 pounds of fuel left, and I did shut it down before getting back to the alert hangar. I learned a lot that night about flying instruments, about the F-86D and myself. I learned mostly that in situations such as this one, and there would be more to come, knowing your aircraft, thinking clearly, making good decisions and not attempting what you are not capable of are important lessons that will help you for a long career of flying.

Life Is Good

I realized how fortunate I was to have joined the 317th FIS. We had such good people from top to bottom who liked each other and strove for excellence. They were a competitive group of guys in everything they attempted whether it was flying or golf. The 317th always broke gunnery records at Yuma besting all other ADC squadrons in a number of successful hits on target. We were the best of the best and we were proud of it. I learned a great deal in the 317th FIS that came in handy later in my career as I moved up in grade. From my time in that unit, my goal became to become a squadron commander of my own fighter unit. That goal was reached, but with unforeseen results.

Those years with the 317th were happy years for Alice and me. We had our little home on Lake Steilacom, and we acquired several wonderful friends and neighbors in the Lakewood area. It was such a great life for me because I was happy in my work, flying a lot, but each night I could come home to my new wife and enjoy home cooked meals and the company of the woman I loved dearly. Alice made the adjustment to military life with not too much difficulty. Despite being far from her home and family and all that she had known and lived with, it was clear that she would be a terrific wife and companion. I now had someone to share and confide with, and our routine included eating out once weekly (mostly at the O club) and hitting the base movies. We owned a TV, but there were not too many great shows on in those years to be enthusiastic about. On weekends that I was off duty, we drove around the region and saw the wondrous sights of the Pacific Northwest. Alice soon became as thrilled as I was with the Puget Sound area, and our lives were full. It was a great time.

In December, Alice told me that she thought that she was going to have a baby. She had missed a period, and that rarely happened she told me. Not being wise in those matters, I was pretty shocked. To be honest, it was not something I had thought about, although, of course, I should have. The doctors would not see her until she had missed more than one period and that happened the next month. I was very happy about it, but it gave me a lot to think about. I was in one of the world's most dangerous professions and I had very little insurance. And, commercial insurance companies would not sell me any because I was a very poor risk. I thought about what would happen to Alice and our child should I die. Where would they go and who would take care of them? It was a concern that haunted me for many years and grew as our family grew in number. Alice finally saw the base doctor who confirmed her pregnancy and predicted arrival of the child in the middle of June 1956.

January 13, 1956

It was Friday the 13th, and I was scheduled for a rocket-firing mission against a target drone dragged on a cable behind a B-26. The firing range was about 25 miles off the Washington coast. All maritime traffic had been warned to stay out of this area. My wingman was John Sewell. It was a cold clear day, and we took off with me in the lead and headed for the Olympic Peninsula. We were under radar control and were directed to the firing range at 24,000 feet. We were carrying 24, two-inch FFAR rockets without explosives. We both made a firing pass on the target and were being positioned for our second pass. As lead ship I entered my final approach on the target, selected my armament (six rockets), and called out contact on the tow aircraft. Then something went drastically wrong. A sudden screeching noise enveloped the cockpit, followed by an explosion in the engine area. The fire warning light illuminated, and the RPM gauge went to zero. I pulled the throttle to off and broke off the run. I called MAYDAY and began a turn back toward the shoreline. Sewell joined up next to me, and I tried to ascertain what my situation was. Clearly, I had a total engine failure, but why? Could I restart the engine? Should I try? I had to maintain glide airspeed and, more than anything else, I wanted to get back over the mainland if I had to bail out. I wanted no part of leaving this aircraft and ending up in the Pacific Ocean on a cold January day. Below I could see lots of whitecaps and high waves down there that made me concentrate on getting this jet back to the beach. Restarting the engine did not seem feasible, and I soon began to concentrate on looking for a place, to either make a dead stick landing (maybe on the beach) or finding a good place to eject. I was in radio contact with GCI and my wingman had me in sight as he followed me back to land. Because the wind was at my back, I was now over land at about nine thousand feet. Landing on the beach did not appeal to me. I saw no stretch that looked safe enough, as there were lots of driftwood and rocks sprinkled about and the wind made a straight in approach tricky without power. I decided that ejection was my only recourse. I announced that I was ejecting, and I went over the procedures in my mind. I cannot remember being panicky, but to say I was completely calm would greatly overstate it. And the worst was still to come. I got ready and pulled the right ejection handle up and the canopy flew off. I could feel the wind blowing into my cockpit and pushing my body back against the seat. Then, taking the proper position, I squeezed the trigger. But nothing happened. I squeezed it again, and still I was in the seat with the wind pushing down against the seat. What the hell do I do now? I checked my airspeed as I could not let the aircraft stall on me, and now I realized my altitude was getting lower and lower. I needed to get out of this aircraft or else I would have to ride it into the ground. I was no longer in the beach area as I had drifted inland over the heavily wooded forested area common on the Olympic Peninsula. I then tried to climb out of the cockpit, but that effort failed as the incoming wind pushed me back down into the seat. I was now at or below four thousand feet. I then, somewhat in desperation, recocked the ejection handle, and looking down at the trigger, gave it a good hard squeeze. My next recollection was that of tumbling backwards up and away from the F-86. The seat had left me, and I cannot recall releasing the lap belt, but I must have done so. I quickly pulled the D-ring and the chute deployed up through my legs and deployed above me halting my free fall. On seeing that chute billowing above me I had my first realization that I was going to survive this experience. But my troubles were not completely over. The shroud lines had become entangled in my legs as they deployed up between them and I was now coming down with my feet above my head. If I continued like this, I would hit the ground on my back. I had my survival parachute knife in my pants leg that I then used to cut the shroud lines releasing my legs to a normal upright posture. This resulted in my coming down faster, but I was at least now able to see where I was and where I was going to land.

And I was going to land in the woods. Below me coming up fast were the huge pines and firs of the Pacific Northwest. I was sure that I would hang up in one of them 100 feet above the ground with no way to descend. But that did not happen. My marvelous luck continued and through some miracle I crashed down through the boughs and hit the forest floor on my feet and tumbled backwards over some logs hitting my head on the ground. But my helmet saved me there and when I pulled myself together, I was amazed and grateful to realize that I had survived the most precarious experience I had yet encountered in my flying career. After getting up, I opened my survival pack and set off a flare that sent smoke up through the trees and marked my place on the ground. My wingman spotted my bailout and now reported my position to home base where rescue procedures were started. He also made a low-level flyover, letting me know that he had spotted my flare. I was now beginning to realize what a lucky guy I was. I ended up sitting in the forest for over two hours until a rescue helicopter arrived and lowered a sling down through the trees and hoisted me out and took me back to McChord. I received a wonderful greeting there from my commanders and squadron buddies. Little did I know that my ejection had been reported on the radio in New York where my brother heard it. Also, some Squadron wives visited Alice to give her support, which was not needed after Sewell reported that my bailout had been successful.

This was my first and only major aircraft accident in my flying career, and one that nearly cost me my life. I had to go through the trauma of an accident investigation board which determines the causes of the accident and what could have prevented it. The F-86D crashed into the forest not too far from where I hit. It was a difficult place for the investigators to get into, but they did. The engine failed due to a failed oil pressure pump. I had flown out to the target and had made one firing pass before lack of oil caused the engine to seize and explode. There was no oil pressure warning light on the F-86 at that time and the oil pressure gauge was down below the console underneath the radar scope where the pilot would not see it during his normal instrument scan. You can believe I made it a point to look at that gauge in the future.

Because I ejected while looking down at the trigger and not in the proper posture, I incurred a back problem. I was taken to Madigan Army Hospital on Ft. Lewis where I was given therapy for my back for several weeks. Eventually, they fitted me with a back brace that I used when not flying. I was also required to perform back exercises daily to strengthen the lower right lumbar muscles that were injured in the ejection. I still do those exercises today. I was also verbally reprimanded for not wearing my flying boots. I had the habit of wearing my low-cut shoes when flying, and those were what I was wearing on my ill-fated mission. The miracle is that they stayed on my feet during the ejection. But when I got off the helicopter back at McChord wearing those dress shoes, I was caught. I have to admit I was pretty dumb not wearing flying boots and that was the last flight I took without them.

The thing that amazed me was how calmly my wife accepted what had happened to me. Her attitude was, and always was, that flying was my business and taking care of the home and the children, etc. was her business. She never seemed to worry or show undue anxiety about my flying. She had my confidence in my skill and that made me feel better. There were many instances of wives interfering in their husbands' business and making things more difficult for them than it should have been. I have never had that kind of trouble from my wonderful wife.

The Spring of 1956

As soon as possible, I was back in the cockpit. Having survived a bailout did not have any effect on my desire to fly nor did it cause any hang-ups. In April, the 317th was again deployed to Yuma MCAS for gunnery. This was a very busy time for me. As the squadron weapons instructor officer, the results of the squadron's firing program would be a reflection of my training program with the pilots. Fortunately, the 317th was a special group of excellent pilots and we soon established a new ADC record for number of hits on target. In addition to flying my own scheduled firing missions, I spent my ground time assessing the gunnery film of the squadron pilots as they returned from their missions. That resulted in long hours on the flight line. But I was also given another duty that entailed much responsibility that gave me much to worry about. Being the lead instructor pilot in the T-33, I was directed to check out the 25th Air Division Commander in the T-33 while we were at Yuma. Brig. Gen. Romulus Puryear had joined us at Yuma and was anxious to become qualified in jet aircraft. I had known the General because I had previously been assigned to check out Division pilots in the T-33, and now he wanted to become checked out too. Every afternoon, following the end of the F-86 firing schedule, I would meet General Puryear in the ops room. I would brief him on our training flight and then we would proceed to the squadron T-33 and fly a two-hour mission, he in the front seat, I in the back. He was very anxious and enthusiastic about becoming a jet pilot and we had no problem about his being a General and me being a Lieutenant. I was the aircraft commander being the instructor and when I felt we were getting into a dangerous situation, I took over the controls, right now. Each day, I gave the General the pilot's T.O. to study, and I would quiz him on every procedure including emergencies. He was a good student. Our flights consisted of his doing the preflight, startup taxi out and take off. We climbed out and then did some airwork. He loved the acrobatics, and we did them all including lazy eights, barrel rolls, aileron rolls, Cuban eights, loops, and Immelmann's. Finally, we would return to base where we performed several touch-and-go's. After our final landing, we would return to ops for a thorough debriefing of the flight. All this was pretty routine, except my student was the Division Commander and a General Officer. Knowing that he wanted to solo the aircraft, I felt pressure to ensure that he was making good progress and that he would be safe. Losing a General Officer as a student would not help my career for sure. Col. Williams, my Squadron Commander, was very interested in Puryear's progress and checked with me every day. He could feel the pressure too.

After seven flights, about 15 hours, and over 22 landings I felt that Puryear was ready to solo. He was running out of time, as he had to return to McChord shortly. We took off, climbed out and did our usual work, but I cut it short and returned to base earlier than normal. After three touch and go landings, I instructed the General that after his next landing to roll out and leave the runway at the mid-field taxiway. After taxiing into the ramp area in front of the tower, I had him open the canopy and I told him I was getting out. I told him to take off and shoot three landings. They could be touch and go or he could taxi in and takeoff in the normal manner as he wished. I would be in the tower. If he had any troubles or questions, he could contact me there. He asked me if I was sure he could land this "dad gum plane safely?" I told him that if I wasn't certain of it, I would not be standing outside on the wing. I told the General that if he needed to, just call me, I would be in the tower. But I was sure he would not have to. "I'll see you in a little while, General," I said and then I jumped down from the wing and waved him off. I hurried up to the tower and watched my star student make his solo flight. I had a lot of my AF career riding on the success of his short flight of three takeoffs and three landings. I was confident that the General could do it as long as nothing went wrong with the T-33 that could upset him. And nothing did. Puryear did me proud. He made three excellent landings and when he taxied in, I was there to meet him and shake his hand and congratulate him. I did not let on to him how relieved I was that he was back on the ground safely. And the General was absolutely thrilled and proud of himself as well he should have been. Of course, the next stop was the O club. Word quickly spread that Puryear had soloed the T-33 and the Club filled up with 317th jockeys. Major Williams was ecstatic. The 317th FIS was again setting a new ADC record for rocket hits at Yuma and now we had checked out the 25th Air Division CO in jets. As the Squadron Weapons Instructor, and also Puryear's instructor pilot, I had carried a heavy load that May. But things had turned out well. That night we had a great party. General Puryear was a popular commander. He loved to party with the squadron pilots and show off his favorite trick of somehow bending his legs so that they ended up vertically behind his head. His aide de camp did not enjoy seeing the General performing this act.

Two days later, General Puryear and I flew back to McChord, nonstop in the T-33. He was in the front seat. As he flew over the 25th Air Division, he took delight in calling each of the radar sites and identifying himself as "This is Mercury One calling!"

When we landed at McChord, we were met by the General's wife, Alice, and also Alice Marie, who came out to pick me up. My Alice was pregnant with Gene Jr. The General invited us over to his house for a drink, but Alice Marie said that she would rather not. So, we declined. It was not until we were driving home that I told her that she had just said no to an invitation from a General and his wife.

But if I had a crystal ball, I would have understood better why Alice wanted to go home and rest. In a short while, she would be giving me the biggest surprise of my life.

CHAPTER SIXTEEN: TWO BECOMES THREE

Alice Marie's pregnancy became more complicated. Further exams seemed to indicate that she might have a difficult time, and when she had some bleeding, she had to stay at home in bed for several days to stabilize the situation. I certainly was no big help and when she returned from a follow up exam and told me the doctor reported we may be having twins, I was totally befuddled. Clearly, the baby was not acting as scheduled and we were a little concerned. Recall, in those times there was no such thing as a sonogram or cat scan or any of the great advances that we know today. The doctors used the touchy-feely method of examining pregnant women and their forecasts were derived from the last time they experienced a menstrual period. Still, Alice and I were optimistic and confident of a good outcome. The baby was due to arrive in the middle of June and we were counting the days.

In the interim in May, I became involved in a 25th Air Division Rocketry meet. I was chosen to represent the 317th FIS. The other squadrons were the 465th FIS at McChord, my old outfit, and the squadrons at Portland, Spokane, Larson in F86s and the one at Paine Field and Namao, B.C. (RCAF) flying F-89 Scorpions. The targets were towed by B-29s off the coast of the Olympic Peninsula, the same place where I had my engine failure and ejection in January. It was a close meet from the very beginning, and I entered the last few days in the lead with a perfect score. On May 30, Memorial Day, the competition took a day off, but I decided to fly my F-86 anyway on a short hop and ensure all the systems worked perfectly before the last day's competition. I arranged a nine o'clock takeoff with my crew chief.

That morning I awoke to see Alice Marie sitting on the bed. She was in obvious pain. She told me that she had a backache. I advised her to take some aspirin and I proceeded to put on my flying suit and head for the base. But Alice was clearly not feeling well at all. I asked her if she wanted to get dressed and I would drop her off at the Base Clinic. She said no and told me to go ahead. Though time was short, I did not feel right about leaving her like that. So, I decided to call the Base Clinic. I got hold of a nurse and told her my wife was having serious back pains and did she have any ideas. She suggested some pain medicine. I then told her that she was pregnant but that the baby was not due for another three weeks. The nurse asked me if the pains were coming and going. I asked Alice, "Are the pains coming and going?" "Yes," she said. I told the nurse, "Yes." "How frequently are they coming and going?" I asked Alice that. "About every five minutes," she said. I told the nurse, "About every five minutes." "Bring her in right away," said the nurse. When I told Alice that, she said no, I'll be okay. No way was I going to leave her there now like that. So, after a little verbal tussle, Alice got into her bathrobe and slippers (I was in a hurry now to make that takeoff time). I dropped her at the Clinic and hurried down to the squadron where my plane and crew chief were waiting.

I only flew a half hour flight. Things looked good on shut down and I felt good about my chances in the last days of the meet. There was a message for me in ops to call the Clinic, but when I did, there was no answer. It was a short drive to the Clinic, but when I entered, I could not find anybody there. It was a holiday, and the Clinic was not fully staffed, but surely there had to be somebody around. And where was my wife? I roamed up and down the hallways looking in rooms and then I found Alice. She was in bed, asleep, and she was flat. My God don't tell me. What had happened? I found the baby's nursery and there I found a nurse. She told me that within minutes that she arrived, Alice delivered a baby boy, and it was a breech birth. Then she showed me my son. Because he came out feet first, he had no bruises on him. He was a most beautiful sight and a truly handsome child. I couldn't believe it. All this happened while I was flying above the Pacific Northwest in my Sabre concerned about its readiness to perform and my wife was giving birth down below. It was some day, a great day and a happy day. But as I learned later, it was not an easy day for Alice. It was not an easy delivery, and the doctor did not do a good job on her, which caused her to have to stay in the Clinic several painful days. She had not yet even seen her child, but when he was brought in, the happiness on her face was something beautiful to behold. Things certainly did not go smoothly but they turned out very well and we were both very happy with that. Our lives now would undergo a complete change from before. We had been married a little over eight months and already we were three. And this from a gal who warned me that she had only one ovary.

We had to buy a bassinette, baby bottles, a bottle sterilizer, diapers, and lots of other things that we had put off until the last minute. Because Alice was still hurting, I took over the duties of changing diapers, sterilizing bottles, and sometimes feeding the little guy when Alice was resting. Alice insisted on naming him Eugene Junior and that was what we did. As Alice regained her strength, she took over the motherly functions and I can tell you no baby received better care. It was a wonderful thing to watch. I now had a very happy wife in my home.

The rocketry meet resumed on the next business day and I went on to capture first place. I received the trophy from my old T-33 student, General Puryear, who was very delighted to award the trophy to me. My first-place finish also pleased no end my Wing Commander and Squadron Commander as nothing pleased them more than beating out their sister squadrons.

Flight Pay

Capt. Reynolds, a Flight Commander, asked me if I would fly him to Los Angeles and drop him off there as he was going to take leave and visit his family. We departed McChord in a T-Bird on a Friday after work and headed south landing at El Toro Marine Corps Air Station, located in Orange County. His family met him there and Reynolds went off with them leaving me to spend a day by myself to see some sights in LA. When I returned to the station on Sunday, I was met by the Marine air ops officer who asked me if I would be willing to take a passenger back to McChord. It seemed that an Army Major had emergency leave orders authorizing him military transport to the Seattle area as his father was critically sick and close to death. This was an unusual situation, as we normally did not take passengers in the T-33, especially non-rated personnel. The Army Major was not a pilot and had never been in a jet aircraft nor had he ever been in a pressure chamber. There was no way of telling what his reactions would be in the tight confines of a jet cockpit, with an oxygen mask on his face at 35,000 feet. The Marine ops officer thought the Major would be OK and the Marines were willing to fit the helmet and mask to the Major's head and brief him on its uses. The Major's father was in bad shape and not predicted to live much longer. I then spoke to the Major. He was a small man but appeared in good physical shape. He was more than willing to fly with me and almost begged me to take him. I finally agreed.

I then filed an IFR flight plan to McChord. The weather was good most of the way, but as usual, McChord had a 500-foot ceiling with rain. After the Major was strapped in by the crew chief, I stood on the wing and briefed him thoroughly on the ejection system and bailout procedures. All the time, I knew that if l lost this guy, my flying career would probably be over for good. We took off in bright sunny weather and climbed up over the San Joaquin Valley and headed north. It was a beautiful day for flying. To the west off our left wing, you could see the blue Pacific Ocean and to our right, the snowcapped Sierra Mountains. As the T-Bird was equipped with both VOR and low frequency radio, I was able to navigate and tune in some listening music at the same time. The Major seemed okay and responded on the intercom that he was feeling fine, and as we rolled along, he became more comfortable with the flight. As we flew high above California in the bright sunshine we began to converse about many subjects, at times comparing our careers in the military. The Major had also served in Korea but had been there before I had. He was primarily an artillery officer and was well educated. He clearly intended to be a career officer. At one point he asked me if l received flight pay. I said "Yes, I did." He asked how much I got. I told him I received $107 a month. "How much do you have to fly to receive that pay?" he asked. I answered that I had to fly a minimum of four hours a month. "That's not much is it?" he said. "Well, I have to fly much more than that a year, including nighttime, etc." "You know," he said, "you're making as much money as I am, and I'm a major. That's not right." I answered that I received this pay because flying was hazardous duty and to get guys to become pilots you had to offer extra money. The Major said, "I don't believe that's right nor fair. Just being in the military is hazardous and Majors should make more money than Lieutenants or Captains."

I have to admit that this last utterance got me a little ticked and I decided not to continue with this conversation. I got busy with my navigational duties. As we continued into northern Oregon, we lost our sunshine. We were now in and out of clouds. A hundred miles south of Tacoma, Seattle Approach Control picked us up on their radar and we began our letdown into the McChord area. The weather at destination had not improved so we would be making a GCA approach to landing. As we descended into the clouds, the turbulence increased and we entered some heavy precipitation mostly rain, but also some hail. We also had some lightning around us and the T-33 really began bouncing around. I turned on the windshield defroster and the pitot tube heat in case of icing. There was a lot of radio static and we were really experiencing a rough ride as we descended down to 2,000 feet and the downwind leg for our landing on runway 16 to the south. The wings were rocking and the nose of the aircraft was swaying from side to side as we encountered some high side winds. On the base leg I lowered the landing gear and flaps, which made things in the cockpit even noisier, and then on the final approach we had some hard rain on the windshield. At about 500 feet we broke out under the overcast. I had the runway in sight, thanked the GCA controller, switched to tower frequency and landed.

As we rolled out, I remembered to check in with my passenger. I asked the Major how he was doing. I could not hear his response. I checked the mirror. I could see his helmet and he seemed slumped over to the canopy a little.

"How you doing Major?" I asked.

"Not so good. I think I am getting sick." He replied.

"Hold on. Use the brown bag if you're going to vomit." I replied.

I then cracked the canopy and told the Major to take off his oxygen mask. I rapidly taxied into the ops area and shut down the engine. I opened the canopy and jumped out on the wing. The Army Major was in bad shape. I undid his mask and took his helmet off. His face was ashen in color, and I could see he was a shaken man. The crew chief helped me get him out of the cockpit and down to the ground where he seemed to recover a little. After a little bit he started to get back to normal. While I was filling out the logbook, he came over to me and asked, "What happened up there, I thought we were gonna die."

"What do you mean?" I asked, "Everything went pretty smoothly, I thought."

"Shit, I thought that plane was coming apart, and the lightning and hail was throwing us all over the place. I didn't know where the hell we were or what was happening, I damn near crapped in my pants."

"Well, actually, that was a pretty normal approach to McChord, maybe a little bumpy in spots, but no big deal, Major."

"Well, damned if I ever want to do that again," he said. With that, the Airdrome Officer arrived and took the Major into the operations building to arrange his transport to Seattle.

I finished my work at the aircraft and about 15 minutes later came into the ops building. The Major was about to leave. He came over to me and thanked me for the ride. I told him it was my pleasure and I wished him well and that his father would be okay. Then as he was about to go out the door, he stopped and came back to me and said: "You know, maybe I was wrong about that flight pay thing. I was never so scared in my life as I was on that last part. I wouldn't do that for ten times the pay." And he left.

We Are Family

The impact of having a family now began to penetrate. Our life was fun and uncomplicated when there was only Alice and I, but now we had the little one and, of course, his presence dominated our lives and our daily plans. Our families back in New York were thrilled, but they were 3,000 miles away and Alice had no support other than me to help out which I did. Another worry entered my mind and that was my concern that what would happen to Alice and little Gene should I not return from one of my missions. Flying jet fighter aircraft was a most hazardous business and the USAF major aircraft accident rate reflected that. I only had the insurance that the Air Force provided and that was pitiably little. My attempts to buy additional commercial insurance failed as no company would issue insurance on jet fighter pilots. It was a worry to me, and I was determined that I would do my very best to fly safely and take no unnecessary risks in the future while still accomplishing the mission. As usual, I never flew without my Rosary beads in my pocket and a silent prayer before takeoff. While I had many more close calls during my flying career, I was able to come back safe and sound. I know that God, through the intercession of the Virgin Mother, permitted me to return to my family intact.

I Become a Headquarters Weenie

In the spring of 1956, Col. Hubert Williams, our CO, received orders and was replaced by Lt. Col. Walter Beane. Some other organizational changes took effect that were not well received by everyone. Both squadrons lost their aircraft maintenance personnel who were combined into a consolidated maintenance organization which serviced the planes for each squadron. The squadrons were now just operational units consisting of pilots and ops schedulers and staff. The squadrons were placed under a new organization, the 325th Fighter Group. This organization, headed by a full colonel, did all the scheduling of flight training and alert for the units. And I was selected to be assigned to the ops staff of the 325th Fighter Group. It was an assignment I did not seek, nor volunteer for, but in the past, I had filled in a few times at Wing Headquarters doing some staff work for temporary periods. Clearly the word was out that in addition to know the fighter business I was also capable of staff work. That is, I could write well and brief clearly and intelligently. However, I was sad leaving squadron life. I knew how squadron pilots felt about group weenies, and under the new organizational structure, their independence was gone.

Still, I had new job to learn, and I put my best efforts into it. Now I was involved in parceling out the training sorties to both squadrons. Naturally, they were watching carefully to ensure that they received their fair share of the flying hours and alert duties. There were plenty of hassles and I did my best to iron them out and display no favoritism. Unfortunately, both squadron commanders despised the new structure, especially losing control of their aircraft maintenance, which now came under the Group. One CO, Lt. Col. Charles Carson, who commanded the 465th, went so far as to fly back to Washington, D.C. and complain to the Chairman of the Joint Chiefs, whom he knew personally, about the new structure. It did him no good. These were interesting times for me, and I learned a great deal about the value of diplomacy and good listening just to keep things going on an even keel. I also began to learn a great deal about how decisions were made in the Air Force. My job required me to attend many staff meetings where current problems and future plans were discussed and then acted upon. Frequently, I would end up being the action officer on any decisions made by the commanders. It would then fall on me to write the orders or draft the plans that would implement the decisions made. This gave me a lot of insight into the thinking processes of officers who were long on experience. Most were veterans of World War II, and I respected their accomplishments in that great enterprise. Gradually, my confidence grew to where I was able to participate in the discussion phases of our meetings and even make recommendations because I had current knowledge of the aircraft we were flying and the mission. Although I missed the daily camaraderie of the squadron and the flying, I knew that I was learning a lot about how the Air Force operates, and that would prepare me for positions of higher responsibility in the future. In April 1957, I was promoted to the grade of Captain.

My new bosses were good to me. Col. Edwin (Chic) Fuller was the Group CO, and my immediate boss was Major Joe Brown, with whom I got along very well. We were busy people overseeing the Group and meeting our air defense missions. Then came along some more challenging issues. Orders came down directing the 317th FIS, my old squadron, to transition into F-102 Delta Daggers. While I was excited about the possibility of my flying the F-102, there was a great deal of work to be done in checking out instructor pilots, bringing the new aircraft online, and getting maintenance personnel up to speed in learning all about this newest model air defense fighter and its weapons system. Fortunately, Joe Brown had been a part of the F-102 checkout program at Edwards AFB and was great help in setting up our transition plan. The arrival of the new aircraft was exciting. I was shaken to see how big the F-102 was and how much more sophisticated it was than the F-86D. The transition went smoothly, and being a staff officer, I was way down the line for a checkout. Still, I studied the tech orders, sat in on the briefings and whenever possible got in some simulator time. Finally, Major Bill Jackson, the 317th ops officer called me and told me to come down to ops right away. He was going to test hop a TF-102, a two-seater trainer version, and he invited me to come along with him. Bill was a great guy, and we were good friends. He was a real character, famous for his trombone playing and his practical sense of humor. We started up and took the runway for takeoff with me in the right seat. Bill was talking all the time going through his checklist and telling me all the things he was doing so I could be learning how to fly this baby. We were cleared to roll. Bill released the brakes, and we started down the runway, with Bill using nose wheel steering to keep us centered. He then moved the throttle outboard to light the afterburner. I waited to feel and hear the afterburner ignite, but I didn't hear anything. We kept rolling and Bill kept talking to me about how fast this aircraft would leave the runway and climb. But we kept rolling down the runway and our wheels were still on the pavement. I finally got through to Bill, “Did the afterburner light?” “Hell no,” he replied, and he brought the throttle inboard and shoved it outboard again to the afterburner position. Still nothing happened. By now we had eaten up a lot of runway, and we were still not in the air. Our drop tanks were full and were carrying a full load. We either had to start flying soon or abort. Jackson let out an expletive and tried the afterburner one more time. Then the RPM began to unwind. The TF-102 had been ready to fly and the wheels were beginning to bounce off the runway, but now there was no chance that we were going to get airborne. I used the intercom again. "Judas, let's abort before it's too late." At this point Bill pulled the throttle to idle and stomped on the brakes. Then he pulled the drogue chute handle and all I could see was the end of the runway coming up fast. We just got it stopped before our nose wheel was about to go off the end of the runway. My heart was about to pop out of my mouth. That was too close. If we had gotten airborne and then had the engine problem, we surely would have ended in the toolies off the end of the runway. As it was, I thought Jackson waited a long time to decide to abort. Still, we came out okay. And that was my first ride in the F-102. It gave me a lot to think about. In this bird, you had to make your decisions a lot faster than you had before.

Not long after the 317th finished its transition, we received word that it would be transferred to Elmendorf AFB, Alaska. In return, we would receive the 64th FIS. The purpose was to position an advanced interceptor squadron in Alaska with the range to intercept the Soviet bomber that frequently approached the Alaskan ADIZ. Our job then became preparing the 317th to deploy en masse to Elmendorf. This included maintenance, parts, equipment, families, everything. It was a huge operation that took weeks of planning and mountains of paperwork. It was the first time I had ever participated in anything of this size and complexity. We also had to prepare for the arrival of the 64th FIS and their transition into the F-102. There was also many accompanying personnel difficulties as many involved wives and others did not want to go to Alaska. Getting all of those problems resolved was handed over to Chaplains and the folks in Personnel. The F-102s were flown to Alaska in flights of six via Edmonton, Ft. Nelson, Whitehorse, and then into Elmendorf. Because the Deltas had no VHF navigation equipment they had to be led by a T-33. I lead one of the first flights from McChord to Alaska. It took us two days. The weather was good but one of the Deltas ran off the short runway at Ft. Nelson. It suffered no serious damage, but we had to leave it there. All in all, the deployment of the 317th to Alaska was successful and while I felt the loss of my old squadron deeply, I was proud of their performance and pleased that I was a part of their accomplishments.

Another Surprise

Little Gene continued to be the light of our lives. He was growing and thriving, thanks to the motherly nurturing rendered on him by his mother. We were visited by my father who came across the country to see his first grandchild. He was very pleased to see his first grandchild and we had a great visit. I took some time off and we toured the Pacific Northwest. Then in the spring, Alice notified me that she was again with child. I was again flabbergasted but pleased. Why not? Two was better than one, right? However, it was clear that our little house on the lake was now growing a lot smaller for our growing family. It was also clear that we needed a bigger car for our growing clan. My cherished Pontiac Catalina Coupe was hit broadside on Highway 99 with Alice at the wheel. It was not her fault as an elderly man ran a red light as she was coming off the base. The car was never the same and now was too small. That spring, I bought a Chevrolet station wagon to be picked up at the factory in Flint, Michigan. Alice and little Gene flew commercial to New York to visit with her family and about a week later I flew a T-Bird back to Mitchel Field to join them. After a short visit Dad, Aunt Anna, Alice, Gene Jr., and I drove to Clearwater, Florida for a week on the water and some sunshine. We had a great trip and then drove back to Michigan where we picked up our new car. We said goodbye to Dad and Anna and drove back to Tacoma. It was wonderful for Alice to show off her son and see her family again and for me also. But now it was back to work.

A Routine Target Mission-Not

It was common to use the T-33s as targets for the interceptors on routine training missions and for larger air defense exercises. The T-bird would sometimes carry a pod that had a chaff dispenser, which would give the interceptor pilots some training in combating electronic warfare devices designed to fool the pilot into locking his radar onto a false target. I was scheduled to fly a night target mission as instructor pilot with a Navy Lieut. Commander, who was on exchange duty with the 25th Air Division, in the front seat. We took off from McChord shortly after dark and headed southwest and proceeded over the Pacific and down the Oregon Coast. The mission went smoothly, and several fighters intercepted us and made runs on us as we proceeded back on a northerly course. We were at about 25,000 feet when I thought I detected a smell of smoke in the cockpit. I removed my oxygen mask and asked the commander if he smelled it. He did and about the same time he called out that he had some flames coming out from behind the right side of instrument panel. I called an emergency situation to ground control and asked for pigeons to the nearest airfield. This turned out to be Salem, which was only about 75 miles to the east; however, the weather there was bad, and the runway was short. Nevertheless, I determined that we had to get this aircraft on the ground as soon as possible so we headed for Salem.

As usual, fog was prevalent in the Pacific Northwest, and it was forming fast this night. We contacted the tower at Salem, and he informed us that the visibility was closing in rapidly. The smoke in the cockpit was also getting heavier and we just had to get on the ground quickly. The Navy guy was great, and he did the flying as I handled the radio and the navigation. I tuned in the Salem radio and informed the tower we were going to try and approach. We descended to 1,500 feet and I took over the stick and headed for the outer marker. At the outer marker we would begin out approach on the runway heading. When the Commander had sight of the runway, he would take over the controls and land. If he did not see the runway at 200 feet, we would take it around and either head for someplace else or try again. We came down the final approach with me just holding the runway heading and controlling the rate of descent to about 400 feet per minute. Down we came through 1,000 feet then 700 feet, 500 feet, 300 feet, then 200 feet. Just as I called out minimum altitude, the Commander yelled, "I got it." And we touched down and quickly hit the brakes. Soon as we stopped, we opened the canopy to vent the smoke, then turned off the runway and shut down the engine. A little flame was still smoldering behind the panel and we both were delighted to get out of that plane and get on terra firma. Salem was a small airport and did not have much activity on it, and it was a long time before they even had a fire truck show up. Needless to say, we were two lucky guys; first that the fire did not spread into the cockpit and that the smoke did not get so bad that we could no longer operate. Second, we were fortunate to find Salem airport and get down, because within minutes of our landing, that airfield went zero, zero, as did the rest of the airfields in Oregon. Sometimes, it pays to be more lucky than good. This incident was written up in my OER as evidence of my flying proficiency and I believe was a factor in my obtaining my next assignment.

Our Family Grows

As Alice's pregnancy proceeded, it became evident that we would need a bigger home to accommodate our next arrival. Sadly, we had to leave our little honeymoon cottage on Lake Steilacom and move to a more conventional home in Lakewood with three bedrooms. This necessitated the purchase of some more furniture as we had been living in a furnished house up to this point. In November 1957, Alice was within a month of the programmed delivery date. This time I was better prepared. Because the Air Force had closed the base clinic to obstetrics, Alice had access to a local civilian doctor for her pregnancy. We were both pleased with this arrangement as the McChord doctors had really botched up little Gene's delivery and caused Alice a great deal of pain and suffering as a result. In the wee hours of the morning of November 10, Alice woke me at about one in the morning with labor pains. I took no chances and did not hesitate. I had her pack her bag. I called the doctor, wrapped young Gene in a blanket and put him in the back of the car, and away we went to St. Joseph's Hospital in downtown Tacoma. It was about a 20-minute drive normally, but at that time of the morning there was little traffic and I set a new speed record. The baby was coming fast, and I was aghast that I might be called on to deliver my own child in the car. We pulled up in front of the hospital and I helped Alice through the front door where she was met by an elderly nun. She was the most beautiful sight I had ever beheld. She put Alice in a wheelchair and whisked her into an elevator. I returned to the car to check on little Gene; he was still sleeping soundly unaware of the historic events of the moment. After about 15 minutes I returned to the hospital and went upstairs to the delivery rooms. Young James had already arrived. I am assured that the Sister had performed the delivery as there was no way the doctor could have made it there in time. I took a quick peek at the new arrival. A handsome lad for sure, and then headed back to the car and returned home with my other charge. It had been an eventful night and a blessed one for our family as we now increased by 25 percent. I was very happy that everything had come to pass successfully, and that mother and child were healthy and sound. God was kind to us. Our new son was named in honor of his great grandfather James Francis, who came from County Mayo, Ireland with his young bride, my grandmother, in the early part of the 19th century to make a new beginning in America.

In the ensuing days I passed the good news to our folks back east about the arrival of our second son. In the short space of just over two years of marriage, Alice and I now had two little boys. We were very happy, and she was very busy. If I had to say so myself, they were two beautiful healthy guys. Young Jimmy was a most delightful baby, always in a happy disposition, which of course earned him the appellation of Sunny Jim. They kept us both busy, but it was a work of love. Young Gene was now walking and riding his tricycle and getting to know his little baby brother. Naturally, our boys put a crimp in our social life, and we took a pass on many of the Air Force socials we used to enjoy, but to tell the truth, we did not miss them at all. I was busy and happy in my job and Alice was busy and happy with hers. Life was good.

I Am a Staff Weenie

My work at Group Operations gave me great experience in Air Force staff work. My day consisted of morning briefings when a staff officer covered the day's schedule of flying, the aircraft maintenance situation, intelligence, flying weather in the area, and other matters affecting the Group. I usually covered the flying and alert schedules for the Group. During the day I read incoming TWX messages (military telegrams) and drafted answers for the Commander, if appropriate. I also prepared letters as necessary and briefings for the Commander as requested for various upcoming VIP's or other occasions. I felt comfortable doing these duties and I found my writing and speaking skills, developed in my high school and college years, were most helpful in performing them. I also kept my hand in flying and tried to fly at least three times weekly. When available, Major Joe Brown, Col. Fuller, or one of the other staff officers and I took some weekend cross-country flights to get in some flight time and gain navigational experience. We of course usually spent the nights at some fairly favorable locations such as Las Vegas, San Francisco, etc. I also took my turn sitting alert and filled in for one of the squadron pilots at least once every two weeks. I always believed that sitting on air defense alert was the most important way I could put my flying skills and training to the best use. But to tell the truth, I missed squadron life and longed to get back to it. Most of all I missed the camaraderie of the squadron plus the fun of the beer calls and intra-flight competitions that were always ongoing.

In early 1958, Alice Marie informed me that she was again pregnant. As Sunny Jim was less than six months old, we were both shaken. I recalled Father Beard's admonition to us the night before our wedding to avoid the sin of artificial birth control. He would have been proud of us.

About this time, Col. Fuller was replaced as Group Commander. He was assigned to ADC Headquarters in Colorado Springs as part of the TACEVAL/ORI Team. This organization was charged with evaluating the operational readiness of all ADC units and determining their capability to perform the mission. Only the best people were assigned to that unit. I liked Col. Fuller very much and we got along very well. He liked to fly with me often and respected my abilities both as a pilot and as a staff officer. The one problem I had with Fuller was he was an alcoholic. He was a tremendous drinker and a Playboy, and I had to watch him to ensure that he did not try to fly when he was under the influence. Our new CO was Col. Tom Schwellenbach. I had now been stationed at McChord since April 1954, meaning that I was coming up on my fourth year there. I was really enjoying my work at the Group and learning a great deal about management and decision-making. I was also flying the F-102 and the F-86D. Our Division and our Group continued to receive outstanding results in their TACEVAL/ORIs conducted by ADC Headquarters and that reflected well on the 325th Group. In June, Col. Fuller called me. He asked me if I would be interested in coming to ADC and becoming a member of the TACEVAL/ORI Team. They were seeking an F-86D/ F-102 pilot with some staff experience. After talking to Alice, I agreed to fly down to Colorado Springs for an interview. The new head of the team was Col. Sam Grashio, who had commanded the Group at Paine Field in the 25th Air Division. I had met him before and knew him slightly. He was a survivor of the Bataan Death March, having escaped from a Japanese prison camp fought with the Filipino guerillas and later picked up by a US submarine. Fuller was my surrogate and I got to meet many of the team members, some of whom I knew. One who really impressed me was RAF Wing Commander Bobby Oxspring, a hero of the Battle of Britain. I spent two days in Colorado Springs and then flew back to McChord. About a week later, I was told that I had been selected: we were going to Colorado. This was an important turn in my professional career. Joining this team was an honor and only the best of the best were chosen. If I could do a good job there, it would serve me well toward my goal of becoming a commander of my own fighter squadron. I was very pleased but also aware that this would be a position full of pressure responsibility. I would now be dealing with very high-ranking people who expected high levels of performance at all times. My old student pilot Gen. Puryear was DCS/OPS at ADC, and I thought that maybe he had a hand in my being selected. I looked forward to this new adventure and, thankfully, so did Alice Marie. This would be our first family move in the Air Force and it would be with two small kids and one more on the way.

That spring I flew down to Colorado Springs and bought a small house on the north side of town. I paid about $14,000 for a three-bedroom house that had a nice big backyard and a garage and small family room. It was the first home I had ever bought. We had not been in it a year before I realized it should have been bigger. I did not foresee that more kids might be coming and the ones I had would be growing faster than I realized. Still, it was a nice comfortable home and our years there were filled with joy. The job I was coming to entailed much travel time away from home, which placed huge burdens on Alice Marie to run the house and take care of our young brood without me. This she did to perfection. She was the epitome of an Air Force wife. I had made a great choice.

So, in late August 1958, it was off to Colorado Springs for the Cunneely Clan. I was a Captain and a Senior Pilot. I had 1,764 hours of flying time, and in 1957, our income was $6,700. Alice was seven months pregnant, and we were anxious to get to our new home, (which she had not seen), find her a good doctor and start my new job. It was an exciting time for us. We drove our new station wagon down through Yellowstone National Park where we enjoyed the sights and fed little James with his food and milk heated on Sterno. Young Gene enjoyed the bears and the geysers, but we could not stay too long. Our new home waited.

CHAPTER SEVENTEEN: AIR DEFENSE COMMAND 1958-1963

Colorado Springs in 1958 was a lovely little town, nearly 6,000 feet above sea level, in the shadow of Pikes Peak and the Rockies. The weather was totally different from the Pacific Northwest where it was mostly rainy, cloudy, and mild. The Springs was mostly sunny and dry, but the nights were cold. The climate was semi-arid, but in wintertime we had our share of snow (which did not stay long), and in the summers, we had afternoon thunderstorms to keep us cool. Our little home was comfortable and our large, enclosed backyard was ideal for the kids to play in a safe environment in the sunshine. I found myself another jalopy to drive back and forth to ADC Headquarters located on Ent AFB. Ent was a base without runways. It was a conglomeration of buildings close to downtown Colorado Springs. On the base was the Headquarters of NORAD, North American Air Defense Command, and its command and control center. Located as it was just off the street, it was most vulnerable to any attack by either enemy saboteurs or to destruction in the first air assault from a Soviet missile or bomber forces. I found my new job to be all that I could expect it would be. I was assigned to the unit that planned and controlled the air defense exercises that were an integral part of the TACEVAL ORI's. These exercises consisted of a large number of target aircraft, mostly T-33s, B-57s, and SAC bomber aircraft that would "attack" the air division that we would be evaluating. To put together such a large-scale exercise required a large amount of calling, coordination, planning and follow-up. This had to begin months in advance. The timing was critical and, of course, FAA had to be involved as huge amounts of airspace had to be reserved. Major Dave Winn was my supervisor. He too had just joined the team so we both were learning the business at the same time. Dave was a cool guy and a brilliant planner, and we got along very well together. We also tried to fly together when we could. To stay current in the F-86D and the F-102, we had to fly a T- 33 from Peterson Field (Colorado Springs) to one of our fighter squadrons and beg for a few flights. By having to do this I soon got to fly with ADC units all over the country and to meet most of the commanders, ops officers, and jocks. From Maine to Florida to Texas to Minnesota to

California, I gradually flew at most all of our ADC bases. I knew the let-downs to each of the bases. Usually when visiting these bases for some flying time, I volunteered to sit alert for one of the squadron pilots That was always a welcome offer and I enjoyed doing it. During my years with the ADC team, I greatly expanded my flying capabilities and through my evaluation work, I learned a great deal what it took to run a good efficient organization and what happens if you fail to apply strong management principles and pay lots of attention to details.

Mary Elizabeth Arrives, Our First Girl

We found an excellent civilian doctor for Alice, and I had to stay close to home as our next child came closer to arrival time. I knew that with Alice, you could not rely on the predicted date. Her history showed that early arrivals were commonplace. In the middle of September, Alice's mother came to stay with us. Kathleen was a great woman, from the same stock as my own grandmother who raised me. Alice seemed to be having more trouble with this pregnancy than with the previous two and I was a bit concerned. On the night of October 12th, Alice complained of oncoming labor pains, and we headed for the hospital located in downtown Colorado Springs. Shortly after our arrival at the hospital, Alice entered the labor room, and I began a long night of waiting. But unlike previous Cunneely childbirths, this one took a long time to come forth. After a long time, the doctor finally came out and told me that he was having lots of trouble bringing the child out safely, and that he might have to opt for a Cesarean procedure. He told me that he would give it one more try and if it failed, he would operate. I was very concerned, and I got out my Rosary. It was now well after midnight, and I knew my wife had to be exhausted and in pain. After what seemed like an eternity, a nurse came running out of the operating room with a baby in her hands. She only gave me a quick look before heading into the baby ICU. "It's a girl," I heard her say as she disappeared. I was not at all content that all was well. The baby I saw was pretty beat up and red with bruises. Finally, the doctor came out and told me that things were okay, and that Alice did a great job in giving birth to a little girl, who, although she was a bit beat up on the way out, was healthy. I was very relieved. He told me that Alice was very tired and needed complete rest. She had asked for a private room and that was granted. It had been a long tough night, but it turned out successfully. We now had a daughter, and I knew Alice was pleased with that. So was I. God had been good to us. Within three years of our wedding, Alice Marie and I now had three children. And this from a wife with only one ovary. For the first time, I would hand out cigars with a band that read: "It's a Girl."

Sooner or later, I had to get to work in my new job. It was soon clear that I was in an organization staffed with very competent and experienced professionals. ADC/TACEVAL had the top guys in each aircraft and in every facet of air defense, including aircraft maintenance, armament (including nuclear weapons), ground radar control, airborne early warning and control (EC-121 aircraft), U.S. Army Nike air defense weapons systems, and air defense tactics at the division and region levels. Our job was to evaluate these units annually, right down to the individual aircrews, controllers, and loading crews and determine their capability to intercept and destroy an enemy force of bombers attacking the North American continent. It was a huge and important task, and we took it seriously. At that time, our air defense fighter inventory included F-86Ds, F-102s, F-106s, F-104s, F-89s, and F-101s. I was current in the F-86D and the F-102 and I flew the T-33, as we all did, in order to fly out to all the bases where we flew with the various squadrons and maintained our currency and sat alert. It was important that we maintained our proficiency in the aircraft in which we would be evaluating ADC squadron pilots. At the same time, I had my staff duties that involved planning the large-scale exercises that we used to evaluate whether the Air Division could detect and repulse a simulated heavy force Soviet bomber attack. We would conduct these exercises about eight times a year, so we were busy people. There was a great deal of traveling and I was away from home many nights. This was the worst feature of the job, and it placed a heavy burden on Alice Marie. Although I had warned her of this beforehand, it was still a tough deal. To this day, I am unable to understand how well she handled the daily burden of three energetic and demanding children by herself. But she did, to her everlasting credit.

ADC, at that time, had over 50 fighter squadrons. Shortly after things calmed down, Dave Winn, who had some hours in the F-106, began to check me out in it. First, we picked up some brand new TF's at the Palmdale factory and ferried them to their home base. Then, I just began flying the A model and by studying the Dash One, I quickly learned the fire control system. During my years at command headquarters I flew with most of the F-86D, F-102, and F-106 units. These included the following.

318th FIS	McChord AFB, WA
64th FIS	Paine Field, WA
2nd FIS	Wurtsmith AFB, MI
5th FIS	Minot AFB, ND
11th FIS	Duluth, MN
27th FIS	Loring AFB, ME
48th FIS	Langley AFB, VA
49th FIS	Griffiss AFB, NY
71st FIS	Richards-Gebaur AFB, MO
94th FIS	Selfridge AFB, MI
84th FIS	Hamilton AFB, CA
95th FIS	Andrews AFB, MD
319th FIS	Bunker Hill AFB, IN
329th FIS	George AFB, CA
438th FIS	Kincheloe AFB, MI
456th FIS	Castle AFB, CA
498th FIS	Geiger Field, WA
439th FIS	McGuire AFB, NJ

In addition, I flew many target missions into other various ANG and RCAF bases in setting up our air defense exercises. All in all, while I was flying fewer hours, I was flying missions all over the command and getting to know most of our bases firsthand. My cross-country navigational skills became so adept that I could look down at the countryside going beneath and tell almost exactly where we were within a few miles. I literally became a geographical expert on the USA. I never was bored flying over our country because it had so many different looks to it from the rocky coasts of Maine to the Badlands of the Dakotas, the Colorado Rockies, the breathtaking Pacific Northwest, the snowcapped Cascades with its series of volcanic peaks, and the green, plush California valleys. What a great country to be able to fly over.

Because I now flew fewer hours in the fighter aircraft, I kept up to date on the Dash One, the pilots' operating manual, and made certain that I memorized the emergency procedures. Evaluating squadron pilots was a major responsibility. I had been a squadron pilot for many years, and I completely understood their life and the pressures they were under. I took that into consideration during my check rides, made them feel at ease and encouraged them to pay close attention to the details of their profession because what they did was so important. I always had good relations with the pilots to whom I gave check rides. The vast majority of them were sharp and well trained, very few flunked. I believe I was always fair in my assessments.

Early in my assignment to the TACEVAL Team, my supervisors discovered my writing abilities. This caused them to assign me, along with Dave Winn, the responsibility for final editing and proofreading our ORI reports and the Team Chiefs briefings. This meant long hours following the end of the exercise when the various team chiefs returned from the units and rendered their reports to Colonel Grashio. We would sit in on these debriefings and assist in finalizing the written reports, proofreading for proper grammar, spelling sentence structure and understandability. As some air divisions would have as many as 20 to 30 units, this made for a long report. Consequently, Dave, I and the secretaries would often have to stay a few days longer than the rest of the team to complete our job. This report was important because it would be read by the ADC Commander and then be forwarded to Headquarters USAF for perusal by the Inspector General. It was tiring work that made for long days, but it was our job and that was that. We also helped the Colonel rehearse his briefing which he would deliver to the inspected unit. This was a suspenseful situation as failing a TACEVAL/ORI was a serious thing and often meant that the commander of the failed unit would be fired, and his career shortened. On the other hand, a rating of Satisfactory or Outstanding would be very helpful and could mean promotion. So, a lot was riding on the outcome. I sat in on both good and bad results. They were very difficult to sit through when the unit failed. We always had to be very sure of our findings and prepared to support our conclusions with solid facts. The commanders would reply in writing to our report but there was not much they could do to mitigate it if it was a bad one. It would also mean that the failed unit would have to repair their deficiencies and be inspected again. That meant we would have to go back and do it again. So, it was no fun for us either to fail a unit.

My years with the ORI Team were very educational for me. I gathered a great deal of information on how decisions are made at the Command level. I met many General Officers, and I was called upon to brief them on various topics of an operational nature in which I was supposed to know what I was talking about. Shortly after arriving at ADC, I was informed that I was required to attend a speech class conducted by the University of Colorado for incoming junior officers. General Partridge who commanded ADC had little patience for officers who did not give their briefings in a concise comprehensive, and thorough manner. I attended these classes twice a week for several weeks, having to prepare and give mock briefings to the instructors each week. The instructors were very hard-nosed and almost none of my offerings met their approval. However, that could be said of their critiques of the other students as well. I had always been confident of my speaking techniques in the past and had always received good evaluations for them, but these guys really tore me apart. I finally began to make some headway and received their approval along with a certificate that permitted me to brief the commander. Later in my career, I better understood how important it was to deliver a brief in a concise and intelligent manner that would give the listener confidence in your judgment. Also, I worked hard to write logically and correctly with a flow of thoughts that made sense to the reader and brought his mind into alignment with what I was trying to convey. I came to grasp clearly that leadership and communications are tightly linked.

My days at ADC turned into years and I enjoyed them. My family had grown, and they too enjoyed the wonderful benefits of living in such a beautiful setting. Alice Marie and I had made some great friends. The children were beginning school. They loved the outdoor life that the sunny climate of Colorado afforded. They were healthy, happy, and thriving. They loved our backyard, and I can still see Gene, Jimmy, and Mary racing around there in play a mile above sea level in the shadow of Pikes Peak and the east slope of the Rockies. Gene Jr. and Jimmy loved the zoo, built into the side of the mountain behind the Broadmoor and we went there often on our days off. We also made great use of the recreational facilities at the new Air Force Academy. I became an early member of the Academy golf course and played there regularly. Young Gene loved to accompany me, and he was always on the lookout for deer and other critters that ranged through the woods there. In the summer, we were members of the Academy Officer's Club using its pool facilities. Alice, I and the kids, spent many wonderful days there, topping off a day's swimming with a buffet meal in the Club. One year, Alice's brother Tom visited us, and I recall his joining us at the Academy pool and playing with his nephews and niece. He was a great favorite of theirs and they loved having him around to play with.

The Delta Dart

As ADC phased out the F-86Ds, I began flying the F-102 more with the squadrons. Then as I began to gain more flying time in the F-106 I started to fly it as well. The Delta Dart was a far superior interceptor in many ways especially in performance and in the MA-1 fire control system. The F-106 was capable of Mach 2 flight (twice the speed of sound), could reach 50,000 feet, and also carried both Infra-Red and radar guided missiles. The "Six" also carried an air-to-air Genie rocket with a nuclear warhead. Thus, ADC was now capable of destroying incoming enemy bombers with atomic weaponry. Arming our fighter squadrons with nuclear weapons brought with it a host of complex problems for commanders involving psychological testing, (human reliability programs, loading training, nuclear storage security, aircrew personal weapons, aircrew and controller alert codes which were used to prevent any unauthorized employment of a nuclear weapon. Nuclear safety and security now became a vital factor in the evaluation of a fighter squadron.

The F-106 cost the USAF about $2 million each. It was the biggest airplane I had ever flown, and it looked fast just sitting on the tarmac. The instrument system was totally different as the gauges were all vertical, not round, as in all the previous aircraft I had flown. The control stick was a yoke, the right side used for aircraft control and the left side for operation of the radar. Down in the center was the TSD (Tactical Situation Display) that projected a map of the area so the pilot could follow his flight path over a map of the ground. The MA-1 system was designed so the pilot could immediately after breaking ground, turn the aircraft over to an automatic flight control system which acted in response to commands received from a digital computer. The information was transmitted from ground stations feeding intelligence to the aircraft's MA-1 system. ADC in the early 1960s brought into operation its Semi-Automatic Ground Environment system (SAGE) which coupled with the MA-1, automatically vectored the F-106 to the target and gave the aircrew directions to fire at the precise second. The pilot only had to handle the throttle settings select the armament and squeeze the trigger. The SAGE system then brought the aircraft back to base, through the letdown and put the aircraft on final approach where the pilot only had to take over the stick for touchdown. It was truly an amazing technological advance. Many pilots disliked using the automatic features, feeling that they took away the prerogatives of the pilot. Others distrusted the trustworthiness of the system, but as time went on, I found it to be a great backup, and I began using the automatic pilot more and more often. I especially liked using it when making weather approaches using the Instrument Landing System (ILS), as I felt that I was not getting sufficient practice in. The F- L06 was a quantum advance in interceptor aircraft. The additional power and performance of its engine permitted interception of high altitude, high speed bombers. It also had additional range and endurance. The Delta Dart had the atomic weapon as well as both IR and radar guided air to air missiles. And the ejection system, after they got rid of the dangerous downward B seat, provided comfort and safety to the pilot. I truly loved flying the F-106.

My years at ADC seemed to fly by. I was very busy on the job and at home. While I traveled a lot, I was able to call home frequently using the USAF phone system. And so, I was able to stay in touch with Alice and our brood. Our children were growing, and they were for the most part healthy and happy children. They had a terrific mother who took care of all their physical and emotional needs on a day to day basis. They loved living in Colorado. In the winters we drove to Camp Hale, an Army Recreation Center in the Rockies, and spent time there sledding, skiing and hiking. We rode the cable car up Manitou Mountain and drove to the top of Pikes Peak. We made countless visits to the Cheyenne Mountain Zoo, a favorite destination for our kids, especially Gene Jr., built into the side of the hills overlooking Colorado Springs. Alice was shocked to learn she needed calling cards printed with her name on them, along with white gloves, to present herself to the wife of the Commander of NORAD who resided at the sumptuous Broadmoor Hotel. Alice was impressed, but definitely not overwhelmed. She thought it was a bit overmuch. We enjoyed watching the construction and grand opening of the USAF Academy located a few miles north of our house. As I mentioned, we began to use the Academy for the recreational facilities it offered, and we even began to utilize its Base Exchange and Commissary. We were in attendance for the first home football game there in its new stadium, and we also began to host cadets in our home on their weekend pass for some home cooked meals. I became one of the earliest members of the AF Academy Golf Course, and along with some other members of the TACEVAL Team played the course almost weekly. I also began my skiing career in Colorado. Deciding I would take up this sport, unbeknownst to Alice, I drove to Denver one day to Dave Cook's Sports store and purchased skis, boots, poles, socks, gloves, hat—the works— for a total cost of $75. I thought Alice would kill me when I told her what I did. She did tell me I was nuts. So, I started skiing with my friends usually at Loveland Pass. I was not very good, but I truly loved being in the mountains and later I tried to pass on the exhilaration of this sport to my kids. We loved the Springs so much, that Alice and I bought a lot there at Falcon Estates, a development north of town very close to the Academy. We planned to build our retirement home on it. But those plans changed later, and we eventually sold that lot for a good profit.

In spite of the fact that I had to be away from home so much, our life in Colorado Springs was a very happy time for our family.

CHAPTER EIGHTEEN: TIME TO MOVE ON

During my tenure at ADC Headquarters, I learned a great deal about many things that would come in handy in my later career. Visiting and evaluating different units, permitted me to observe both good and bad leadership at command levels. I took note of the things that effective commanders did and did not do to make their units excel. And I also saw the other side, where commanders failed to look after the details that were vital in making their organizations fully prepared to fulfill the mission under pressure. Soon, I was able to sense these factors very shortly after visiting an organization. The aura of a fully effective squadron hung in the air, and you could 'feel' it when you first arrived. You could detect it in the way the pilots and crew chiefs spoke and acted. Esprit was alive and almost visible. In failed units, the opposite was also detectable. I made note of those factors and studied them carefully. You can learn so much in just paying attention when you observe what is essential for success in forming a solid team with everyone on the same page and geared to achieving mission success. That's what I someday wanted to do.

I also had the opportunity of working side-by-side with some very outstanding people. Major Dave Winn was someone I truly admired and respected. He was a very cool guy and knew how to get things done. I loved working with him and flying with him. Our families became quite close. In 1968, when I was in Korea with the 318th FIS, Dave was shot down in North Vietnam and imprisoned in the Hanoi Hilton until released in 1974. He was a Colonel then and later was promoted to Brigadier General. He retired in Colorado Springs where Alice and I visited him in 1985. Lt. Col Howard (Scrappy) Johnson was another terrific guy. An F-104 pilot, Scrappy took an F-104 Starfighter to over 95,000 feet, a record altitude for a jet aircraft at that time. Col. Bill Moore, an F-106 jock had been a POW in Korea and had undergone the brainwashing techniques introduced by the Chinese and North Koreans. Col. Sam Grashio, who headed up the team after the departure of Col. Vic Milner, had been a survivor of the Bataan Death March in World War II. He, and several others including William Ed Dyess, successfully escaped from the Davao Prison Colony,

Philippines, and after hooking up with some Filipino guerillas, were evacuated to Australia by submarine. Grashio and I flew a lot together and he used me to edit his post ORI briefings. After our retirements from the service, the Grashio Family and the Cunneely's were close friends and neighbors on Lake Pend Oreille. All in all, my tour at ADC was very worthwhile for me professionally as well as for my family. We made many friends there and we have many fond memories of the springs. In 1960, Alice and I bought a piece of property on Falcon Estates, just north of our house and fairly close to the Air Force Academy. Our plans were to someday build a retirement home there. The lot was on an elevated piece of land that provided terrific views of the mountains and beyond. Often we would visit the site, picnic there and talk about the future. It was a great time in our lives and the future still looked bright, but little did we realize what was coming over the horizon.

But before all of that, Alice Marie and I were given another blessing. In 1960, Alice informed me she was with child again. This time she was able to go the AF Academy for her pre-natal visits and, as the hospital there was scheduled to open that fall, we expected our next addition to be one of the first babies to be born at that beautiful new installation. Everything proceeded well and Alice was very pleased with her doctor at the Academy, but again, Alice's propensity for early deliveries again foiled our plans. A few weeks before the hospital opened, Alice went into labor. And this time, I was away on an ORI/TAC in Arizona. Poor Alice had to call on her friend Rosemary Drew to drive her to the hospital at Fort Carson in a mad dash to beat the stork. They barely made it and not much before they were to run out of gas at two o'clock in the morning. The folks at Fort Carson had never heard nor seen Alice Cunneely before, but they had no choice, as a baby was coming and coming quickly. Thus did Michael Joseph arrive in November 1960, a beautiful red headed lad, healthy and content, unaware of the confusion and anxiety that his early forthcoming had caused. I learned all about this the next morning in Tucson, grabbed a T-Bird and headed home to see my new dependent. I was also given some heat over my failure to be where I should have been when needed the most. I pleaded guilty as charged.

In 1961, I was selected to attend Air Staff and Command School at Maxwell AFB, Alabama. It would be a one-year assignment beginning in the summer of 1962 and considered a must for field grade officers if they wished to attain the grade of full colonel. I let it be known that I did not want that assignment. What I did want was to return to flight duties in a squadron, preferably as an ops officer. However, it certainly looked as though Alabama was the next stop for us. I even flew down to Maxwell and spent a few days with my friends searching for a house for our family. But then a miracle happened. A general officer I knew, told me that he could get me off of the school list without prejudice if that's what I wanted. I told him that's what I wanted, and he did it. So, I was off the hook and my tour at ADC was extended another year. I was very happy about that. A year at Maxwell was not my cup of tea.

I continued my work in evaluation, but there were some major changes. Headquarters USAF told ADC that the ORI function had to be broken away from the TACEVAL section and had to be run by the Inspector General's Office. This required a major overhaul, and our old organization was split in two. I was sent to the IG section that would run the ORI's. I now worked for Col. Vern Stinson, a World War II pilot and former POW, Col. Grashio having left for Air War College. ADC now had two evaluation teams harassing the units, the TACEVAL team from OPS, then the ORI team from the IG. The IG's reports were sent to USAF and were used to estimate the operational readiness of the AF's combat units. Thus, they were important in assuring that the AF was able to complete its mission. Thus, we took our job seriously. As a team member I kept myself prepared to evaluate the unit I was assigned to and to do it fairly and meticulously. Finding a unit unsatisfactory was a serious thing and was not done without a great deal of discussion. The team captain and his team members had to show beyond doubt that the unit did not meet AF standards of performance in combat readiness. One of the most depressing experiences of my career with the ORI Team was my last evaluation. General Bill Elder was now the Command IG and he assigned me as team captain at the 27th FIS at Loring AFB Maine. This squadron had failed a previous ORI, and this was a repeat inspection. Thus, a great deal was on the line for the commander and his staff. I thought things had gone well, at least on the flying end, but when I met with my armament evaluators, they informed me of serious violations of safety in the breaking out and loading of the nuclear weapons. This constituted a must failure of the unit. The evaluators were experienced people in their field and were unanimous and positive in their findings. This was not what I wanted to hear. When I called

Gen. Elder to tell him of our findings, I had hoped he would find a way other than to flunk this outfit, but he did not. He left it strictly up to me. I thought it over and concluded I had no choice. My meeting with the squadron commander was unforgettable and I truly sympathized with him. His people had let him down and they had ruined his professional career. He was devastated and angry, especially with me. On that day, I did not like my job. Later Elder told me, he was waiting to see what my decision would be. He told me that if I had not flunked the 217th FIS, he would have had my butt. I believed him.

That was my last ORI with ADC before I was reassigned. Col. Grashio, after completing the Canadian War College was assigned to the Ottawa SAGE Sector at RCAF station North Bay, Ontario, Canada. The Sector was collocated with the Northern NORAD Region, and it was looking for an officer qualified as an air defense specialist to work in its evaluation branch. As the Sector and the Region were not yet operationally ready, it appeared that there was a lot of work to be done there. Also, I was assured that I would be able to continue flying the F-106 with the Loring squadron, (the one I had just recently failed on its second ORI). North Bay was a small town about 200 miles northwest of Toronto. It was a beautiful landscape, hilly and dimpled with hundreds of lakes surrounded by forests. I flew up there and met with the Canadians, and Grashio, who was very popular with the Canadians, introduced me around. I liked what I saw. I believed that this would be a great place for the family. There was a good Catholic school for the kids, and we would be fairly close to New York. At last, we could see our families again with more regularity. I had a chance to meet the RCAF officers I would be working with, and I liked them. I had had previous experiences with the Canadians at ADC setting up large-scale exercises that originated in Canadian airspace and utilized RCAF aircraft and aircrews. I always found them to be helpful and effective. It would be fun working with them, I was sure. I decided to take the assignment.

And so, in the summer of 1963, we sold our little home in Colorado Springs and headed east. We said farewell to all of our remaining friends and neighbors and began a new adventure. It had been a great five years for us. We had added two new children and now had two of them in school. I was now a field grade officer and I had accumulated a wealth of experience in both flying and staff work at Command level. I had expanded my coterie of friends to include many influential officers who had attained high grades and important positions in the Air Force. I was a Senior Pilot with a total of 2,806 hours of flying time, almost all in single seat fighter aircraft. I was still alive and very happy. God had been good to me. And, as we left Colorado, just as we had left Tacoma five years previous, Alice Marie was pregnant.

Our trip east was difficult. Young Gene had come down with mumps a few weeks before our departure date. As I had never had that ailment I had to stay away from Gene and take some shots. As we were packing and moving, this made things more difficult than normal. We set off on a hot sunny day packed into our 1957 Chevrolet station wagon, sans air conditioning. At about six in the evening we arrived in Hiawatha, Kansas. We checked into a motel with a pool that helped cool the kids off after a long tough day in the car. Alice and I were tired, hot and we both had headaches. After dinner, we hit the sack early in preparation for another long day on the morrow. We had just fallen asleep when the sounds of sirens awakened us with a jolt. Tornadoes were in the area, and we all had to evacuate the motel and take shelter in a National Guard Armory nearby. There we stayed until about three in the morning when the all-clear was signaled. By this time, we were a mess. Returning to the motel, we tried our best to recapture some rest before it was time to hit the road again. It was a bad first day on our long trip to New York where we planned to visit with family before reporting into RCAF Station North Bay. Alice, being pregnant, was not in any good shape for this trip. And that would prove disastrous when we arrived in New York.

CHAPTER NINETEEN: RCAF NORTHERN NORAD REGION—I GO UNDERGROUND

Per ardua ad astra
Through adversity to the stars.

-Motto of the RCAF

It's customary that when you leave an organization after several years of service, they give you a going away party and a gift. Such was the case when I left ADC after a five year tour. I still have the beautiful silver bowl inscribed to Major Eugene P. Cunneely, from the ADC ORI Team. Ed Cleary, who was a gifted artist, did a sketch of me in a SPAD, chasing a German WWI bi-plane. This commemorated an ORI in which in the confusion of air traffic, I lost the flight I was chasing and ended up returning with another flight of the same squadron. We also have a silver tray from the 317th FIS, and other gifts including a beer mug from the RCAF, a beautifully done portrait of my F-106 when we left the 95th FIS, a complete Saudi Arabian clothes outfit when I departed USMTM, and so forth. All of these represent great memories of my days of service and of those with whom I served. That's why I still retain and cherish them. I also had some great memories of flying adventures, too numerous to insert into my narrative. (I may include some at the end of my story.) In retrospect, I believe that my tour of duty at Headquarters ADC was pivotal in my career. There I had close personal contact with some of the finest professionals ever assembled from whom I learned so much about leadership and management. I also worked for some intelligent, capable, and proven General Officers, most with combat experience, whom I studied carefully and from whom I learned much about how to handle difficult problems in personnel, operations, maintenance, and simulated combat. I carefully watched how these experienced and successful people reacted to seemingly insolvable problems under pressure, with poise and brilliance. Now, not all of them did it right every time, but I learned from their mistakes as well. It was a terrific educational experience that prepared me for what was to come in my own career. And I put what I learned to good use when I needed to.

Our visit to New York was great. We split time between my dad's place in Patchogue and with Alice's mother in Manhattan. But Alice began to have problems with her pregnancy and eventually had to be hospitalized in the Bronx. This was bad news. All efforts were being made to stabilize the situation, but it became clear that she could not drive with us the rest of the way to North Bay. As the time arrived for me to leave, we decided to leave Alice. Sadly, the boys, Mary and I departed on the last leg of our trip to Canada. We did make a great stopover in Niagara Falls, which mesmerized the children with its forceful energy and overwhelming booming beauty. But leaving Alice behind was a real setback and a great worry for us. Arriving in North Bay, we rented a small cottage on the shores of Lake Nipissing. There we enjoyed great days swimming, playing in the sand, and hiking. But the children and I missed their Mother Alice who was still recuperating in New York. I spent several hours each day searching for a permanent home for us. There was a great influx of personnel into North Bay with the forthcoming operational readiness of both the collocated Ottawa Sector and the Northern NORAD Region, which was putting a great deal of pressure on the local housing market. The pickings were few and I was beginning to become a little discouraged. While reporting in for a new assignment and then searching for housing, I hired babysitters for the kids.

Living in the same court with us was the family of Brig. Gen. Harrison Thyng. Gen. Thyng was a double Ace, having shot down more than five enemy aircraft in both World War II and the Korean War. Gen. Thyng had two teenage daughters who were my babysitters. They were a great family, and we became close friends. I later had the privilege of flying with Gen. Thyng and that was quite an experience in itself, which I will get into later.

After a few weeks, Alice was cleared to fly to North Bay. Her brother Tom drove her to the airport in New York, and she flew in landing right on the RCAF Station where the commercial aircraft also came in. It was a happy reunion. But Alice was not over her problems. A few days after her arrival, she suffered a miscarriage, and we lost what would have been our fifth child. It was a very sad day for our family.

North Bay was a small town approximately 200 air miles north of Toronto and west of Ottawa. It was situated on the north shore of Lake Nipissing, a large lake, about 40 miles in width. It was not a deep lake but noted for its sandy beaches and excellent fishing. It was on this lake that I had my first experience with ice fishing, which turned out to be both successful and pleasurable. North Bay was a confluence of rail lines and a hub for both lumber and mining. The RCAF Station was on the north side of town and was home to air defense CF 101 Voodoos. On the station were also some Canadian T-33s similar to our American versions, except that they had a more powerful Rolls Royce engine. I would be flying these aircraft during my three-year tour with the Canadians. I had also received permission to continue flying the F-106 with the Loring, ME squadron which was under the operational control of the NNR. An interesting aspect of the Canadian T-33s was that they could be put into a spin, a maneuver that was prohibited in the American version. So, for the first time in my flying career I was able to spin the T-bird. Fortunately, the plane came out of the spin as advertised.

The Canadian people were friendly to all Americans and the North Bay folk perceived the influx of military personnel as a great boon to their economy, which had depended a lot on tourism. My job was Chief of the Evaluation Branch and I worked for a Colonel Jim Wood. Col. Wood was a true character and we got along right from the start. His knowledge of air defense was very scant, and he left it totally up to me to set up the shop and put things in motion to do what had to be done as both the NNR and the Ottawa Sector, which was co-located with the Region became operational. Both the Region and the Sector were built in a building underground. To enter the building, you had to take a bus from either the RCAF Station or a south entrance. The site was hewn out of a mountain of basalt rock and took over four years to build. It had to be located near a lake to provide the water and cooling necessary. The underground building was in the shape of an H and was positioned on huge springs to withstand a nuclear blast. A great vault-like door about two feet thick sealed the entrance. When there was an alert called, that door was closed, and no one could enter or leave until an all clear was sounded. The inside of the building was comfortable and when at work, you easily forgot that you were in a subterranean setting. When you exited, you did not ascertain what the weather was, or whether it was night or day. It took some getting used to.

NNR was the last SAGE region to come online. Under NNR was the Bangor Sector, located in Brunswick, ME and the Ottawa Sector. Bangor was an operational SAGE sector and was commanded by RCAF Air Commodore Michael Pollard. The Ottawa Sector was also commander by a Canadian officer, Air Commodore Sharp. RCAF Air Vice Marshal Harvey commanded NNR, and his vice-commander was Gen. Thyng. So, the region was a mixture of USAF and USAF personnel with RCAF holding the command positions. I found it easy to work with the RCAF people with only a few of them displaying any animosity to Americans. However, one of them was the CO of the Ottawa Sector and that lead to some interesting interplay as we got into bringing the Ottawa Sector into operational status.

My family continued to enjoy the cottage on Lake Nipissing while I frantically searched for a house. Several leads failed to pan out, but one day someone told me that a house on Queensway Road might be coming up for rent. I immediately headed over there and luckily caught the owner, Mr. (Swede) Swenson, who was at the house doing some repair work. I introduced myself and told him of my needs. He was a slow moving, slow talking man, but very affable. I looked over the house and told him I would truly like to rent it. He said, “Well, dat vould be alright vith me.” And the deed was done. We were so lucky. It was a good house, large and comfortable, with a garage, a backyard and a basement, that eventually Mr. Swenson would finish as a family room. It was also convenient for me for work, the kids for school, and we ended up with wonderful neighbors. Getting that house made things so much better for us during our three years in Canada and we still remember fondly our time there. Shortly after, we moved in and our tour of duty with the RCAF got underway.

North Bay was a small town that grew as a hub for the timber and mining industries in northern Ontario. The Canadian people were friendly and gracious to all of the Americans who were now coming into their neighborhoods in increasing numbers. This placed some pressure on the housing situation, traffic and schools. But it also favorably helped the economy. Gene Jr. and Jimmy were enrolled at the local Catholic school to which they could walk to and from except in the most inclement weather. Typically, they asked to be driven to school, but I asked, “What did the Canadian kids do?” When they admitted that most walked, I decided that if the Canadian kids could walk to school, then the American kids could join them. Mary, we enrolled in a kindergarten at a new Catholic school that was further away, and so we arranged for her to commute by taxi each day, which made her feel pretty special. The kids seemed to enjoy their new environment and soon made friends with their new neighborhood kids. Alice had her work cut out for her settling in our new digs which was a roomy comfortable home and convenient for my work, schools and shopping. One problem was we had only one car that Alice and I shared with her often driving me to work each morning. Things soon fell into place as we began our new life in Canada.

My new job required me to work closely with my RCAF compatriots whom I liked and appreciated. However, this required diplomacy and careful use of language. The NNR and the Ottawa Sector were still not operational and had a long way to go to attain that status. The Canadians, while as bright and knowledgeable as the Americans were not as aggressive and much more laid back in their approach to meeting deadlines. This caused frequent tensions between the RCAF commanders and the USAF underlings. Nowhere was this more apparent than the impatience displayed by Gen. Harrison Thyng, the NNR Deputy Commander. He was a true fighter pilot, a double ace from World War II and Korea, and a man who did not suffer fools gladly. It was not long before his reputation for letting staff officers know his unhappiness with their poor performance and failure to accomplish objectives at the established timelines became widely known. He was the exact opposite of AVM Harvey, who was a delightful man, charming and always prepared to understand the reasons provided why things were not going as they were planned and programmed. Staff officers shuttered when AVM Harvey was away from station and Gen. Thyng would be chairing the staff meeting. One of the best ass chewings I ever received came from Gen. Thyng. Just prior to Ottawa Sector's achieving operationally ready status, I ran a small no-notice exercise. I had ten "enemy" target aircraft penetrate the Sector from the north. The Sector was provided early warning intelligence data and their DEFCON status was raised gradually signaling the beginning of an attack. However, the senior controller failed to notify the units and only at my urging did he do so and summon the Sector Commander to the Control Sector. This delayed their response, and the Sector failed to intercept the target aircraft until they had penetrated deep into Sector airspace. In my debriefing, I covered their problems in depth and made important conclusions and recommendations. Thyng was furious that I did not come down harder on the Sector and really blister them. He thought that I had been a wimp and had gone too easy on the Canadians. I thought that it was better to start easy on them and that way encourage better performance in the future. While I believe I was right on this, it perhaps showed why Gen. Thyng was a fighter Ace, and I was not.

November 22, 1963

In November 1960, America voted in its first Catholic president, John F. Kennedy. He was a popular choice because of his youthful looks and his gift for eloquence. He also had an attractive wife and delightful children who made the White House an interesting place and more like an American household. While off to a rocky start, Kennedy was generally liked by the American people and admired and approved abroad. But on Friday, November 22, 1963, Kennedy was assassinated in Dallas by Lee Harvey Oswald. His murder rocked and saddened the world. The Canadians were very sympathetic and in the late President's honor, they held a memorial service in their local Cathedral in North Bay with full military honors. It was a moving ceremony and again displayed the friendship of the Canadian citizenry for its American neighbors.

During my assignment in Canada, I flew the RCAF T-33 and, when possible, I flew the F-106 with the 27th FIS at Loring AFB, in Maine which was in the Ottawa Sector and under NNR. Gen. Thyng, also used his influence to procure a USAF T-33 from Kincheloe AFB for our use at North Bay. So, I was able to keep up my flying although my currency in the Delta Dart was a problem. Whenever I could, I flew into Loring and flew the F-106 and sat alert. But most of my duties entailed helping the Region improve its air defense capabilities derived from my experience with the ADC TAC/EYAL team. Although I was only a Major, I had compiled a great deal of firsthand experience in all facets of air defense including ground control tactics and strategies. It just took some time for me to impress the RCAF staff enough so that they would pay attention to my suggestions. Gradually, we began to make some headway, and in due course, both NNR and the Ottawa Sector attained operational readiness and passed their NORAD Readiness Evaluations. My standing with the RCAF staff rose.

My years with the RCAF in North Bay flew by. My family also enjoyed the tour. The kids thrived in school and in their new surroundings and with their new friends. Being close to New York, we were blessed with visits by Alice's mother and brother Tommy. Alice's mother, thank God, was with us when Kathleen Anne was born in January 1965. We have fond memories of her playing hockey with Gene Jr. and Jimmy in our basement. My father was also a frequent visitor, arriving and departing by train. We were also able to spend some vacation time in New York, where we visited both families. My father had by now retired from the railroad and owned a home in Patchogue near my brother and his wife. So, we had some good times with them, and they all got to meet our children, some for the first time. The kids especially loved Uncle Tommy and swimming in the pool in Patchogue. I also took them to Fire Island where they were able to swim in the Atlantic Ocean for the first time. They had a marvelous time frolicking in the surf and being knocked down by the ceaseless waves crashing over them. Back in North Bay, our family enjoyed the many lakes that were sprinkled around the countryside. Lake Nipissing, of course, was a fabulous lake for swimming and fishing. Also, Trout Lake, adjacent to the RCAF station was another beautiful lake with several islands that we hiked on and explored.

Hockey

As a kid, I loved the game of ice hockey. Dad and I frequently attended the Sunday afternoon games at Madison Square Garden. The New York Rovers was the home team and they played in a triple A league that included the Hershey Bears, and teams from Atlantic City, Boston, Brooklyn and others. We also would stay and watch the New York Rangers of the NHL. There were only six teams in the NHL in those days and the competition was fierce. All the players, with one or two exceptions, were Canadians. They were terrific skaters and puck handlers, and we were enthralled with how they were able to accomplish so much while going at such speed up and down the ice, stopping on a dime, reversing directions and skating backwards. I never got tired of watching hockey and still love the game. Gene Jr. was now six years old and seeing the other Canadian kids skating, he had to have ice skates too and he was out there with them. It seemed that every other home in North Bay had a backyard ice rink and there were ice rinks all over town. North Bay also had a semi-pro hockey team that played in the local arena, and each high school had JV and varsity hockey teams. Gene was soon playing with a peewee team. They practiced and played their games outdoors, and on most days, it was very cold out there. The boys had to be very well dressed from head to toe, and that went for the mothers and fathers who came to watch them. Gene had a lot of catching up to do in his skating, but with every day he got better and better on his blades. I recall his first game. It was a typically cold afternoon, but I was determined to see how Gene would do. I dearly hoped that he would not be embarrassed out there and come away feeling badly.

However, in the first period he scored a goal on a very well executed shot from the right side. I was very proud, especially when I heard some of the fathers say, "The Yank scored." I did not have to worry about Gene from then on. He progressed nicely and seemed to have a knack for being in the right place for goal scoring. The other Canadian kids liked him, and I could see that he was having fun. Later in our tour, Jimmy also began playing hockey. Jimmy was an excellent skater from the beginning and was selected for an all-star team. Both boys received a good start in their love for hockey at North Bay, a love they carried for the rest of their lives. It served them well.

Scollard Hall

Early in my tour in North Bay, I established a relationship with Scollard Hall College. Col. Sam Grashio's son, Sam, was attending that school which was an equivalent to our high school. It was a Catholic high school, all boys, which also provided dormitories for boys from out of town. It was a school with an excellent academic reputation, and I enjoyed meeting Father Pat Mackan, the football and basketball coach and athletic director, and some of the other priests. Somewhere in our conversations, either Sam or I mentioned my background in basketball. The school was having some problems in its basketball programs and Father Pat began questioning me for some ideas about how to shape up the program. Little did I realize what Father had in mind. In the middle of the 1963-64 season, he called me and invited me over to the school. There he asked me if I would take over coaching the freshman team for the rest of the season. The team was below 50 percent and floundering. I told him, I would have to get permission from my boss as my real job was Major, USAF. Coaching would require my having to leave work early and may interfere or conflict with my AF duties. I spoke with Col. Woods, who had to speak with AVM Harvey, and I got an okay. I took over the team and I could see right away there was a great deal of work to be done. The team needed a lot of work in fundamentals and that's what we did ad nauseam, even on Saturday mornings. But I also worked them hard on their foul shooting and rebounding, which paid off in their very next game. In fact, we began winning and the next thing we were in the playoffs. Unfortunately, we lost in the championship game by a close score, but it surely was a great turnaround and I felt good about what I had accomplished. It was fun being back in the game, I had to admit.

But I did not expect what followed. After the season, Father Mackan asked me to take the varsity. This was a large escalation in commitment, and I had to, yet again, go back to Col. Woods and get permission. This would mean that during the season, I would be leaving work at about three o'clock each day, and head for Scollard Hall for two hours of practice. I was surprised when I received the okay, the AVM believing that there was a large personal relations benefit from a USAF Major coaching a Canadian high school basketball team. Perhaps he was correct. For the next two seasons I coached Scollard Hall to two North Bay NOSSA senior championships. I remember well our final championship game against Algonquin when we had to come from behind in the fourth quarter. Winning those championships were great for the school in that, being primarily a hockey school, they put basketball back into prominence. I thoroughly enjoyed my relationship with Scollard Hall, and it reinforced my inner belief that, if things had turned out differently, I could have been successful in my first goals of becoming a teacher and coach at the high school level.

Father Pat Mackan was the man who brought me back into high school sports at a later stage in my life. He was a great priest and educator. His first love was sports, and he was an excellent football coach and fierce competitor. When he turned over the basketball team to me in 1964, it hurt him a little inside and he always wanted to travel with the team and sit on the bench during the games. This was okay with me, but one night I had to put an end to it. We were playing Chippewa and it was going to be a close contest for sure. We were both tied for first place and playing on our home court, we needed to win badly. But in the first quarter, things did not go so well for us, and we dropped behind. On top of that, some referee calls went against us. Father Mackan, never one to restrain his emotions let the officials know his displeasure by jumping off the bench and stating so in no uncertain terms. A technical foul immediately ensued. Not five minutes later, the same thing happened, with another technical foul against us. Going into the locker room at half time we were behind by seven. After my half time talk, I sent the team out, but as Father started out, I stopped him. I said, "Pat, you stay here." Father was shocked, "What do you mean?" "Pat, you're killing us out there. I can't afford any more technicals. You stay in the locker room, and I'll have Mulligan give you the score every five minutes." I can't describe the look on the face of that priest when I ordered him to stay off the bench, but he obeyed. He must have smoked a pack of cigarettes during that second half, but the outcome was worth it as we came from behind and won. I loved Father Pat Mackan. We had some great times together and we shared some good times and some tough ones. He died from heart problems not too long after I left North Bay, but I won't ever forget him.

In 1972, I was asked to return to North Bay and address the Senior Athletes' Banquet at Scollard Hall. I took a T-33 and flew from Dover AFB to RCAF Station North Bay. I had a memorable reunion with several old neighbors and friends. It was an honor speaking to the seniors, the faculty and the parents at the school where I had three wonderful years (1963-66) when I was a Major. Now as a Colonel, I returned and I attempted to say something to inspire these young men to take what they had learned at their school, in the classroom and on the playing field, and put it to use in their lives. I hope I succeeded.

Living in Canada

Living in Canada was a great experience for our family. Our children received a good Catholic education and the Canadians, being wiser than we Americans in this area, permitted us to direct the property taxes that came from our rent to be directed to Holy Cross Parish School. We should do the same in the USA. Another benefit we received was the strength of the US dollar versus the Canadian dollar. As we were paid in US dollars, we could purchase more with our money in Canada than we could living in

America. Because my position was viewed as diplomatic in nature, I, and all the Americans received privileges at the Provincial liquor stores. Each month I was able to purchase a case of liquor at about one-third the retail price. Naturally, I took advantage of that privilege as it was a great savings, but at the same time all the Canadians, knowing we had this privilege, did not hesitate to impose on our hospitality. The Canucks were mostly a party crowd, and we had an active social life in North Bay. Although the ladies were not permitted in the Officer's Club, they made up for that with a plethora of dinner parties and holiday parties. I enjoyed flying the RCAF T-33, but I had a difficult time keeping current in the F-106, as I had to rely on the 27th FIS at Loring to provide me some flying time in exchange for alert duties. I also received my first experience in ice fishing on Lake Nipissing. With Father Mackan, Sam Grashio and a few others, we rented a shack on the lake and went out on a very cold day. The shack, thank God, had a small potbelly stove, but most of our fishing was done outside through small holes drilled in the ice. I mean it was cold and when a little wind came up, the chill factor dropped through the floor. In addition to some stew that we had to keep our bellies warm, someone thoughtfully brought along a bottle or two of booze to maintain our spirits. All in all, it was quite an experience. We did catch some fish, and we had lots of good cheer and laughs. It was not an experience I won't ever forget, nor will I repeat.

We Buy More Property

Alice and I had bought some property in Colorado Springs and our intentions were to build a retirement home there some day. That's how much we loved our days in Colorado. But we also loved the Pacific Northwest and I had told some friends to be on the lookout for some good property there, preferably waterfront, and let me know about it. One day I received a call from my friend Sam Grashio, who had now retired and was living in Spokane and working for Gonzaga University. Sam told me about some lake property in northern Idaho that he thought I should look at. So, I put in for a T-33 and with a friend, Major Bill Davenport, we flew from North Bay to Spokane. Sam and Bob Jones, a longtime friend of Sam's and a fellow alumnus of Gonzaga, picked us up and we drove to Sandpoint, Idaho. Bob was a Realtor and was handling the sale of waterfront lots on Lake Pend Oreille. It was a beautiful, warm sunny day and as we hiked the wooded areas on which these lots were located, all of us were impressed at the beauty of this place. To be brief, I ended up purchasing two of these lots, each with 100 feet of lake frontage. Sam also bought one. Flying back to North Bay, I wondered how I was going to explain my impulsiveness in buying two lots in northern Idaho on a lake Alice had never seen and which we may never use or see again. Thank God, Alice trusted my judgment when I told her what I had done. I hoped it would work out somehow.

Next Assignment

During my years in North Bay, the skies darkened on the far side of the world. President Kennedy, foolishly as it turned out, got us involved in the small country of Vietnam. Perhaps in an attempt to make up for his failures in Cuba, this intervention with American troops, resulted in the years to come in the worst military and diplomatic failure in American history. Where we were stationed, we just began to hear more and more of the growing participation of American troops and air forces. Lyndon Johnson, who succeeded Kennedy, appeared to be escalating our participation with growing frustration from the military about the restrictions placed upon it by the administration. I was hoping to return to a flying assignment in ADC after my assignment in North Bay with my goal of becoming a squadron commander within the next few years. I requested a flight assignment and, thanks I believe to some friends at ADC Headquarters., I was assigned as operations officer for the 318th FIS, McChord AFB. I could not have been more pleased. An F-106 squadron at my old base, a base I loved, what more could I have asked for?

CHAPTER TWENTY: I AM BACK TO MY FIRST LOVE—FLYING FIGHTERS

The Fighter Pilot:
Say what you will about him... Arrogant, cocky, boisterous, and a fun-loving fool to boot—he has earned his place in the sun. Across the span of 50 years, he has given his country some of its proudest moments and most cherished military traditions.

-Friar Tuck

In June 1966, we bade farewell to our friends in Canada and left for our new assignment. First, we headed for New York to visit with our families before heading west again. We had a great visit. A year previous, we had bought my brother Joe's 17-foot Thompson boat and trailer. Alice, Mary, and Katie were going to stay behind in New York and attend the wedding of Alice's brother Tom and Mary Lou. The boys and I were going to drive across the country in our new 1966 Mercury station wagon, trailering the boat and camping out in various state and national parks. It was well planned, and I hoped would be a great educational adventure for the boys. It turned out to be quite successful. Some of the highlights were our stops in the Badlands, Mount Rushmore, Yellowstone Park, and finally Lake Pend Oreille, where the boys first saw our new property and stayed a couple of nights in Jack Chambers' cabin. There we listened to the critters scurry across the beams in the attic and lived by the light of Jack's kerosene lamps. At the camp sites across America, we stayed in our tent, slept in our bags, fished on Yellowstone Lake, saw bears and more other wildlife than we expected and truly had a great trip. The boys and I had a great time together and they really saw what a great nation we have from coast to coast. It was a memorable trip that all of us look back on with fondness and wonderment. The boys were truly excited about Lake Pend Oreille. They swam and fished it from morning until night. They also water skied behind the Thompson and took in the beauty of Northern Idaho. They were thrilled to learn that we owned some property on this most beautiful lake. Finally, arriving at McChord, we moved into our house on the base, without furniture; tired but happy to be there. And I was ready for my new job. Back to flying again!

We had to wait a few weeks for Alice and the two girls to arrive by air, and we were truly happy to see them come in safely. We had a nice house on base with three bedrooms and a wide backyard. Being the Ops Officer of the fighter squadron, I had to live on base, and I enjoyed doing so for the convenience of it. Alice had a lot of work setting up everything in the house and arranging for schools for the kids. Thankfully, Gene, Jim, and Mary, would be attending a Catholic school, and the base would provide bus transportation for them. As ever, Alice Marie was terrific. In no time she had the house arranged into a warm and comfortable home.

Our kids soon met other kids in the base housing and made fast friends. The boys and Mary enrolled at St. Francis Cabrini School in Lakewood. This was a fine school, and this we knew from our former assignment at McChord and our days in Lakewood on Lake Steilacoom. We both loved the Pacific Northwest and were delighted to be back. I was also pleased to be with the 318th FIS. I knew many of the pilots from my days with ADC TACEVAL and I knew the flying area like the back of my hand. The 318th was a small squadron of pilots and ops personnel only, as it was when I left McChord in 1958, when maintenance became an organization of itself. The commander was Lt. Col. Merle Schuder, whom I knew and liked a great deal. He was easy to work for and would depend on me to run the flying operations for the squadron. I first left on a TDY to Tyndall for a few weeks of refresher training in the F-106, because, in fact, I had not been able to fly as much as I wanted while at North Bay. I really enjoyed being back in the cockpit on a daily basis and I learned a great deal about many of the newer systems and tactics that had been brought in since my ADC days. On my return, I took over the reins of the operations officer duties. The ops officer is the head pilot and is responsible for all flying operations including training and air defense alert. Safety of flight is a lead responsibility, and he must ensure that every pilot in his squadron is capable of flying the mission he is scheduled for safely and successfully. The ops officer uses his flight commanders and his aircrew evaluation and flight safety officers to assist him in meeting these responsibilities. Any mistakes in my judgment can be life or death decisions. I made it a personal duty of mine to fly with every pilot in the squadron and personally evaluate his performance in the air. I tried to never allow personal friendships interfere or cloud my professional judgment. To do so would be callous and perhaps fatal. This was a 24-hour, seven day a week job, that I had to take seriously, but that I loved. Being with and around fighter pilots and fighter aircraft was a terrific way of life. We had fun in everything we did, no matter how treacherous much of it was. We partied a lot of the time, but we also were businesslike most of the time. A majority of the guys were married men with families, and they well understood the importance of knowing their job and being prepared for whatever exigency might happen and often did. The more I worked with my squadron pilots, the more I admired and appreciated their skill and professionalism. That made my job a little easier.

Things had changed at McChord since I had left in 1958. The air base had its own Approach Control that reached out a few hundred miles and was capable of vectoring our F-106s out to and back from our training flights without having to adhere to the standard published departure and letdown routes. It was a great system, well ahead of its time and one that other bases were trying to establish. In a region with lots of bad flying weather and high mountains close by McChord Approach Control was a terrific asset to safety.

The 318th had some experienced flight commanders and I used them to oversee the training of their pilots and to ensure that they never put one of their pilots in a position he was not ready for. But I also used my flight examiner to back up the flight commanders and when there was a doubt, I myself flew with the pilot in question. Of course, even as Ops Officer, I was subjected as everyone else to annual flight checks and annual instrument flight checks. I tried to fly three times a week and sit alert with each flight once monthly. The 318th had an additional mission nicknamed White Shoes (later College Shoes) given to it in 1963. When the Soviets began to penetrate the Alaskan airspace on a regular basis, the F-102s of the 318th FIS stationed in Alaska did not have the range necessary to: first intercept them, then stay with them until they left Alaskan airspace, and finally return home. The 318th FIS and the 498th FIS out of Geiger AFB, Spokane were deployed, nine aircraft in all, to handle this mission. The F-106s sat alert at Eilson AFB, Fairbanks, Galena AFS and King Salmon. The 318th bore the brunt of this additional mission and I had to rotate pilots to Alaska on a regular basis as well as handle our own McChord alert commitment. The Alaska mission provided many problems personnel wise, but we also had some exciting intercept missions with Soviet Bear bombers in the Arctic circle frequently testing our defense systems.

I remember my days in the 318th FIS from 1966 through 1970 as some of the happiest of my career. I was flying an aircraft I loved, and I was supervising an organization that I saw as truly top notch in effectiveness and esprit. Our inspection results were always Outstanding and the reputation of the 318th was now command-wide. At home, the family was thriving. Alice Marie and the kids were loving the Northwest. The boys were into Little League on base and hockey at the Lakewood Ice Rink. Mary had begun swimming, thanks to the Dean Family influence and we all enjoyed boating and picnicking at American Lake. Living in the beautiful Northwest, with Mount Rainier ever present in the southeast, standing majestically penetrating high above the Cascade Range, was a joy despite the rainy weather. We skied in the winter, often at White Pass and the kids gradually improved their skills and broadened their love of the outdoors and the natural wonders of God's creation. Flying over the Northwest was never boring. On any given sortie, you could look down and see the blue Pacific Ocean, the Olympic Mountains, the straits of Juan de Fuqua, four or five volcanic peaks of the Cascade mountain range and the farms and cattle ranches of western Washington state and British Columbia. Alice and I enjoyed an active life as parents and as an integral part of a vibrant organization which had its share of interpersonal problems that looked to us for guidance and leadership. My major goals were to ensure that the 318th aircrews were well trained and prepared to fulfill the vital mission given them, i.e., to defend the country against any invading bomber force, and secondly, to ensure that they flew their missions and met their alert requirements safely. My goal was to never lose a pilot. And toward that goal, I dedicated my total efforts every day.

Flying the F-106

As ops officer, I had to control my own flying schedule. I tried to fly three times weekly and sit alert once monthly. Flying the F-106 on a regular basis was great and my proficiency in the aircraft and with the MA-1 fire control system became better and better. We began to obtain some aircrews returning from Vietnam with combat experience in the F-105, F-4 and other combat aircraft. They were on the whole excellent pilots and within a short time became combat ready. Doug Brenner, an ex-Thunderbird pilot came to us after a tour in Vietnam in F-105s. Doug became my assistant ops officer and was a tremendous addition to our squadron. Dan Barry, who was our Flight Evaluation Officer, and Jon Dean, the Weapons Instructor Officer, formed a great team of supervisory aircrews that kept our aircrews in top condition. Whenever we received an evaluation from ADC, the 25th Air Division or other higher headquarters team, the 318th FIS received the highest ratings. They made my job easier. Doug and I pushed our aircrews to become proficient in formation flying techniques, including formation approaches and landings day, night and in weather. In 1967, I felt that the 318th FIS by far was the best squadron in ADC.

318th FIS

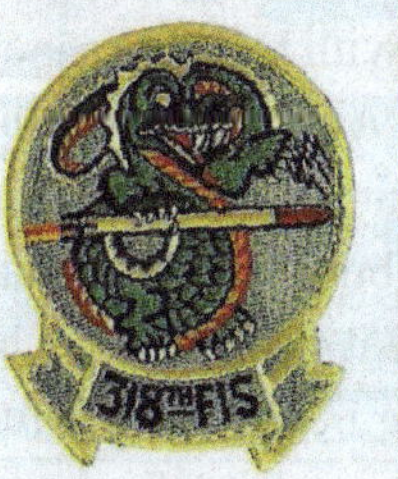

Left: The last markings worn on 318th FIS Darts were the full-size North Star as seen on these aircraft at Tyndall AFB.

Right: Earlier versions of the unit's marking were in the form of this two-tone blue, three-tailed dart shape. Note the '0' serial number prefix, indicating that the aircraft was over 10 years old.

The 318th FIS at McChord AFB, Washington was assigned its first F-106A (59-0055) on 21 March 1960. Like so many users of the 'Six', the 318th had flown the F-102 Delta Dagger. The 318th took the brunt of the first White Shoes deployments to Alaska. Five F-106s of the 318th arrived at Elmendorf AFB, Alaska, on 15 July 1963 along with four F-106s from the 498th FIS. In all, the 318th made seven deployments to Alaska.

The squadron also deployed to Tyndall AFB, Florida five times to participate in the William Tell competition, achieving first place in 1963, fourth in 1965, third in 1970, third in 1972 and third again in 1974. The squadron also initiated the Project College Cadence deployments to Korea during the *Pueblo* crisis. On 20 November 1967, 10 F-106As of the 318th flew non-stop from McChord to Tyndall with air-to-air refuelling. This was the first test of College Cadence.

In the spring of 1974, squadron F-106s deployed to Howard AFB in the Panama Canal Zone for Exercise Blackhawk '74.

With other surviving 'Six' squadrons, the 318th was transferred to TAC on 1 October 1979. On 2 November 1983, the squadron ended its Delta Dart era as the last F-106 departed McChord. One of the last active-duty air defence outfits, the 318th was an operator of the F-15A/B Eagle before being inactivated on 7 December 1989.

Air Refueling and IRST

Gen. Arthur Agan, the Commander of ADC always felt strongly that the F-106 would be a great air-to-air fighter because of its turning capabilities and strong thrust to energy ratio. Though it was too late to make it into the Vietnam conflict, in 1967, the F-106s were retrofitted with in-flight-refueling capabilities, 360-gallon supersonic external fuel tanks, and an infrared seeker device, IRST, located on the nose centerline in front of the windshield. The IRST gave the pilot the capability to locate a heat source from another aircraft in a situation where the enemy might be dumping chaff to confuse air-to-air radar. These new capabilities gave the F-106 some added potential that now made it possible for long overseas deployments into areas where air-to-air fighter capability was needed as well as air defense. These new capabilities also required us to put our aircrews through new rigorous training programs. The 318th was the second squadron to receive aircraft with these modifications and I ordered Capt. Larry Haight, our training officer, to get our pilot training program underway immediately. Our pilots had to receive air-to-air refueling training, then aerial combat training (ACT). Col. Schuder, the CO, was in the first group to train and then Doug Brenner and myself were next. The training was intense, but truly enjoyable. Air refueling was not more than simple formation flying, except when it was at night and in the clouds. But ACT with Doug Brenner as my daily opponent was a blast. I had not flown any ACT since Korea, but I was not a neophyte in it, and we had some great battles. Brenner was a terrific fighter pilot, and he won more battles than I, but I did get him a few times. Overall, the squadron's training program went without a hitch. Only when the Wing Commander, Col. Harry Downing in his aerial refueling tries tore up the boom twice, evoking a serious complaint from the Geiger refueling squadron, did we have any problems. Because we got down to work, did not complain and showed a positive attitude from start to finish, the 318th passed the squadron that had received the modifications ahead of us. I then sent a message to ADC declaring that the 318th FIS was combat ready and prepared for deployment if needed. Little did I know that my message would have such eventual repercussions. But we were given a mission and we fulfilled it. I was never prouder. November 1967 marked the first air-to-air refueling of the F-106 when the 318th FIS deployed to Tyndall with ten birds. I lead that mission.

Shortly after, Col. Schuder was reassigned. I had been promoted to Lt. Colonel and I secretly hoped I would get the job. I was disappointed. The squadron commander at Griffiss AFB, Lt. Col. Fred O'Connor was named as our new CO. O'Connor did not have a great reputation and many wondered why he was relieved of his command at Griffiss to come to McChord. He had just been divorced and his previous squadron did not have the best reputation. He did not have many hours in the F-106 and would need lots of catch up training. The 318th aircrews were not enthused about his arrival, and it was up to Doug and me to set them straight and remind them that the mission takes priority over personal preferences. Not getting the job as CO was another lesson that I learned about life and how you have to take its ups and downs. My goal of becoming a squadron commander was still attainable, but I would have to wait and keep working. This was a tough adjustment. The aircrews did not like O'Connor, and he did not earn their respect. Brenner did not like him at all and when O'Connor assigned him as Executive Officer, Doug was flabbergasted. Brenner was a fighter pilot, first and foremost and he detested desk jobs. But O'Connor was adamant, and Brenner was hogtied to the orderly room. Ironically, Doug replaced a man whom all of the pilots actually respected deeply. Lt. Col. Bill Shomo was a Medal of Honor recipient and an Ace in World War II with eight kills. He was no longer on flying status and the 318th was his last tour of duty before retirement. I loved talking to Bill, but he rarely spoke about his accomplishments. He truly was an American hero whom few actually realized was right there in our midst. When he retired, we in the 318th gave him the best sendoff we could. He deserved it.

My office in the ops building overlooked the tarmac where the 318th F-106s were parked. The Delta Darts were beautiful aircraft and the 318th tail design was especially striking. From my desk, I could watch our aircrafts taxi in and out, and I could see the end of runway 34, which was the active alert runway. At any moment, I was able to look up from my desk and keep a close eye on our flying operations for which I was responsible. I also could keep a close eye on weather conditions that played a vital role in safety. McChord AFB, often referred sarcastically to as McFog, was prone to sudden changes in weather conditions that could bring visibility down to minimums and below in a hurry. As a flight commander and the supervisor of flying, it was a continuing responsibility of mine to make sure that only our most experienced pilots were airborne when deteriorating conditions were forecasted. Despite our hectic schedule that included training, air defense alert, Alaskan deployments, additional duties, paperwork and an unpopular CO, the morale of our squadron was high, and our unit performance was consistently rated well above average. We completed our ACT and air refueling training in record time and aircrews that had not been to Aircrew Survival School were being scheduled as quickly as we could get the slots. This was a requirement in the event our squadron was deployed overseas to a combat area, and many of our aircrews, who had returned from Vietnam, had been through this training. I had received this training prior to my Korean tour but the training I received was obsolete now and so I placed myself on this list to attend the survival school at Fairchild AFB in Spokane, WA. In October 1967 I put in my three weeks that included ground school, four days of prison camp that included what some people now call torture and some very realistic, hardnosed interrogation tactics, followed by week-long escape and survival training in the Colville National Forest. It was three weeks no one ever forgets. The prison camp experiences and the evasion and survival training in the woods without much food and little rest, brought us out exhausted and very hungry. The course was designed to counter somewhat what our guys were experiencing in North Vietnam where the enemy paid no attention to the Geneva Convention and treated our prisoners like animals. I was never certain that anything could prepare a man for that kind of treatment. He either had it in him or he didn't. In any event, I was now fully prepared for an overseas deployment should it be ordered and if not, I had completed a training event I would need prior to an assignment to Vietnam which I was sure would be next. As I had a combat tour on my record in 1953 and just completed an overseas tour to Canada in 1966, I felt that I had another year or two before being tapped again. I was not overly concerned about a Vietnam tour, but I had hoped to first complete a Squadron CO position in ADC. The Vietnam War did not look like it would soon be ending, and I felt my chances for a command position in ADC were at their best now. But you could never be sure, so I just kept my focus on my job and doing the best I could and enjoying what might be the best assignment I would ever have.

The *Pueblo* Incident

At the same time that 200 Americans a week were returning from Vietnam in body bags, tensions were getting hot in Korea. The South Korean President Park Chung-Hee was nearly assassinated on January 21, 1968, and two days later, the North Koreans seized an American spy ship, the *USS Pueblo*. This was an embarrassment for the USA, which had always bragged that it could fight two wars simultaneously. The truth was different. The only tactical airpower in Korea were F-4s and they were unable to alter their operations to cap the Pueblo. The 318th FIS, being the first deployable air defense F-106 unit under the College Cadence program, was alerted to deploy. Although we were not told where we were headed, it was not difficult guessing that the destination would be Korea. In February 1968, we received deployment orders. Because Col. O'Connor was barely trained, I was designated the squadron leader. After tearful goodbyes, I lead the first deployment of 18 F-106s overseas departing McChord on the morning of February 9, 1968. After rendezvousing with our tanker force and refueling just east of the Farallon Islands, we landed some five hours later at Hickam AFB, Honolulu. All 18 aircraft landed safely and in good flight condition.

This was the first overseas deployment of the F-106, and much was riding on its being a success. ADC was now invading mission territory that only Tactical Air Command possessed, and the reputation of the Command and its aircrews and maintenance personnel was at stake. Doug Brenner and I carefully selected the aircrews who would fly on this deployment, and we attempted to select those whom we could rely on to come through with top performance under pressure. One aircrew that we differed on was a relative newcomer, Lieutenant Rodman. Rodman was a West Point grad and had graduated from F-106 school at the top of his class. There was little question about his abilities. However, lately he had shown a tendency to find questionable excuses to abort missions especially where there were difficult flying conditions in the area. Brenner, and a few others thought he should not man a cockpit. I held out to include him. My judgment was found faulty the next morning as we prepared to depart on the second leg to Naha, Okinawa. This would be a ten and a half hour flight with at least five refuelings. Breakfast was at 0500 followed by briefing and 0700 take-off. All went as scheduled until we headed out to the aircraft. Then I was intercepted by our Flight Surgeon Dr. Gordon who informed me that Rodman could not fly due to problem with one of his testicles. I was dumbstruck. Engine startup was minutes away. I could not believe that one of my pilots would refuse to fly this mission with such a problem. I called for our backup pilot, Major Harold (Yogi) Van Daele. He had already headed for the KC-135 in which he was scheduled to ride to Okinawa. He had sat through the briefing, but I had noted he was a little under the weather from his previous night on the town in Honolulu. No matter, I sent someone to get him and get him to Rodman's plane in a hurry. I told Rodman I would see him in Oki. I had my own problem and that was to get to my aircraft, get it started and get my flight to the end of the runway in time to meet the tankers. It was a miracle that it all came to pass in good fashion and 18 F-106s of the 318th FIS flew to Okinawa and arrived safely nearly 11 hours later. It was a joyous arrival and we met by many VIPs, and I know that when the word was received back at ADC Headquarters and at the Pentagon, there was much relief self-congratulation. Looking back at that flight. I can only remember that despite knowing beforehand how long this flight would last, I ran out of water early. I let out the rudder pedals so my legs could extend, and I tried hard to relax and stay alert during the many hours that droned by so slowly. Keeping track of my flight and staying close to the tankers who did the navigating for us was crucial. Although there were rescue ships stationed along the route, looking down on the Pacific Ocean, which seemed to be endless in all directions hour after hour, I felt that if my engine ceased to operate or some other calamity occurred that sent me down into it, all would end there in its cold and unfriendly churnings. Our arrival at Naha Air Base on the historic island of Okinawa was a great feeling, but to be truthful, most of us were completely exhausted. Then the word came down from PACAF, it wanted two F-106s on alert as soon as possible. I did not have the heart to assign that duty to any of my aircrews, who had performed so well and only wanted to relax with a beer before hitting the sack. Yogi Van Daele and I sat alert that night and protected Naha and Okinawa from enemy attack.

The SR-71

The 318th remained at Naha for a week before it received orders to deploy to Korea. The squadron's aircraft arrived in good shape, and we were soon up and ready to go. While at Naha, one of our missions was to provide support to the Blackbirds flying out of Kadena Air Base that was just north of us. The SR-71s flew their Top Secret reconnaissance sorties out of Kadena. Their flights took them over Red China and the Soviet Union and God knows where else. We were taken through the Blackbird squadron, met some of their pilots and visited the alert stalls from where they were launched. Our mission was only to assist in the event an SR-71 should experience troubles in the air by intercepting the aircraft, looking it over, communicating with the aircrew and escorting it back to base. The SR-71 was an aircraft unlike anything I had ever seen before or since. A product of the fertile mind and energetic imagination of Lockheed's Kelly Johnson, the SR-71's ebony color gave it stealth attributes, while its 2,100 mph airspeed and 80,000 foot altitude capability gave it immunity to interception. Although the aircraft was easy to operate, the crews had to wear astronaut-like pressure suits and the missions could last from two to twelve hours. While flying out of Naha, I had one occasion to meet up with an SR-71 in the air. Flying next to this fabulous plane was a great experience. The pilot and I talked, and he challenged me to a race. He generously gave me a 30 second head start. After hitting the afterburner on my Delta Dart, I broke through Mach 1 and was still accelerating when I felt this sudden whoosh and my F-106 suffered a heavy thump as the Blackbird went by me with its afterburners glowing. It then went vertical and was still separating itself from my sight as if I had suddenly gone into reverse. It was an incredible display of air power.

Korea and the Most Controversial Flight of my Career

We finally received an ops order to deploy the squadron to Korea. This time, Col. O'Connor decided to lead the squadron into Osan, our deployment base. The order called for a deployment of three flights of six aircraft, the first flight departing in the morning, the second later in the morning and the third in the afternoon. I would be leading the third flight. The flight to Osan was relatively short, about an hour a half for the F-106, and it was a choice whether or not to load the drop tanks. We decided to carry full fuel tanks, and as things turned out, that was a fortunate decision. Our flight north out of Okinawa was uneventful and the weather was clear as we approached the southern coast of Korea. However, the weather at Osan was deteriorating and the base now had low cloud cover and rain. As we passed over central Korea, I noticed a base I had flown into many times during my Korean tour in 1953. Taegu, or K-2, as it was called then, was an F-84 base, and I landed there several times as it was an alternate landing base, and I visited some old friends there from flying school. Taegu was wide-open weather wise, and I surprised myself at how quickly I recognized it after 15 years. As we proceeded further north, I noticed that I was unable to receive any of the TACAN signals. I notified my wingmen and asked if they were receiving them. The response was mixed, a few said they were, and some said they were not. I then used my element leader's navigation directions as we approached Osan. The weather now at Osan was ceiling 1,000 feet, visibility 2 miles and rain. I was totally unable to receive the Osan TACAN, so I was now relying on my wingmen to do the navigation. We then broke into elements for approach, and having the most fuel on board, my element was the third and last to descend. My wingman was Lt. T.J. Brown, a young, bright and aggressive officer who had much potential. I checked with Brown to be sure that his TACAN was receiving the Osan station, and he assured me it was. After the second element made its approach, we began our approach, but it was only now that Brown informed me that he was not receiving the Osan TACAN. Thus, we were not sure where we were when we began our descent. I passed this information to Approach Control as we entered cloud cover. I was familiar with Osan from my former tour, and I reconfirmed this from my letdown chart. To the east of the base there were hills rising approximately 800 feet that approaching aircraft had to stay above before beginning final approach. Usually by that time you were under radar control as I was hoping we would be also. However, as we descended, Osan Approach Control was unable to pick us up. The lower we descended, the more this worried me, as without TACAN reception, I could not confirm our exact position. As we began what should have been our final approach, radar control still did not have us on their scope and neither Brown nor I had TACAN positioning. I decided then to descend no lower than 1,000 feet unless they picked us up. And when they didn't, I declared a missed approach and we began our climb back up.

I was not particularly upset at this point; we had plenty of fuel, thanks to the drop tanks, and surely, we were in vicinity of the field and would be picked up by radar and vectored back to the landing approach.

Brown was a great pilot and was right on my wing. I did not worry about his flying formation in the weather, so I was confident that we would get down all right. But then things took a strange turn. Ground radar now claimed they had picked us up and directed us to turn north. I followed their directions and took up a northerly course. But after a while, I questioned ground control whether they were sure they had positive identification on us. They then gave us a change of course, but after complying with their course change, ground control did not assure me they had positive ID. Then they again directed us to turn north. I complied with this request, but now I was worried about where in the world we were and where we may be heading. Osan was not very far south of the DMZ and entering the demilitarized military zone would be a violation of airspace. Now I was getting worried. While when we made our missed approach, I was fairly certain that we were close to Osan Air Base, now I didn't have the faintest idea where we were in relation to our destination. After questioning ground radar again whether they had us on radar and receiving a negative reply, I made a command decision. We would head back south and find the base at Taegu that was wide open when we passed over it an hour ago. So, Brownie and I did just that. We headed for K-2. After about 20 minutes we could see Taegu ahead of us, still in the clear. We would be on the ground soon. But then I got a call from Osan. They stated that they had now positively picked us up on their radar. When I asked them for our position, their response verified that they had indeed finally had positive ID. With that, we turned around and flew back to Osan, made our approach and landed safely. But that did not end the confusion. As soon as we shut down our engines, Brown and I were directed to report to Division Headquarters and write out individual statements giving the details of our flight. It was clear that the Division Commander and his staff were very upset with what had happened. The details of our flight had been reported all the way back to the Pentagon and the State Department. As I had suspected, our flight had indeed penetrated the DMZ and had gotten the North Koreans all up tight. The Division Commander, Maj. Gen. Bob Burns, was an old ADC friend from my days at McChord and ADC. He was in a tight spot, and the statements provided by Brown and me only made things worse for him. Unfortunately for the Division weenies, it was their ground radars that had vectored us north into the DMZ. If not for my decision to break off their directions and turn south, we would have penetrated deep into North Korean airspace. While the Division would have preferred to blame the problem on the pilots, the evidence was clear and on the tapes that the ground controllers were the ones who directed us to fly north.

But the flight was still an embarrassment all around and I felt badly that it happened. I did a lot of second-guessing on myself about it. I did make a mistake in trusting Brown when he stated that he had good reception on the Osan TACAN, which he didn't. I should not have relied on the Lieutenant's assurances on something that important. Still, after reviewing all of my decisions, I concluded that I made the best decisions a pilot could make when the unexpected happens and, most importantly, I got my flight back on the ground safely. The 318th took an awful lot of ribbing from the F-105 squadron pilots on base about our navigational abilities. However, they soon learned not to get into any simulated aerial combat (ACT) with the F-106s. And, as it turned out, it was soon discovered that the TACANs in our F-106s were not compatible with the ground installations in Korea. Modifications had to made.

A footnote to this story: T.J. Brown, after leaving the USAF, was active in an organization that flew F-86 Sabres (the plane I flew combat in Korea), at air shows around the country. He was subsequently killed in an accident.

Osan

It had been almost 15 years since I had left Korea at the completion of my wartime tour. The difference was startling. The South Korea I had left was war torn and destitute. Its cities had been shattered and the population, poverty stricken and heartbroken from its wartime losses of young men and the destruction and separation of families. But this newer Korea was thriving. Seoul was a cosmopolitan city with modern hotels and skyscrapers. There were several posh resorts and even golf courses and the infrastructure of roads, bridges, and railroads had been replaced. This was a new Korea, and you could see it in the faces of the people. I was pleased that the USA had played a role in saving this country from the throes of communism that was choking the Korean people unfortunately imprisoned in the People's Republic to the north.

Our mission was to take over air defense alert duties and this we did immediately after our arrival. There were no alert hangars. Our F-106s were set up and parked on hard stands far from the end of the runways.

There was a small shack adjacent where the alert pilots and ground crews were housed during their alert duty. Our operations building was composed of a few tent shacks and a Quonset hut far from the tarmac. As ops officer, I was issued a jeep of World War II vintage, which I used for transportation around the base. This was the envy of all the guys, and I was asked frequently for the use of it by almost everyone, which I generously consented to. During our stay at Osan, I was involved in the design of a new ops building for the deployed F-106 squadrons and we moved into it a few months before we were relieved by the 48th FIS out of Langley AFB, VA. If I must say so myself, the ops building which I designed was quite functional and took care of our needs very well during our stay in Korea. Another mission I undertook in my first month at Osan was to visit all of the GCI radar sites in S. Korea and brief the controllers on the F-106. This led me to some very interesting places, some of which I had to reach by helicopter. Most of the controllers at these sites had no experience vectoring F-106s and they appreciated my presentation and had lots of questions. They also enjoyed a visit from a pilot, and someone newly arrived from the States. One of the sites I visited was on a small island off the western coast of Korea. The island actually was north of the DMZ and the North Koreans had their guns trained on it all the time. On my way down from the radar site located on the top of the island's highest point, I visited a Catholic Church where I said a few prayers and marveled at the interface of Korean culture and traditional Catholic architecture and art. I then met the Pastor who was a Maryknoll priest from New York. We conversed for a long time and reminisced about the USA and our old hometown. I surely admired the work he was doing in this far off outpost where his parish could be attacked at any time and destroyed. Thirty-nine years later I met this priest again in Sandpoint, Idaho where he was temporarily assigned to St. Joseph Parish. I reminded him of our first meeting, which, I believe he just barely remembered in such starkly different circumstances for the both of us.

Our deployment quickly began to settle into a routine of alert duty and training missions over the Korean countryside. Occasionally we were able to mix it up with the F-105s, but they usually dove for the deck and poured the coal on, and we could not catch them. But the F-105s were soon relieved by F-4s from Yokota Air Base and then we really had some fun. The F-106 could out turn the Phantoms quite easily so whenever we got them into a horizontal air battle we prevailed easily. The F-4s were more difficult in vertical air battles that usually ended in stalemates. The F-4s with their two engines often ran low in fuel early and had to call off the fight before a conclusion was reached. We noted that they often called the breakoff for low fuel just as the F-106s were closing in on their six o'clock position. The F-4s had as one of their prime missions escorting our spy aircraft that flew along the DMZ and up and down the coasts of North Korea. I believed that that mission could be better served by our F-106s which had greater range and endurance. When I suggested this to the people at Division, they quickly shot it down. The Commander of the F-4 Wing was none other than Colonel Chuck Yeager, the man who first flew through the speed of sound and was one of the USAF's top flying heroes. I got to meet and work with him at Osan, and I approached him with my idea of relieving his F-4s of this tedious escort mission and giving it over to the interceptors. Yeager thought it was a good idea and when he told the Division people that, they quickly changed their minds too. So, in addition to our air defense alert role, the 318th FIS, now added to its mission the role of escorting our airborne spy aircraft on its secret and dangerous sorties close to the North Korean coastlines and within range of enemy aircraft and antiaircraft installations. This turned out to be a real coup for our unit and for ADC.

Our deployment to Korea was a huge success, but it was also a hardship in that it caused the separation of fathers and husbands from their families. This was especially difficult for some of my pilots who had only recently returned from combat tours in Vietnam. I had several of these and I made it a priority to rotate these pilots home first when we were able to do so. For the most part, I found morale high. Most of the aircrews responded well and did their job without complaint. I had great Flight Commanders and I had Doug Brenner as my Asst Ops who took care of any problems early before they became too hard to handle. I had two problems though which I had to personally step in and decide and not to the liking of the persons involved. Lt. Rodman, the pilot who came up unable to fly the leg from Hawaii to Okinawa at the last minute, still displayed all the signs of fear of flying syndrome. Though cleared to fly from a medical standpoint, he showed all the symptoms of a pilot that wanted no part of becoming airborne. I could no longer assure his safety, and therefore I had him grounded and returned to the States for psychological evaluations.

Rodman left us and never flew again. The USAF had spent a lot of money training this man and his return on that investment was almost zero.

Another pilot I had, Eynon Hughes became more trouble than he was worth. Recently returned from Vietnam and now finding himself in Korea, he complained unendingly about the unfairness of it all. There was nothing I, or anyone could do about changing his situation, but I did promise him that I would ensure he would be among the first to rotate home when that time came. This failed to satisfy Hughes, who made a pest of himself on this subject, and it was beginning to affect aircrew morale. Doug Brenner was a great intermediary for me with the aircrews and kept me abreast of what was going on with them. For instance, I always knew which of our pilots were hitting the Osan nightclubs and partaking of the enjoyments of what the ladies of the evening had to offer. On this I did not care that much as long as the pilots were discrete and smart. But Hughes came to see me one day in my office and again complained bitterly about his situation and how he missed his wife and family, and this deployment so soon after his Vietnam tour was intolerable and unfair, etc., etc. I let him finish and then asked him how a man who missed his wife and kids so much could explain his actions of visiting Osan bordellos so frequently. Was this the way in which he displayed his loneliness for his wife and his children? Hughes was stunned. He called me a bad name and walked out. He did not realize that I knew what he was doing with his off evenings, and he also knew that I knew his wife Jane very well. That was the last time I heard from him about his situation. Nevertheless, he was one of the first I sent home. I was glad to be rid of him. His Evaluation Report would reflect my thoughts about he handled an unfair situation under stress in a national emergency.

When our aircraft needed periodic inspections, these were performed at Naha Air Base, Okinawa. So, every aircrew was able to fly back to Oki and spend a few days there where the weather was great, the food was good, and you could relax and do some sightseeing. I took advantage of that myself and spent some time on this historic island where one of the bloodiest battles of World War II was fought. It is a tropical island with beautiful beaches and heavy rain forests. Surrounded by the blue-green Pacific, it is hard to imagine what it was like in 1945 when the Marines stormed those beaches, and the Japanese defenders fought them to the death. At the Marine Museum one can see a replica of that awful battle. I also had the opportunity to do some shopping. There I purchased some jewelry pieces for Alice Marie including a string of pearls by Miki Moto. She has that piece of jewelry today and it is still a lovely string of pearls.

My days at Osan became pretty routine. Each morning, I had early breakfast at the Club and then aircrew briefing at Ops about 0800. Then the flying day began, interspersed with meetings and my own flying schedule. At 1700, I attended Mass in the Base Chapel followed by dinner, usually in the O club, but sometimes in the mess hall. At night, I tried to write a letter home or maybe hit a movie in the base theater, although they were usually pretty old and bad. I tried not to think about home too much as that only made things worse and me feel worse. I gave myself 15 minutes each day to think about Alice and the kids, and that's all. This was not easy and many times I did not adhere to that rule. But I knew it was not good to brood about being away from home and family too much. I stayed busy doing my job as best as I could, and I knew Alice Marie was doing her job the best as she could. I had all the confidence in the world in my wife. I knew she would be taking great care of our five children, and nobody could do it better than she. And I was absolutely right about that.

One of my major responsibilities was the morale of the men who worked for me. Keeping them active and out of things that could cause trouble for them was important. An opportunity arose when our squadron took up sponsorship of an orphanage located near the air base. The sadness of this place was the children there were sons and daughters of Korean mothers and American serviceman. Because they were not "pure Koreans", they were not accepted as citizens of South Korea and had no future in their own country. It was a particularly depressing situation for these youngsters, who for the most part were bright and beautiful. The men of the 318th FIS took these children to their hearts and every day we had our guys at the orphanage helping out in many important ways, building, repairing, bringing goodies, and interacting with the children. The word quickly spread and soon Alice Marie and the wives back at McChord began to gather toys, clothes, food, and medical supplies for the orphanage. Despite AF rules forbidding such things, the MAC C-141s began smuggling these items from the USA to Osan, where we ensured that they were given over to the kids and their caretakers. These orphans took the place of our own children, from whom we were temporarily separated. We knew our children back home had so much more than these orphans ever had.

Doctor Bob Gordon, our Flight Surgeon, also chipped in by visiting the orphanage, checking the children, and treating them for whatever problems they had. It was a truly great relationship and again, I was truly proud of our American servicemen who always display humanitarian qualities wherever they go in the world and under the worst conditions. At the opening of our new ops building, the children of the orphanage came and sang songs for us, which they had memorized in English. We fed them hot dogs and soda pop and that's how they said thank you. When the 318th FIS rotated after its six-month deployment, the 48th FIS that replaced us, and the following ADC squadrons, continued to support the orphanage. Sponsoring the orphanage and helping those unfortunate children and those that took care of them, was one of the best things we did during our deployment.

The Pueblo incident had pretty much faded from the news stories. The North Koreans continued to hold the ship's captain and crew accusing them of spying and they milked the incident for all it was worth accusing the USA of provoking all the trouble. But it became clear shortly after American forces arrived in South Korea, that any thoughts of invasion were dispelled. American forces were still deeply engaged in Vietnam defeating the North Vietnamese in their Tet offensive, falsely perceived by American citizens as a setback for our side due to biased reporting by anti-war media. Thus, the Korean story became a back page issue. The North Koreans clearly had hoped that America would not react as strongly as it did with its heavy involvement in Vietnam. The Pueblo crew was finally released in December 1968. North Korean MiG-17 fighters shot down a US Navy EC-121 Constellation on April 15, 1970, killing all 31 aboard, so the F-106 mission in Korea continued until May 1, 1970, when the 95th FIS redeployed to Dover AFB, DE to meet its new Commander. Thus ended College Cadence, the F-106 deployment to Korea beginning with the 318th FIS. Over 27 months more than a half dozen ADC squadrons contributed aircraft and airmen to the Delta Dart presence in South Korea. At times, our F-106s were within minutes flying time of North Korea's MiG-17s and MiG-21s, although neither side's fighters ever crossed the line. I was proud that our squadron led the way. We were the first ADC squadron to deploy overseas into a potential war zone and take up alert duties. We did it without incident and on schedule. We broke new ground, and I was honored to have been in the lead aircraft.

Another anecdote relating to my six month's tour in Korea was my experiment with a 'stache. It was fairly common for fighter jocks to grow various forms of facial hair while in combat areas. They did so because the regulations against them in those locales were relaxed and being apart from home, wives, girlfriends and society in general lent itself to such nonsense. I had not partaken of that foolishness in my first Korean tour, but this time I decided to grow a mustache. At first it started out to be a thin version, but it then grew and ended up as a rather healthy handlebar with curled ends which I waxed to keep them symmetrical and stylish, rather like the old barbershop quartets used to have. Surprisingly, my mustache came out slightly reddish with some gray and, to be perfectly truthful, I did not like it at all. It was, in fact, a pain in the neck because I had to trim it, keep it waxed, and also, I ended up fingering it too much by habit. Many days I thought I would shave it off, but then I wanted to keep it until I returned home and surprise Alice Marie and the kids. However, my plans were thwarted by Jack Ryan. General Ryan was the Commander of Pacific Air Forces, and he came to visit the 318th at OSAN and I had to actually show him the cockpit and explain its capabilities. I knew I did an impressive job because he told me so, but he also told me to get rid of the lip hair. When a four-star general tells a Lieutenant Colonel to do something, he usually does it. And I did, but only after I had a photo taken of my facial masterpiece for posterity. To be honest, I was glad to get rid of it.

As we approached the end of our deployment, I was able to release some of our aircrews back to the states. As promised, I selected the pilots who had recently completed combat tours in southeast Asia. (Including the pesty Hughes). Here I encountered trouble stemming from working under too many commands. Although at Osan, the 318th was under the operational control of Fifth Air Force and PACAF, we still were under the administrative command of 25th Air Division back at McChord and ADC. PACAF and Fifth AF both agreed that we could release eight pilots to return home. I selected the pilots and notified them that they would be rotating. But then ADC overruled that decision and ordered that we could only release seven pilots. We informed ADC of PACAF's approval to release eight and asked for their okay to do so as PACAF was the operational command. The response was horrendous. The squadron commander, Col. Fred O'Connor, was back in the states at this time and being the acting CO, I received a blistering TWX, plus a phone call from the Wing Commander at McChord, threatening that I would be summarily fired if I released the eighth pilot. I had no choice but to inform one of the pilots, Ron Bernd, that he would not be returning early. Ron was a class act and took the news with no complaints. Doug Brenner, my ops assistant, volunteered to stay and let Bernd go home, but I said no to that. Doug, even though I would miss him badly, was in line to return, but I wanted him back at McChord to prepare things for the return of the full squadron.

We were now preparing to turn our Korean alert and other mission duties over to the Langley squadron. Col. Jack Bailey who was a close friend of mine from my days with the ADC TACEVAL Team in Colorado Springs commanded the 48th FIS. ADC had decided that we would leave our aircraft at Osan for the 48th to fly and we would fly the 48th's aircraft at McChord until they returned. This would save the costs of a full-scale overseas deployment of F-106s from the mainland to Korea. Naturally, we did not like that decision as we felt that our aircraft were the best maintained and best looking in the Command. But orders are orders, as they say. And that's the way it went down.

As we neared the end of our six-month tour, more of the 48th FIS aircrews came in until they finally took over the mission. I was ordered to report to Headquarters PACAF, Hickam AFB, Hawaii for an out briefing on my return to the states. I wrote to Alice Marie, and I suggested that we meet in Hawaii for an R&R and then return to McChord together. She sent a prompt reply that it was a great idea, and she was looking forward to it. I started to make arrangements and reservations, including a rental car for a week.

However, shortly after, Alice wrote that on afterthought, she decided it would be better that I come directly home, as the kids were anxious to see me. To put it mildly, I was furious at this change of decision. I had things pretty much arranged; I would arrive in Honolulu a few days before Alice and stay at Hickam. I would give my briefing to the PACAF people and then rent the car and check in at the Fort DeRussey, the R&R hotel for servicemen. Now Alice was backing out. After a few drinks at the O club I wrote her a letter. In it I told her that I was sorry that she could not meet me in Honolulu and that I was going to miss her. However, I had heard that an Air Force Lt. Col. should not have any trouble getting into mischief in Honolulu if he tried. I sealed the envelope and mailed it. About a week later, one of the maintenance sergeants came into my office and told me they had received a message from my wife via the AF communications system. She said she would meet me as originally planned. He asked me if I understood the message. I told the sergeant, yes, I understood exactly what the message meant. Alice, thank God, had her Mother come out and babysit our kids for a week and she flew to Hawaii. We had a great week that we both needed after a long hard six months. I had just completed a pressure filled tour, a first for ADC and the F-106. We deployed halfway around the world to show the North Koreans we meant business and we did it successfully and faultlessly. Alice also had a tough six months taking care of five children without help and also being the Mother Hen to a bunch of wives, many of whom were in a state of shock and helplessness without their husbands. She brought an aura of common sense and reason, providing the more bewildered of them comfort and a helping hand. (One spoke of suicide, no less.) I could not have been prouder.

We had a great time in Hawaii. Not surprisingly, the weather was superb. After two days we moved to the Royal Hawaiian Hotel, which was right next door to Fort DeRussey. We spent lots of time on the beach, which we both loved, saw several shows, hit a few luaus, drove around Oahu spending a day at Bellows Beach, and downed lots of Mai Tais. I even tried my hand at surfing that ended in near disaster. We had great time and then flew home together to a reunion with our five offspring. What a joy that was for me. They looked so good, healthy, and happy—a grand tribute to their mother and my mother-in-law who watched over them in our absence. What a lucky guy I was, and I knew it. I thanked God.

College Cadence Summary

At the time of our deployment from McChord to Osan, I had accumulated approximately 3,850 hours of flying time and was a Command Pilot. Nevertheless, this was the most exciting and important mission in which I had participated. To have led my unit across the vast (and seemingly endless) Pacific Ocean, involving many tanker aircraft join-ups, aerial refuelings, weather formation penetrations, and safe arrivals at strange airfields, was a major responsibility. Thank God it came out well. The 318th FIS received many kudos for its pioneer accomplishment, especially breaking new ground for the F-106 and Air Defense Command. We also demonstrated that ADC could deliver air defense capabilities to anywhere in the world on short notice. For this our aircrews and I received The Air Force Commendation Medal, and I received an Air Medal with Oak Leaf Cluster. I now hoped that I might be in position for an assignment as a squadron commander. I would have to wait and see.

Following is the Comments Section of the Evaluation Report I received in November 1968 written by Colonel Fred O Connor, the 318th FIS Commander and endorsed by Brig. Gen John Rouse, the 25th Air Division Commander:

> Lt. Col. Cunneely is one of the most competent Operations Officers I have known. Under his supervision the operations section of the squadron has flourished to the extent that it has achieved singular success in every test to which it has been subjected. Lt.Col. Cunneely combines diplomacy and firmness in the skillful management of the aircrews, the quality of which is emphasized by the absence of flying accidents or significant incidents. He is meticulous and thorough in preparing a new aircrew for the perfection the mission demands. The ease with which the squadron passed its ORI was largely the result of the training, briefing, and general preparation Lt. Col. Cunneely fostered among the aircrews. His skill in this regard was proven further in the non-stop deployment to the Command firing range, in which a record was established for firing by F-106 aircraft. The ultimate confirmation of the efficiency with which he manages the operations section came in an extended over-water deployment of the squadron to a hostile zone in Asia. All aircrews and aircraft arrived on schedule. In the most dangerous mission confronting the Command, that of Air Combat Maneuvers, Lt. Col. Cunneely, by meticulous and diligent personal supervision of the program, has insured the utmost in safety and perfection in the execution of this mission. Lt. Col. Cunneely,while extracting a high level of performance from the aircrews, elicits enthusiasm and dedication to the mission from them.

STRENGTHS: He is diplomatic but firm and relentless in attaining the utmost in performance from the personnel under his supervision.

SUGGESTED ASSIGNMENTS: His operational experience and overall capability should make him adept at command of a fighter interceptor squadron.

This ER truly bolstered me. My goal had been to become a squadron commander and this evaluation plus the reputation I enjoyed at ADC Headquarters gave me encouragement that I might soon achieve that goal. One worry was that my next assignment could well be Vietnam as that conflict was still raging. While I was in Korea, my good friend Dave Winn, whom I worked closely with and flew with, was shot down over there and became a POW. If l did not get that squadron soon, I may have to wait a long time. I would just have to be patient and do the best job I could.

CHAPTER TWENTY-ONE: I LEARN PATIENCE, FIND DISAPPOINTMENT, THEN A SURPRISE

God grant me the serenity to accept the things I cannot change.

-Reinhold Niebuhr

The 318th's return was triumphal and it was certainly great to be home with families reunited. We resumed our NORAD duties with a mix of our 318th and 48th aircraft. The 48th FIS having replaced us at Osan. Maj. Larry Haight ran ops at McChord in our absence and displayed maturity and an excellent sense of command and control. He kept the wives informed about what was taking place in Korea as well as run flying operations with the aircrews and planes left at home. In short time, we were back to routine flying operations and alert duties. And of course, we were back to our Alaska deployment and also the opening of our deployment base at Walla Walla. That did present us new aircrew rotation problems. For me, getting back to family life was terrific. The older boys were in Little League baseball, and we spent lots of time at American Lake at Fort Lewis for swimming and water skiing behind the Thompson boat I had purchased from my brother Joe when we were in North Bay. In the fall, Gene and Jimmy played hockey at the Lakewood Ice Rink. Mary was now into competitive swimming with the Dean's, so Alice and I were on the go a lot. Thank goodness, much of their activities were on base or close by. The kids had lots of friends and they were good students. They were kept busy with their schoolwork and their sports. Mike was attending Kindergarten on base and would soon join the others at St. Francis Cabrini for first grade. All in all, the children loved McChord, and were coming along fabulously. The boys loved watching the F-106s and I brought them down to the flight line often. Thanks to their mom, they were happy healthy, children.

We Buy a Vacation Home

While I was in Korea, I thought a great deal about the property I owned on Lake Pend Oreille. I was not satisfied with the lot I had, and I was determined to buy the piece that Bob Jones owned with the log cabin on it. This was the one the boys and I stayed in on our camping trip. So, I called Bob and told him I wanted to buy that lot and asked him what he wanted for it. How lucky I was. Jones was just beginning development on an RV campsite south of Hope, Idaho, and was in need of some cash. He pegged the price at $13,500. I immediately snapped it up, not being sure how I was going to pay for it. I did it by selling my other lot for a nice profit and borrowing the rest. We still own that lot today as I write. It was a great investment, but more than that, it became a wonderful second home for our family. In the following years, the kids, Alice and I, and many of our family and friends have spent beautiful days and nights together at our lake home on the shores of that most lovely, pristine lake. The boys, Alice and I have put in many days working on our cabin retreat, updating it, improving it, building decks, adding rooms, painting, chinking, working on our dock, and doing so many chores on was once Jack Chambers' home built cabin. Sam and Dee Grashio had their vacation home next door and for many years we shared the water system and a floating dock built by the infamous Freddie Fanin. But we also shared many days and hours together having lots of parties, meals, adult beverages and laughs. Sam and Dee were our best friends for so many years, a relationship begun during our Air Force days. When I bought the lot with the cabin, the family approved of the lake as our vacation home and wanted to spend as much time there as possible. But it needed lots of fixing up to be livable.

And so... That's what we did.

In the next year, I began to work on the log cabin. Jack Chambers, the original tenant and an interesting historical character, had built the cabin with the help of Bob Nesbitt's uncle who owned the property. They used the trees on the land surrounding the site and performed this feat without the use of electrically powered tools. Jack lived alone in that cabin, which was only about 20 feet by 30 feet until his death. Heated by a wood stove on which he cooked and baked, Chambers lived off the land and the lake. He was a popular figure in the area and many people visited him at the cabin, enjoying his company, and partaking of his cookies baked on his wood stove. But the cabin I found was filthy dirty and so I flew from McChord to Geiger and spent a few weekends cleaning it out and getting rid of Jack's old furniture that was no longer usable. I did keep some kerosene lamps and the wood stove, which I moved outside. I had our McChord C-47 transport a carpet I had bought in Tacoma over to Spokane and I installed it, which gave the cabin a little warmth and color. In the summer of 1969, we had a few carpenters from Spokane come up and build a kitchen and a bathroom. Bob Dawson, Sam Grashio's son-in-law, installed the bathroom plumbing and the infamous Shell put in the septic tank and the drain field. We now had indoor facilities to cook eat, shower and go to the potty, something that pleased Alice Marie to no end. The boys and I built a new door, a dining room table with benches, and an outside picnic table we fashioned from Jack's original door. We then built a 12 foot by 12-foot deck so that we when we exited the cabin we did not step down into the mud. We were really in business. I will write more about the cabin in later chapters.

Decision Time

I was now completing my third year as Operations Officer of the 318th FIS. Clearly, I was approaching the time when I would be up for reassignment. I was still in hopes that I would receive a command position. My ER's all provided strong recommendations that I be considered for one. My leadership in the 318th had been highly praised all the way to USAF level and I felt fairly optimistic. However, my hopes were suddenly dashed. Fred O'Connor was promoted to Colonel and assigned to 25th Air Division opening the 318th Commander's position. But Lt. Col. Larry Pool was given the assignment. Larry was an old friend of mine and a very able officer. I was pleased to receive him as our new CO, but I had to admit, I was disappointed. It appeared that my goal of becoming a squadron commander would be delayed for an indefinite period. I had no choice but to welcome Larry into our squadron and to provide him my best support and utmost loyalty. Larry recognized this and was most understanding and appreciative.

I then began to consider volunteering for an assignment to Vietnam. The war had taken its toll of fighter pilots, and even though I had recently returned from a foreign duty assignment (Canada), my last remote (unaccompanied) tour was Korea in 1954. I decided to get some counseling on this matter. Through Larry Pool, I put the question up the line to ADC on whether I should wait for a squadron commander assignment or should I volunteer for Vietnam duty. A few weeks later I got my answer. Gen. Archie Burke, the Commander of 25th Air Division called me. He told me that Gen. McGhee, Commanding General of ADC told him that I should go ahead and get my Vietnam tour in and that when I returned, I would be given an ADC squadron to command. That was a promise that came from the highest level in the Command and so I had little recourse other than to accept it. My next step was to talk it over with Alice Marie. This was a tough decision as an assignment to Vietnam could mean as long as 18 months away from home depending on how long it would take to complete training into a new aircraft and then fly my 100 missions. Leaving Alice with five kids, for that long a period was a tough deal for both of us, especially for her. Alice decided that she would remain in the Lakewood area, and we even began to search for home sites. In January 1970, I put my papers in volunteering for a tour of duty in Vietnam.

College Shoes - I Get My First Command (interim)

The College Shoe program that evolved from the White Shoes program was still continuing, but now was rotated among the Command's F-106 squadron. It began in 1963 with only the 318th FIS providing the aircraft and the aircrew. The program has begun because the F-102s stationed in Alaska did not have the range and endurance to intercept the Soviet bombers that frequently flew close to the ADIZ. The 318th's turn again came in January 1970 and this time I volunteered as one of the aircrews. This would be a two week TDY. I enjoyed flying in Alaska and, frankly, getting away from the routine at McChord. But about 10 days into the TDY, Beaver Thiel and I just returned from alert duty at Eielson when I received a call from Gen. Burke. He informed me that he had just fired the commander of the College Shoes detachment and that I should immediately assume that position until further notice. To say that I was little shocked is to understate the case. The detachment at Elmendorf came under the control of the 25th

Air Division. The detachment commander was dissatisfied with the support he was receiving from the people at Elmendorf and said so in a letter to the commander of Alaska Air Command, who forwarded the letter to Gen. Burke. And that ended his career. I knew this Major, and I thought he was an effective officer, but he made a serious mistake in writing that letter. I visited him and he handed over command to me. I then met with all of the supervisory officers and NCO's and looked at all the facilities. In the next few days, I visited with the supervisors at Eielson, Galena, and King Salmon, which were the alert bases at which the F-106s performed their alert duties. I found the facilities to be in good condition and the personnel well qualified and effective. All in all, l felt that the unit was operating in good order, and I reported the same to Gen. Burke. I also reported the same when I met with the Commander of AAC. So, things calmed down and I settled in as acting detachment commander until a replacement was found. I hoped that that would be soon as this was not the command position l was hoping for plus, l was also missing Alice Marie and my kids very much. Finally, Gen. Burke informed me that he had selected a new commander and that he himself would be coming to Elmendorf for a change of command. That was great news for me.

I was the College Shoes Detachment Commander for only a few weeks before I was relieved. As I had not packed my bags for staying this length of time, I was glad to get the news of my return to McChord. All had gone well under my command and while I learned a lot from the experience, it did not fulfill my real ambition of attaining command of a fighter squadron. Word came through that Gen. Burke would be flying back to McChord with me in an F-106B (a two-seater) after the change of command and that was also good news. I liked General Burke and found him to be a straight shooter and very operational minded. He always treated me well and seemed to appreciate the accomplishments of the 318th FIS that was considered to be the best in ADC.

When all the affairs were done, Gen. Burke showed up at Elmendorf ops and we were ready to go. As usual the weather in Anchorage was bad, low clouds, poor visibility and light snow. General Burke was clearly a little nervous about the flight home, as we would be flying a long flight, most of it over very rugged terrain, lots of it over the cold Pacific, and over half of it outside of radar control at night. We took off and climbed up through clouds, finally breaking out on top at Flight Level 300 (30,000 feet). It was a moonless night, and I was on instruments much of the time and paying lots of attention to my map display that provided me a visual picture of where we were as we proceeded down the bleak and rugged Alaskan and northwest Canadian coasts. It was a great relief for the General when we were finally picked up on the radar of Seattle Center who proceeded to vector us to a direct letdown and approach to McChord. When we broke out under the clouds and saw the lights of Seattle and Tacoma, he was a happy warrior for sure. He truly enjoyed that flight from Anchorage to McChord that took us about an hour and 45 minutes.

Getting home was special for me too. I had been away nearly a month on a TDY that was to last only two weeks. I recall my little daughter Mary had been ill and I was anxious to see her. I had bought her an Alaskan doll (an Ootpic), and I remember how delighted she was to get it. It made her feel better.

An Assignment I Did Not Expect

My duties as ops officer continued to be busy and enjoyable. I had Doug Brenner as my sidekick, who was a terrific pilot and a conscientious officer. But we were beginning to lose some of our older and better heads to Vietnam assignment. People like Dan Barry and Jon Dean got their orders and left. Their replacements were fine men but had less all-weather fighter experience. Vietnam also caused problems in maintenance as it too lost experienced men to the war area. We now had a dispersal-operating base (DOB) at Walla Walla airport in southeastern Washington. There we maintained six F-106s, nuclear armed, on one hour alert. This placed an additional workload on our aircrews. The pilots would fly training missions out of McChord and recover at Walla Walla. There they would remain on alert for a few days and the aircrews at Walla Walla would take their planes and fly a training mission, then recover back at McChord. As you can guess, this took a lot of sophisticated scheduling on the part of both maintenance and operations to get pilots and planes in the right place at the right time. Walla Walla was a nice little city, and the aircrews enjoyed it. Because they were on one-hour alert status, they were permitted to eat downtown and even go to the movies with their walkie-talkies with them at all times. More than a few times I sat alert at Walla Walla where I enjoyed being away from my office and the constant phone interruptions. I could then work on ER'S and other things. Flying from McChord and landing at Walla Walla brought us over another interesting part of the northwest, the magnificent wheat country of the Palouse. We flew many low level intercept missions in this area and I marveled at the unending miles of fertile farm country that produced crops that fed the world. It was a sight to behold.

I was returning to McChord from a night refueling and intercept mission in early February 1970. I was surprised to see Col. Pool, my CO, waiting at my airplane as I shut down. Oh, oh, I thought, something's up. What could it be now?

"Okay, Larry, let me have it," I said.

"You've got an assignment." Larry answered. Immediately, I knew I was headed to the war in Vietnam at last. "What aircraft did they give me?" I asked. That was my first question. Larry answered, "You aren't going to Nam; you are going to Dover."

"What the hell are you talking about," I answered. "Come on in Gene and let me tell you all about it," said Col. Pool.

Larry Pool and I went into ops, and he gave me the story. It wasn't pretty. The 95th FIS was still in Korea on College Cadence. It would be returning the end of April as that deployment was ending. The Squadron Commander, Lt. Col. Roy White, had been fired and sent home. The squadron performance had been horrific. Their pre-deployment ORI was unsatisfactory, but ADC whitewashed them, and they were sent anyway. Their performance in Korea had been terrible and PACAF flunked them on their Flight Safety Inspection. In the past two years they had five major aircraft accidents, and their aircraft maintenance is pathetic. Two other officers had been offered the job of commander and have refused to take it. Larry said, "Maybe you should too. You can if you want."

I was in a state of confusion. The Commanding General of ADC had told me to first get in my Vietnam tour, but now he offers me this job. Why? Nobody else wants it. Is that why? However, if I don't take it, will I ever get another chance at being a commander? One never knows for sure. "Larry, what would you do?" I asked. "Take it," he said. "This squadron is so bad you can't make it worse. And if you make it better, you'll be a full colonel." Larry had a good point there. "I'll have to think about it and talk it over with Alice Marie," I told Larry. "I will give them my answer tomorrow."

When I told Alice the news she was as stunned as I. It was so unexpected. As always, my wife deferred to my decision, but she clearly liked the idea that at least I would not be on my way to the war zone, at least for a while. She also liked the thoughts of being stationed on the east coast closer to her mother and family. The following morning, I made a few calls to some old ADC friends who were closer to the situation than I. Many told me what a mess I would be walking into. But most of them thought I should take the job if I felt that I wanted to take up the challenge. I finally decided to say yes. This was a huge decision. I knew I was placing my career future on the line. It would be a job where I had to succeed; failure would not be accepted. And if I failed, I would pay the price.

Report I received in March 1970 written by Lt. Col. Lawrence Pool, Commander, 318th FIS and endorsed by Brig. Gen. Archie Burke, Commander, 25th Air Division:
Lt. Col. Cunneely is an absolutely superior officer with the keen insight and managerial expertise required in the effective operation of a fighter squadron. His consummate skill in scheduling resources was responsible for this unit's attainment of a high state of combat readiness and fulfillment of all operational commitments. Recognized by his superiors and subordinates as expert in every facet of the air defense mission, he was able to inspire squadron pilots to a high degree and insure a complete support of higher headquarters policies and directives. He has applied vigorous attention to pilot write-ups of aircraft/weapon system malfunctions to ensure that complete and logical maintenance actions were possible. The result has been a minimum recurring discrepancies with a significant improvement in systems reliability. Lt. Col. Cunneely speaks and writes with crystal clarity. He has the enviable knack for addressing his remarks to the basic problem and he exercises firm control over his subordinates without stultifying their efforts. The superb rapport he has established with host base personnel has created a harmonious relationship with all support agencies. He has consistently stimulated close cooperation between his squadron and higher headquarters as well. STRENGTHS: He is an outstanding organizer with the ability to select the right man for any task. He is able to assimilate the myriad detail attendant to his job while maintaining complete awareness of the complete picture. ASSIGNMENTS: LT. Col. Cunneely has unlimited potential as a leader. He should command an ADC

fighter squadron then be selected for attendance at a Senior Office Service School. He would be a welcome addition as a staff officer at any echelon. On 17 February 1970, Headquarters ADC issued an order designating me for assignment as commander of the 95th FIS, Dover AFB, Delaware, reporting for duty on April 20th. My goal to become a fighter squadron commander was achieved in an unexpected way. The unit was certainly not my first choice, and probably would not have been on my top ten lists. But it was my opportunity to show whether I was commander material or not. The command had given me a very tough assignment. It was giving me a unit that was in dire shape and in need of new leadership. They wanted me to clean up a mess that others made and should not have been allowed to exist. I knew that I would have to reach down and pull out the very best I had within me to succeed. I was 39 years old, a Command Pilot with 4,200 hours of flying time and over 720 hours in the F-106. I had served in several fighter squadrons, some good, some not so good. I had served at Group, Wing, Major Command, and Joint Command levels. In my years on the ADC ORI/TACEVAL Teams, I learned much about what it took to operate outstanding organizations. Now I would be given a chance to show what I had learned. This is what I had prayed for. As the old saying goes: Be careful of what you pray for; God may grant you your prayer.

CHAPTER TWENTY-TWO: MY FAMILY

Your wife shall be like a fruitful vine
in the recesses of your home;
Your children like olive plants around your table.

-Psalm 128

By February 1970, Alice Marie and I had been married over 14 years. We now had five children and were about to embark on our fifth move. The years had literally flown by, and I attribute that because they had been very busy and happy years. We had been blessed in so many ways.

First, our marriage was secure. Both of us had married when we were in our mid-twenties and had both decided that we wanted to begin married life and have a family. I had briefed Alice about the disadvantages of military life, and she was well aware that frequent moves and separations were part of the deal. Yet I was surprised how quickly she adjusted to it and how, when orders came, she did not complain but got busy and organized to make the change as easy as it could be for all of us. Even the kids got caught up in the excitement of traveling to a new home, although as they got older, this became more of a hardship on them. When I had to go on extended TDYs, Alice kept the house and the kids on routine. Even my six-month deployment to Korea, when the burden of being the squadron "Mother Hen" at McChord also fell on her shoulders, Alice never missed a beat. She was exceptionally popular with the squadron wives. One reason was that Alice never indulged in gossip sessions nor ever said an unkind word about anyone. When a wife had a problem, she would call Alice. By just talking, Alice usually helped the troubled one come away feeling better. Alice had that talent because she was more mature than most of the wives and had a deeper understanding. I was very proud of her. In so many ways, my wife was a tremendous asset to me in my career.

Also, the kids were coming along in great fashion. Gene Jr. was a handsome, athletic boy who was an excellent student at St. Francis Cabrini School in Lakewood. He continued his hockey at the Lakewood Ice Arena and was a high scorer on each team he played on. He also played Little League baseball, and one year played football. Jimmy, a year behind Gene, was an excellent skater and hockey player and a good baseball player as well. The Air Base provided great facilities for their sports, and they always had solid coaching mostly from NCOs who loved working with the kids. Mary, thanks to the Jon and Helen Dean Family got into competitive swimming and became quite competitive herself winning many ribbons. It was great to see her become interested and active in sports as I always thought it a productive thing to have the girls involved in athletics along with the boys. It kept them busy and left them less time to think too much about boys. Michael, although still only in his early years, began to play sports. Being the youngest boy, he was always the one being teased, picked on and injured. I made it a point not to give him too much sympathy, but to send him back out there and battle through his adversities. Which he did with good results as in later years he bested his brothers in all of the competitions they had e.g., tennis, golf, skiing. Michael was a blonde and rather small for his age, but in his teen years grew to be the biggest of our sons. Little Kathleen was still too young for sports, but not too young to be heard from as she was noted for providing vocal input on all family occasions. She was a delightful, beautiful child, very bright and happy. Each Sunday after Mass, our family would eat breakfast at the Officer's Club. Our children were always dressed handsomely, and their conduct was impeccable. Alice and I always received compliments from the servers on their table manners, always requesting permission to be excused before leaving the table.

At McChord, we enjoyed what the Pacific Northwest offered in the form of recreation and entertainment. We skied at White Pass each year, usually at Easter time, where the views of Mt. Rainier were truly glorious. The boys and I enjoyed watching the Seattle Totems hockey team, about an hour drive from our house on McChord. And in the summers, we swam and waterskied on American Lake on Fort Lewis, always in sight of the majestic Mt. Rainier in the background. At Christmas time, we would select a Douglas fir grown on the airbase, cut it and bring it home to be erected and decorated for the holidays. (There was always an argument about which was the best tree, as each child had a different opinion.) During our years at McChord, we were visited by my father, my brother and his wife Pat, Alice's mother and brother Tom. They were all impressed by the beauty of the Northwest and enjoyed close hand looks at the F-106s taking off and landing. All in all, we look back at our years at McChord as very good years, years in which our family flourished. Our children grew in stature and their educations went forward. They were secure and Alice and I continued to grow in love and respect for each other. I realized with each passing day what a lucky guy I had been to find a wife like her. God had blessed us.

But now, with this assignment to the 95th FIS at Dover AFB, Delaware, I had a premonition that things were going to be more difficult, not only for me but for the entire family as well. We had such a good life at McChord in so many ways, and I had such a great job in an outstanding unit. But this assignment was going to move us across the country. The kids would change schools and leave their friends behind. Alice would have to supervise the moving and that would be an awful task. And that situation was further complicated by a decision by the MAC Base Commander that truly angered me. My orders came for me to report to Dover in April as that was the month the 95th would be rotating back from Korea. The kids would not finish their spring semester until June, so I requested the Base to permit us to remain in Base Housing until school ended in June. But the request was denied. I was flabbergasted. I could not understand why the Base could not let us stay in quarters an extra month and a half. But the Base CO, flat refused my appeals and the Division Commander could not help me. This meant I had to move out of our quarters in early April and find a place for Alice and the kids to stay until school ended. This became a real Chinese fire drill. We had to hold out clothes for Alice and the kids from the packing and then find an apartment for them. We finally found a motel apartment on Highway 99 just off the base and moved the family in just before I left for Dover. I did not feel good about this arrangement, and it placed a terrible burden on Alice. This was a terrible disruption to my family as the apartment we rented was not luxurious, nor spacious. Alice had to drive the kids to school and pick them up. The kitchen and dining facility was small. And the kids were now separated from their friends on base housing. I felt awful about the situation that I was leaving Alice Marie and my kids in, but I had no choice. I was furious with the Base Commander for his heartless decision, but I lost that battle as well. In addition, the magnitude of the task awaiting me at Dover became more apparent as details of the 95th FIS's decline in mission performance came filtering in as well as calls from ADC friends informing me that I had accepted an assignment that was doomed to fail. The data they passed on to me about the condition of the 95th was both unbelievable and frightening. My own morale began to plummet.

After four great years as Operations Officer of the 318th FIS, I was given a farewell party to end all farewell parties. I was deeply touched by some of the nice things said about me, and also chagrined with the many playful thrusts aimed at me covering many incidents that took place during my tenure. I realized that I would probably never again be associated with a flying unit as good as the 318th FIS. I was privileged and fortunate to have been a member of this unit when it had such great people, pilots, airmen and staff, coming together at the same time. My job was to take advantage of their knowledge, skills, and professionalism and form an unbeatable team. The accomplishments of the 318th FIS during the years I was with it are still legendary. Foremost, it was the first ADC squadron to take over air defense alert in Alaska to intercept the Soviet Bears that approached the Alaskan ADIZ. The 318th was the first F-106 squadron to complete air refueling and air combat training and thus become operationally ready for overseas deployment. This came to pass in 1968 when I lead the 318th on the first overseas deployment of a F-106 squadron when we deployed successfully to Korea following the abduction of the USS Pueblo. During my time with the 318th we opened a dispersal base at Walla Walla, Washington and I also lead the first flight of F-106s from McChord AFB, WA, employing aerial refueling, that ended on the firing range over the Gulf of Mexico, south of Tyndall AFB, Florida where our aircraft successfully intercepted and fired on the drone targets. All in all, I look back on my time as Operations Officer of the 318th as some of my most effective and productive year. There l honed my skills as a decision maker, a keen judge of men, a manager, and a leader. These skills would all be required in my next assignment and in those following. I prayed that God would help me do my best.

The 95th FIS was due to return from Korea to Dover on or around May 1, 1970. Sadly, I departed McChord, leaving Alice Marie and the children holed up in a motel on April 10. I left with great sorrow and trepidation.

PART IV:
I BECOME A COMMANDER, A MAJOR CHALLENGE

CHAPTER TWENTY-THREE: LEADERSHIP TESTED

I slept and dreamed that life was beauty,

I woke and found that life was duty.

My first stop was Air Defense Command Headquarters at Colorado Springs. I flew there in an F-106B with my B-4 bag packed with personal belongings for a month's stay or longer. I met with several high-ranking officers including General Thomas McGhee Commander of ADC and his deputy. He was most explicit in what he expected me to do... clean up the mess. And the mess he described was a bit frightening. In the past few years, the 95th had suffered several major aircraft accidents and incidents. The result of its pre-deployment ORI was actually unsatisfactory but had been whitewashed so that the squadron could meet its deployment schedule. That was a mistake as the unit flunked a Flight Safety inspection in Korea conducted by PACAF, which embarrassed ADC. The unit's aircraft in commission rate was unsatisfactory and they were barely able to meet alert commitments. In February, ADC relieved the squadron commander Lt Col. Roy White and sent him home from Korea. Gen. McGhee told me that I would be given full support from his headquarters to get the 95th FIS back in shape, but he wanted the job done quickly. I also spoke with Col. Lou LaSalle, Deputy for Personnel who told me to call him personally if I needed to make some changes in any management positions in the squadron. I told him that I was sure that I would be talking to him...soon. I even visited with the Command IG, General Robinson. I told them that I knew that they would be visiting the 95th as soon as they could arrange it. But I asked they give me a little time, at least three months. He would not promise.

From Colorado Springs, I flew into Dover AFB and reported in to take command of the 95th FIS. Lt Col. John Mancus, the Executive Officer met me and took care of the administrative details. I also toured the squadron facilities which were located on the east side of the base somewhat distant from the main base facilities and base housing. The squadron facilities were old but adequate in most respects. They certainly did not compare to the 318th's which were fairly modern and efficient. But I immediately put it out of my mind to compare everything I saw to my previous squadron. They were alpha and omega, and it was a useless exercise. While the unit was in Korea there was left only a small force of airmen to keep the facilities in shape, take care of transients, and be ready for the return of the F-106s. Mancus was helpful and provided much valuable information. My office was on the second floor of the Ops building and I met my secretary Ruby who gave me lots of material to read.

Later I met Roy White. He was still on base and invited me to his home for dinner. As this house would be the house my family would move into, I was anxious to look at it and report back to Alice Marie. That evening I met White and had dinner with him and his family. Mrs. White cooked a nice meal, and I enjoyed her showing me the house and hearing her comments about it which I would report to Alice. But it was an awkward evening. White had been fired from his command position and his career was at an end. His bitterness came through and he laid most of the blame on others, mostly on the Chief of Maintenance and his people. He went on about this at length and did not mention any actions that he took which might have contributed to the situation. I certainly did not take issue with his comments but what he said was in great contradiction with what I had been told by others at ADC. I thanked the White's for their hospitality and returned to my quarters in the BOQ where I would be living for the next month or more. I left the White's house somewhat disturbed.

It was becoming clearer to me that the job I was facing was greater than I at first blush thought it was. Almost everyone I spoke with, whom I trusted, presented a dark bleak picture of the problems I would be facing. And as I tried to bring together all that I was told, analyzing where to focus my energy to turn around this very sick organization, I came back to one recurring thought. This squadron suffered from the lack of leadership—strong leadership. And if it's ever going to reverse course and become a combat ready air defense interceptor squadron, strong leadership is what it was going to take. That is just what Gen. McGhee, and others, who entrusted me with this job expected me to bring.

Hollywood Helps

During my career to date, I had attended numerous training schools. These included schools on survival, electronics, the SAGE system, maintenance, intelligence, et al and of course management and leadership. One I remember, and I don't recall exactly where it was or who ran it, had to do with leadership. The instructor was a civilian and our class spent a whole day reviewing a movie that was produced in 1949 called 12 O'clock High. The instructor would run the film for 20 or 30 minutes, then the class would analyze and discuss it. The star was Gregory Peck, who as General Frank Savage, was ordered to take over a B-17 Bomber Group in England at the height of World War II. Peck was replacing his good friend Col. Keith Davenport, played by Gary Merrill, who was a good guy and well-liked by all of his men in the Group. But the Group was in terrible shape operationally, suffering many losses and unable to neither get its bombers over the target on time nor hit the targets when they arrived. Peck felt terrible about taking over from his friend, but the General who sent him to do the job pressured him because it was critical for the defeat of Nazi Germany that daylight precision bombing succeeded.

As Peck is about to arrive at the Base to take over, he stops the staff car, gets out, lights a cigarette, and silently thinks for a few minutes. Then he gets in and the car enters the base. Starting at the Air Police gate shack, where the AP just waves Peck through without checking his ID, to the orderly room where the sergeant who greets Peck is improperly dressed, Peck lets everyone know new leadership has arrived. Before long, everyone in the Group hates Peck and every pilot asks for a transfer. But Peck presses on. He now is demanding that the aircrews meet higher standards in their flying, formation, bombing, navigation, etc. Those who abort combat missions without due cause, he denounces, and he does not hesitate to lead himself on the most dangerous missions. Peck pushes his men unmercifully. Naturally, the Group comes through with flying colors on a vital bombing mission against a German ball bearing factory and the pilots realize what a difference Peck's leadership meant.

I thought about that movie and what our class said about Gen. Savage's strategy. Most of us felt that under those circumstances, he had no other choice. Being popular was not what was needed in such a situation. Leadership did not call for popularity when a huge change in attitude across the board had to take effect. The 95th FIS would have to undergo a large scale change in attitude if it were to regain operational readiness in the near future. And if I were the one to lead it there, my popularity rating would no doubt end up in the gutter. So be it. I knew what my job was; that had been made clear to me. And my career was on the line. This was what I had been preparing for. This was my first big test. I would refuse to fail.

General Jack Gamble

The 95th FIS was one of two interceptor squadrons under the 20th Air Division. The Division was commanded by Brig. Gen. Jack Gamble. I knew General Gamble from my days with the ADC TAC/EVL Team and I admired him greatly. I was pleased to have him as my commander in these circumstances. His Director for Operations was Col. Joe Joiner. Joiner was a terrific fighter pilot and a true friend whom I knew going back several years. With these two behind me, I felt confident that I would get great support from command. I flew down to Langley AFB and drove over to Fort Lee, the location of the 20th Air Division. There I spent time with Gen. Gamble and his staff. The General and his staff were candid about the situation and the problems I faced. The trouble was compounded by the fact that they wanted the 95th to resume air defense alert as soon as possible following their return from Korea. I had no way of assessing whether this was a feasible possibility, and I did not want to commit to this before knowing for sure if it could be done safely. The 20th Air Division had the nation's capital Washington, D.C. within its region, and it was important to have both the 48th FIS at Langley and the 95th FIS at Dover up on alert. I understood this and promised to do the best I could to meet this commitment. Gen. Gamble also warned that the ADC ORI teamed was anxious to return and inspect the 95th because they knew that the conclusions of its pre-deployment inspection had been overruled and the 95th should never have been sent to Korea. All of these thoughts I carried away with me as I returned to Dover and awaited the return of the 95th to Dover. I needed to be prepared and ready when that happened because I would have to enact many changes which would not be well received and against which there would be much resistance. This will be a situation new to me and for which I would have to be totally mentally and psychologically ready. I increased my daily prayers and attendance at five o'clock Mass in the Base Chapel. I would need all the help I could get.

Family Concerns

During this period, I tried to stay in touch with Alice Marie and the family. I was able to call them through the Air Force tactical phone system from the combat alert center (CAC) in the squadron to the 318th CAC. There my old sergeants would patch me through to Alice on an off base line. It was heartbreaking to me that Alice had to cope with five kids in such a cramped situation. She would drive four of them to St. Frances Cabrini School each morning after feeding them breakfast and pick them up after school. Katie was still too young for school and stayed with her mom all day. The boys and Mary often had sports after school which Mom had to chauffer them to and get them back for supper. Then it was homework, baths, and bedtime. It was a tough time for her, and she had little help. On my end, 3,000 miles and a three hour time difference away, I was living in the BOQ across from the Officer's Club. I did have a staff car which permitted me transport around the base and into Dover on the weekends. It was a lonely existence for both of us but busy. I could not wait until the squadron arrived back, I could start my real work, and as soon as possible, I could get my family back with me in our new home.

Dover AFB was owned by Military Airlift Command (MAC) and the 95th was a tenant unit. The MAC Wing Commander was a Brigadier General, and he was a careerist with an ego the size of the big outdoors. On top of that, he wore a toupee with a beautiful wave in front. He demanded that I attend all of his staff weekly meetings that were endless and fruitless. I also had to wear my class A blues at these meetings that served only for the General to go on and on about many things that were of no concern to the 95th FIS.

The 95th FIS Redeploys

Reports coming into me about the squadron were not good. Both PCAF and ADC were concerned whether the 95th was operationally ready to redeploy the 6,000 miles across the Pacific safely. My former commander Fred O'Connor was sent to Osan to look at the squadron and determine whether they could go or not. After some delay, the squadron was released to redeploy. O'Connor flew in one of the lead tankers too oversee the flights back. The squadron had to divert into Wurtsmith AFB, Michigan due to weather, but Col. O'Connor flew on into Dover where I met him, and we spent a long evening together as he briefed me about his observations of the 95th. His observations were disturbing to say the least. O'Connor was an experienced ex-commander, and I trusted his knowledge. Up until the end he was concerned about the unit's capability to make the redeployment. He spoke about their lack of air discipline, and he mentioned that even on the redeployment the aircrews violated safety regulations on several occasions which would be in his report and in the SAC tanker crew's report. O'Connor provided me with names of aircrews and officers that he thought I should have transferred out. Some of these were high ranking people. His major criticism was failed leadership at the command level and low morale in the maintenance organization. O'Connor's briefing was valuable, and I thanked him for it. It confirmed some of the ideas I had been harboring myself and strengthened my plans to completely shake up and remake the 95th FIS.

A few days later, the 95th FIS aircraft finally made it into Dover. The maintenance crews had already arrived and were awaiting them. On the tarmac were the Base Commander Col. Harry Odren, Col. Ken Swisher, the Deputy for Ops, me and Lt. Col. Walt Newton, the interim 95th Commander. Also, there were the wives and kids of the returning pilots and airmen. The welcome banners were out, and much cheering was heard as each of the flights entered the flight pattern, pitched out and made their landings. (I noted that some of the flight patterns were non-standard.) I shook hands with all of the returning aircrews and then the raced over to greet their families after a six month separation. I knew how they felt. This was the first time that I met most of these pilots and the first time they had met me. I would have to get to know them as quickly as I could as there was much ahead to be done that would affect all of us. But at this point they were all happy to be home and I would do nothing to spoil this festive occasion. I was now pleased that the 95th FIS, my new command, was all in place and that the work that needed to be done would commence soon. I was anxious to get started. It was best that the personnel of the 95th be unaware of what lay ahead of them for the time being.

CHAPTER TWENTY-FOUR: 12 O'CLOCK HIGH TIME

The good Lord never gives you more than you can handle. Unless you die of something.

-Guindon cartoon caption

After the welcome ceremonies died down and the returning aircrews and maintenance people had some time off to regroup, it was time to get back to work. Joe Joiner passed on to me on the day the aircraft returned, that Gamble wanted the 95th back on alert in two weeks. I was surprised at that early date. I still did not know the condition of the aircraft and whether or not we could provide sufficient OR (operationally ready) aircraft to both meet the alert requirements and our training schedules. The following week I met with our senior officers from operations and maintenance to assess the situation. I presented to them what was expected from us. Both sides were skeptical. The Chief of Maintenance was a young Major who came across to me as weak and unsure of himself. Lt.Col. Newton, who had been the interim squadron commander after White was relieved, was now the Ops Officer, although he would soon be leaving the 95th for an assignment to Vietnam. One thing the meeting verified was there were no good feelings between ops and maintenance, and this confirmed what I had been told beforehand. This was a serious problem which I would have to face immediately. After this, I had my first meeting with the aircrews. Some of the pilots I knew, but the majority was new to me. Several were Vietnam returnees and had excellent combat records. My talk to them began with my commending them for their successful redeployment from Korea. (I did not mention the flight violation that were submitted by the SAC tanker pilots and Air Traffic Control). But then I spoke to them in more grave tones of what I expected, and that included professionalism in every way, including not just their adherence to flight safety and mission accomplishment but also to how they conducted themselves on the ground, in their relations with enlisted personnel, their conduct in the O' Club, and in their grooming. I ended by telling them I expect to raise the 95th FIS to where it would be considered the best interceptor squadron in ADC and that nothing less would be acceptable.

I did not know how that initial briefing was received but some feedback I heard was many of the pilots did not think much of it. They thought I was talking down to them and of course, they felt I was an outsider sent in to clean up their very own squadron which they felt did not need any cleansing. They also still liked their former CO Roy White better than this bald-headed guy that headquarters had sent them. Well, they had only just gotten a nibble of what was to come.

After that meeting I had a Commander's Call in the Base Theater. There I spoke to the entire squadron. I again commended the squadron for its performance in its successful deployment. I then pinned medal on some officers and airmen who had been awarded them prior to and during the deployment. I then addressed the squadron concerning what I expected of them. I warned that they would see some changes, but they would also be told why the changes were needed and what the goals of the squadron were. One of the things I mentioned in this first address to the whole squadron was my dedication to create a better relationship between operations and maintenance. The aircrews cannot perform their mission without the full support of our crew chiefs and those who fix our planes and keep them OR. And to do this, our pilots must do everything to help identify the problems and assist maintenance people in the fixing. Again, I don't know what reaction my words had. But from the First Sergeant, I got the impression that some of the senior maintenance chiefs were hopeful.

The 95th FIS History

The 95th was a full squadron unlike the 318th and many others in ADC. The 318th was composed of pilots, a combat alert center, a scheduling and records section, and a personal equipment section. All the maintenance on the aircraft was performed by an organizational maintenance unit under a Chief of Maintenance who reported to a Group Commander. In the case of the 95th as squadron commander everything, including the aircraft, maintenance, and weapons storage came under my responsibility. In addition, the 95th had a Dispersal Operating Base at Atlantic City International Airport that also was under my command. All in all, the 95th had over 700 personnel, 25 F-106s, three T-33s, two nuclear weapons storage sites, and maintenance facilities spread over two bases. In truth, I was commanding one of the largest F-106 squadrons in ADC.

The 95th Aero Squadron was organized at Kelly Field, Texas in August 1917. In October they sailed to France and in July traded in their Nieuport 28s for SPAD 13. The 95th was credited with bringing down 16 enemy planes. After deactivation, the 95th was again organized in September 1942 at Harding Field, Baton Rouge, Louisiana. It was then deployed to Africa where, with its P-28s, it participated in the Africa Campaign. The squadron had a splendid record operating in Tunisia, and in Italy. The 95th was awarded three Presidential Unit Citations including one for its participation at Ploesti in June 1944. The 95th was again activated in 1947 at Grenier Field, New Hampshire and in 1949 was redesigned the 95th Fighter Interceptor Squadron and assigned to Andrews AFB, Maryland as a unit of the 85th Air Division and equipped with F-86L aircraft. In March 1958, the unit converted to F-102 Delta Daggers, and in February 1960 the 95th converted to F-106 Delta Darts. In July 1963, the 95th moved from Andrews to Dover AFB. The 95th FIS came under the operational control of the 20th Air Division (SAGE) located at Fort Lee, Virginia. The current emblem of the 95th was the "Mr. Bones" patch which is worn on the flying suits of the pilots and is seen on the vertical stabilizers (tails) of its aircraft. The patch shows a skull wearing a top hat, wearing a bow tie and its right hand wearing a white glove holding a cane. There is also a yellow lightning flash coming into the picture over the left shoulder of Mr. Bones. The death's head, emanating from a cloud, is symbolic of the unit's dauntless capabilities for accomplishing its mission in any weather, day or night. The 95th FIS certainly had an illustrious history going back to World War I, producing many victories in combat and several legendary heroes of the air. But the squadron was now at one of its lowest points and clearly not ready to enter any serious wartime engagement. It was my job to get it back on course more in line with its illustrious past history.

My next step was to begin flying the 95th's aircraft and get a taste of what shape our birds were in. When I did, my worst fears were fulfilled. The 95th possessed some of the oldest aircraft in the inventory. They had the "round eye" instrument panel where the gauges were the old style clock like ones with hands pointing to digits that told you your altitude, airspeed, rate of climb, etc. I had been used to the vertical instruments that we had in the 318th aircraft, but I had no trouble adjusting. My first few flights in the 95th aircraft were unsuccessful as the MA-1 fire control systems were bad on both aircraft. I soon learned that this was prevalent throughout the squadron. I found this out by discussing this problem with the Electronics Officer, a Captain, and the Hughes tech rep. After some prodding by me, they both advanced that the fire control systems of the squadron's aircraft had been badly neglected during the deployment. Most of the aircraft were behind in periodic inspections, and in order to meet flight schedules in Korea, fire control system write-ups were usually not corrected. Aircrews training flights in Korea concentrated mostly on air combat training and patrols. Both the electronics officer and the tech rep informed me that we had a real problem on our hands returning the unit's aircraft back into condition to perform the air defense mission. This was worse than I what I had ever expected.

'Mr Bones' F-106s originally featured the squadron badge in a large disc on the tail, as seen on these examples photographed in 1963.

Ops versus Maintenance

I began to meet the aircrews, study them and make my judgments. There was a fair mix of experience in the group with several Vietnam returnees. But I sensed there was a streak of arrogance present, and I could feel that the pilots, in general, did not like me. And they knew that I had come to make some wholesale changes in the way the organization had been doing business. They were certainly right about that. Like most people, they did not like or want change.

The first real confrontation came at our first scheduling meeting. I instructed Ops and Maintenance to put together a flying schedule and meet with me in my office on a Thursday afternoon. Maintenance was represented by the Chief of Maintenance and his senior NCO's, and Ops by Lt. Col. Newton and Major Cardwell, the ops officers. Looking at the flying schedule they had come up with, I could see that it was totally unrealistic. Our daily in-commission rate had been pathetically low since the unit's return. We were unable to reach a 50 percent OR rate, but this proposed schedule called for 20 sorties a day. I asked the maintenance chief if he could provide those 20 sorties, and he answered that they would try. I said I expected that they would try, but my question was, could they do it? He looked at his NCO and finally said no, they could not do it. Why did you come to me with a flying schedule that was doomed to fail, I asked. Why did you sign off on a flight schedule that you knew Maintenance was unable to fulfill? Maybe this was what the 95th was about in the past but not now, I told them. I then directed that they meet again and come up with a revised schedule, one that Maintenance would be able to meet and that Ops could depend on, even if it did not meet their needs. I directed that when they had a revised schedule, one that was realistic, they should call me and we would meet again and make it final. And the final schedule, I will expect will be flown as agreed on. Further, I will expect Maintenance to improve its performance by gradually increasing its output of daily sorties until it meets the operational needs of our aircrews for alert and training. Then I dismissed them.

The impact of this meeting was felt throughout the squadron. The final schedule they came up with provided only 10 sorties a day for the pilots, half of what Maintenance had been browbeaten to agree to on in the initial submission. It was clear to me that in the 95th FIS, prior to my arrival, Operations ran the show and Aircraft Maintenance had to swallow the pill. From my own sorties in 95th aircraft, I could discern that these planes were in poor condition, especially the fire control systems which had been badly neglected. The morale in the maintenance ranks was very low and I saw that in that area I would have to put a lot of my attention. I called in the Chief of Maintenance the next day. He was a fairly young Major and was inexperienced for such a large job. I informed him that Ops would no longer be calling the shots around here, but that I needed him to shake up his troops and get them working harder than ever. He told me that his people were tired and that they felt that the aircrews did not respect them. He said the pilots talked down to the crew chiefs and told me that there was poor esprit in the squadron because of that. I told the Major that from that day on I would attend the daily 4 o'clock maintenance meetings he held, that I will do everything I can to resolve this problem and that I wanted his full support. The Major thanked me and left. I immediately called Col. Lou LaSalle at ADC Headquarters and asked him to send me a new Chief of Maintenance as soon as possible. Lou said he would look for one.

I Concentrate on Aircraft Maintenance

The repercussions that vibrated out from this meeting went out quickly and widely. A new day had arrived in the 95th FIS and it was clear that things were going to be different from here on out. From this point I did the following:

I moved my office out of the operations building into a building that was halfway between ops and the maintenance complex. The symbolism of this move was clear.

I began attending the daily maintenance meetings at 4 o'clock in the maintenance conference room. At this meeting, I reviewed the status of each aircraft that was out of commission and its estimated time in commission (ETIC). By my attendance I assured the maintenance chiefs that I would support their efforts to attain equal status in the squadron and I would reward and recognize their efforts.

I authorized the construction of a new maintenance control center, similar to the operation's combat control center. This would be a 24-hour center, with full communications equipment that would be more efficient in coordinating and controlling aircraft maintenance activities.

I made it a practice to be seen on the tarmac when there were aircraft departures and arrivals. I also began to sit in on aircrew debriefings with maintenance personnel, as I had received reports of pilots writing up aircraft malfunctions that were erroneous and incomplete. On one occasion, I was called to the debriefing shack by maintenance when two pilots were competing with each other to see which one of them could come up with the most write-ups. When I walked in on this competition, I emptied the shack and gave two pilots, one a Major, the biggest ass-chewing either had ever received in their careers to date.

I began to carry a small notebook in which to keep track of things that came up that I had to deal with. It seemed that during the day I was deluged by reports by phone and in person with problems that required my personal attention, except that each one took time that I had to find.

I also began to get up in the middle of the night and drive down to the maintenance area and inspect to see what was going on there. I started to do this because maintenance kept failing to meet its ETIC's. (Estimated time in-commission). What I found provided some answers. I walked into some hangars and found that maintenance personnel, who were supposed to be working on aircraft, were not. In some cases, they were not even there at all. Some I found asleep. At three o'clock one night, I called the Chief of Maintenance out of bed and his NCO and got them down to the squadron to show them what was going on and why the aircraft were not ready to fly on the next day as scheduled. This resulted in some Article 15 actions.

In one of my squadron walkarounds, I found a door in the maintenance facility locked. I asked the Chief what was in there and he gave me a vague answer. I insisted on his opening the door and looking in this room. After fumbling around for a while looking for the key, I finally got into this room. It was a room with a long bar, fully stocked and adorned with Playboy centerfolds. The Chief insisted that this room was only used on special occasions and only with his approval. I took the key and told him this room was closed, for good.

My Flying Hours Dwindle

I tried to fly at least twice weekly. It wasn't always possible with the amount of commitments I had, including those with the host base. I was not pleased with the way the aircrews approached their work, but I always attended the operations morning briefing which included many items such as flight safety bulletins, intelligence, weather forecasts, and the daily flight schedule. One morning, a day on which I was on the flight schedule but not the flight leader, I was dissatisfied with the briefing as the briefing officer acted as if he was a comedian making snide comments about the mission, the aircraft and the proficiency of the SAGE Controllers. I stopped the briefing and addressed the Ops Officer. I told him that I would be returning to my office. I told him that when his aircrews were ready to provide a professional briefing for the mission that was coming up that he could call me. But that I will not accept that the briefing I just sat through was anything but a professional briefing in preparation for a safe and successful air defense mission. I then left the room. The pilots who jumped to attention when I left the room, I'm sure did not like what they heard, but I believe that they realized a new day had dawned. I made another call to Lou LaSalle. I needed a strong Ops Officer and soon.

I began to crack down on the number of airmen who failed to meet Air Force grooming standards with regard to haircuts, facial hair, and uniforms. When my exhortations did not result in improvements, I began to schedule monthly squadron Commander Calls where I and the First Sergeant conducted personal inspections on the enlisted men who were formed on the flight line. Business in the Base barber shops experienced a significant increase. But then I brought this problem to the Ops Officer concerning the appearance of the aircrews. When my words here failed to improve the situation, I required the officers to stand inspection with the enlisted men. You can imagine the reaction this received. However, I began to get results. The word finally was getting out, this Commander meant business. My popularity rating was in the toilet. But, the ORI Team was champing at the bit to inspect the 95th and my squadron was a mess. I had never been so discouraged, I needed help. I needed a strong ops officer, a maintenance chief, and I had several other officers and senior NCOs I had to have replaced to turn this squadron around and time was running out.

My days were long and tedious. I missed Alice Marie and my children. With all that was going on I had little time to think about them. Only at night or on the weekends did I have a chance to call. I did not always get through, but when I did, I caught up on the many problems my wife was having in running a household of kids out of a motel room. It was clear that Alice Marie was just as busy as I was 3,000 miles away. I felt terrible about how this assignment worked out for her and the kids. Maybe I should have turned down this assignment and gone to Vietnam. Some days and nights I began to believe that the 95th was a hopeless case, as many smart people told me it was. We made a plan that when school ended in June, Alice would drive the kids over to Sandpoint and vacation at the lake cabin. I would ask for some leave, fly into Spokane, and spend as much time as I could with the family, then we would drive back to Dover. It would be a short summer at the lake and a long drive across the country to a hot and humid region, but Dover had a pool and a golf course. I also set Gene and Jimmy up at a hockey clinic in Wilmington, so they could continue their hockey aspirations there. I hoped that they would find Dover a good place to live. The house was okay, roomier than the one we had at McChord and there was a good Catholic High School nearby. Alice would enjoy living in the East where she could see her family more than before and the same was true for me. So, I hoped the best for my family as I wrestled with the many problems of my new job. We made our plans and both of us continued to perform our responsibilities as well as we could, thinking of each other and missing each other badly. Neither of us had much help.

I know that I felt quite alone in what I was doing. The one person I could talk to and ask advice from was my commander, General Jack Gamble. I always kept him informed of what I was doing and where I wanted to go with the squadron. He most always supported me and offered me whatever he could in the way of backing. My biggest problems were personnel and I was working with Col. LaSalle at ADC to get me some strong officers for the important supervisory positions, especially maintenance and ops. I also had named some aircrews that I wanted reassigned and Lou was working on that too. The other problem was aircraft parts. The Vietnam War was still in progress and the major portion of the defense budget was going there. When we had an aircraft out for a part, it was often weeks before we could get the replacement. If we took that part from another aircraft (cannibalize), it did not completely solve the problem and it only hurt your tail number scheduling program. Our aircraft OR rate was still below satisfactory levels and we were not meeting the training requirements of our pilots. The fire control systems were our worst problem. At Dover, the MAC Wing flew the largest aircraft in the world at that time the C-5 Galaxy. The MA-1 should pick up a C-5 at 40 miles. Our F-106s could hardly pick them up on radar at five miles. We needed to do something about this or fail the ORI for sure. I was applying a great deal of pressure on both maintenance and the aircrews, maintenance to provide the pilots more sorties per day and in better aircraft that would meet combat standards. The pilots also were pressured to meet their training requirements and also to fulfill their air defense alert commitments at both Dover, where they sat five minute alert and at our Atlantic City Airport dispersal base where they sat one hour alert in nuclear armed F-106s. Morale was not very high throughout the squadron and this permeated into the ranks of the wives, many of whom let their feelings be known to me. I took my turn sitting alert, but I noted that whenever I was in the alert hangar, the squadron pilots chose not to converse with me. They wore their feelings on their sleeves and then made it clear that they did not appreciate the severe changes I had brought to the 95th. Living alone in the BOQ, I usually ate alone in the Officer's Club or hit an off base restaurant. My work week was seven days and from Monday to Friday, it was 12-hour days. My mind never ceased mulling over the problems of the 95th FIS and what I could come up with to turn things around I called on my experiences on the ADC TAC/ EVAL Team where I flew with almost all of the ADC units, good and bad, and met with the best commanders that were around at that time. I learned so much from them and now I tried to remember some of the things they did and how they approached and solved the problems they faced. One thing I knew to be a successful commander of a troubled outfit, you had to be strong minded. You could not waver. You had to keep your eye on the mission and surround yourself with the best people you can get, tell them what you expect of them, and let them do their job. In this respect, I was in trouble. I needed a strong ops officer badly and a chief of maintenance who could ride herd on the maintenance NCOs and get the best out of them. I needed to be rid of some bad apples who were hurting the unit by back biting and talking down the squadron's leadership. I now knew who these people were, and I had given their names to ADC Personnel. They would soon be gone. What I needed also was a little time for things to fall in place. If I got some time, I might just make it. Prayer would help.

I Get an Ops Officer-First Class

At the end of May, I got some good news finally. Lou LaSalle called and told me he had a name for me as a possible ops officer. The name was Harold Van Daele. "Yogi" Van Daele was returning from an RF-4 tour in Nam. I knew Yogi well from my ADC days and on the 318th's deployment to Korea, Yogi was a spare pilot. When we were departing Hickam for Okinawa, I had to roust Yogi out of the KC-135 tanker and put him in one of our F-106s at the very last minute prior to takeoff. Van Daele was just the man I was looking for. He was one tough son of a gun, a no-nonsense guy, who would make things happen by example. I could not have asked for a better choice. I jumped at this opportunity to obtain an officer who would share my values and put into effect the standards I expected of the 95th aircrews to meet. At last, I received some good news. I hoped for some more.

In mid-June, I flew an F-106B to Colorado Springs and met with some of the ADC Headquarters people. I gave them a realistic appraisal of the situation and again asked for some help in obtaining some new people. I spoke to the IG folks who would not give any hints as to when they would arrive to inspect the 95th but I could tell we were on the schedule. From there, I flew into Geiger Field, Spokane, but only after I gave Alice and the kids a buzz job on Lake Pend Oreille. I'm told that my tailpipe left a nice wake on the pristine waters of that beautiful lake. Capt. Norm Komnick was my back seat pilot, and he then took the aircraft back to Dover. After the kids finished their school, Alice packed up their things and they drove over to the lake cabin. And that's where I would spend about 10 days with them and then we would all drive back across the country to our new home in Dover. It was a great reunion. Although we no longer owned a boat, (I had sold the Thompson before leaving McChord), we still had a great time. The weather was super, and we enjoyed the lake knowing not when we would be able to return to it from our new assignment to the east coast. Alice and the kids had grown to love McChord dearly. It was their home for so many years and that is here the kids had attained their first real friends. Then there was the lake cabin, which they loved dearly. It was their second home where they piled up many adventures in the surrounding forests and back roads. They loved Sandpoint, the St. Joseph Parish, and the many of projects we had working on the cabin—building decks, stairs, and painting just about anything that didn't move. They loved the wildlife in the area and fun on the water: water skiing, fishing, and attempting every convoluted manner of dive or leap from the dock into the cool, clear refreshing waters of Lake Pend Oreille. They loved the sunsets over Mt. Schweitzer and the great night skies that were often percolated by northern lights and fantastic and frightening lightning displays. The days went quickly, and we were soon packing for our cross country trip.

All the time we were getting in our last licks at the lake, I could not help but think about my problems with the 95th FIS that were still facing me. I often sat on the dock or on the deck and tried to arrive at some answers, but few came to mind. I frequently tossed and turned in bed at night truly concerned whether or not I would be able to turn this outfit around before my time ran out. I was especially troubled about the poor condition of our aircraft's fire control systems. One day I came up with an idea. As I tried to flesh it out, I became less sure that it was feasible. And yet, I also concluded that it would require a drastic concept, requiring total squadron dedication, and approval and support by Division Headquarters if we were going to bring the unit's aircraft into operational condition sufficiently adequate to pass an ADC ORI that was surely coming. After several days of thought and many pencil drafts, I inked a letter to General Gamble outlining my plan. I gave it a code name "Better Bones" and I suggested that it begin as soon as I return to Dover. I warned that unless something as irregular as this plan is put into effect, I felt that our chances of success were extremely poor. Shortly after mailing that letter from Sandpoint, the Cunneely Family, plus Brandy the dog piled into our Mercury Marquis station wagon and headed east to Dover. Not too many who got into that car was happy about it and I, the driver, was the most worried of all. It's at times in your life like these, that you have to place your faith in the Lord. and hope for the best. And that's exactly what Alice and I did.

Our drive across America was uneventful. We had no time to sight see and I kept mostly on the interstate systems that had been built since the days when I first made my first lonesome trip from New York to McChord in 1954. After we left Idaho and Montana, the scenery became rather dull. Thank goodness for air conditioning and we made it a point to stop at motels with a pool. The kids were good but Alice and I could see that they were worried about this move. Gene Jr. was 14, Jimmy 13, Mary 12, Michael 9, and Katie 5. The three older ones especially were concerned about leaving their friends and changing schools. To help out a little, I had signed Gene and Jimmy up for a hockey clinic in Wilmington for the summer. I knew that this was a big transition for all of them and I hoped to make it come off as smoothly as possible. There was an excellent Catholic school in Dover and I had made inquiries for their enrollment. My concern was, I knew that this job would be a temporary one for me, two years at most. That is, if I didn't get fired sooner. And then there would be another family move. And my next assignment would likely be Vietnam which would be a family separation for as much as a year and a half. I kept these thoughts in the back of my mind because Alice Marie had enough on her plate for the time being. And so, did I. First things first…

Driving back to Dover, I was anxious to get back and take up my work in repairing a badly wounded squadron. I was also anxious to learn if General Gamble approved of my plan to pull our aircraft out of the schedule, four at a time, and fly and repair them until they met operationally ready standards. If he turned down my idea, I did not know what else to do. I was out of ideas. There were so many problems facing me on so many fronts, I often felt overwhelmed. I carried a small notebook in my pocket in which I recorded each problem brought to my attention. The notebook was crammed. It was impossible for me to keep track of the multitude of problems begging for my personal attention. At the beginning of each day, I perused my problem book and determined to correct 10 of those discrepancies. I rarely met that goal. Some days I solved none of them. On a good day, I would correct maybe five. But even on a good day, my notebook would pick up more problems to be solved than I had solved. My biggest problem was a lack of leadership and experience in the maintenance area. I was still pressing ADC for help there. At least I was getting a new strong ops officer in Van Daele. I was confident that Yogi would lift that burden from me in short order so that I could concentrate on the aircraft situation. I knew that the aircrews were in for a shock when they came under the scrutiny of Col. Van Daele who would brook no nonsense. I had fired the Electronics Officer, who was hopeless, and I had asked Hughes Aircraft for a new Tech Rep to assist in bucking up the fire control systems. I was also able to get a new First Sergeant who would help me bring under control disciplinary problems in the enlisted ranks mostly dealing with pot smoking, grooming, homosexuals, and AWOLs.

So, having received a few breaks, maybe things would pick up when I got back to Dover. I hoped so. At least I would no longer be fighting the problems alone. No more BOQ living and back to some home cooking. Alice Marie was with me again along with my children whom I missed so much. I would have their moral support to buck me up when things were bleak as they were sure to be in the coming days ahead. I knew that they would not like Dover AFB as much as they did McChord, but Dover wasn't all that bad. There were some good things involved about being on the east coast and we would try to enjoy as much of them as we could as a family, come what may. The next year or two would be very interesting indeed.

CHAPTER TWENTY-FIVE: IT'S TIME TO PUT UP OR SHUT UP

Perseverance, my dear Lord, keeps honor bright.

-Shakespeare

We arrived at our new home on Dover AFB in good time and in good order. It had been a long hot drive, and much of it had been rather dull and uninteresting scenery-wise. But Alice Marie and the kids seemed to like the house. It was somewhat larger than the one we lived in at McChord. This was a two-story home composed of brick and it was situated on a large corner lot in the officer's housing area. The home had three bedrooms upstairs and one down, with a nicer kitchen and a comfortable dining room. The family immediately noted the higher humidity than they were used to and the need for air conditioning. The officer's club had a nice sized pool and there was a golf course on base that interested the two older boys. The job of unpacking and finishing moving in was still to be done and that chore is never enjoyable. But Alice started in soon and got after the kids, once it was settled who had which bedroom, to unpack and put away their clothes and other belongings. While this was going on, I was anxious to get back to my responsibilities.

Two good things awaited my return. One was the approval of General Gamble to my proposal to begin pulling F-106s from the schedule, four at a time, and test hopping them until they met OR standards. This plan, officially titled "Better Bones" although conceived by me and approved by General Gamble, was still unknown to the rest of the squadron.

The second good news was the arrival of Harold Van Daele, my new ops officer. Lt. Col. Van Daele, or "Yogi" as he was fondly called, was just what the doctor ordered for me and the 95th. We had an early closed-door meeting during which I brought him up to speed on the status of the operations section. I also told him what I needed him to do and what I expected from him in the shortest possible time. Van Daele had already been alerted to the problems of the 95th and what I told him came as no surprise. As I expected, he assured me that he would fulfill my expectations to the hilt and could not care less about the sensitivities of the pilots. Somehow, I knew that I had found the right man. I could now concentrate on the squadron's aircraft maintenance problems and restoring unit pride and morale.

I now unveiled my plan to revitalize the unit's aircraft to operationally ready condition. I brought in the Chief of Maintenance, the Electronics Officer, the Hughes tech rep, Van Daele, and our Weapons Instructor Officer, Capt. Komnick. I laid out my plan and I then asked that the maintenance personnel select the aircraft and put together the schedule. The schedule would operate seven days a week and selected aircrews would be on duty to fly these missions. The sorties would be about one half hour in length. Two aircraft would take off, enter the training area and conduct intercepts on each other with WSEMs (simulated radar missiles) on board. On return, they would be met by a team of aircraft mechanics and radar specialists selected by the electronics section and the Hughes tech rep. Any discrepancies would be immediately given priority and go to the head of the line to be fixed. As soon as these aircraft were deemed repaired and ready, they were rescheduled and the mission would be flown again. This was repeated until each tail number met OR standards and could then leave the program. It would then be replaced by another aircraft as we worked our way through the entire fleet. I originally thought that we could complete the full program in 30 days' time. I was totally wrong. Despite hard work and faithful adherence to the concepts of my plan, we still had not completed the program after 60 days. Some of the aircraft were in such terrible shape, that it took weeks and as many as 10 to 15 flights to get them out of the program successfully. This kept four aircraft out of the regular daily training schedule for the pilots. To keep them up to speed, we received permission to fly some of the alert aircraft on the daily flight schedule. Van Daele and I participated in the flight testing, usually taking the weekend flights. Yogi and I were two of the more experienced F-106 pilots in the squadron and taking the weekend flights did not bother us. It actually gave me more opportunities to get in some flying time.

Gen. Gamble and the 20th Air Division staff were beside themselves at how slow the program was progressing. They could not understand why it was taking so long and so many flights to bring these aircraft back into shape. I had to admit, I too was amazed. Some of the aircraft were so badly maintained that it took the very best electronic minds in the squadron many hours to find the basic problems that were the cause. But gradually, I began to see improvement. And so did the aircrews. When the aircraft came out of the program and returned to the daily flight and alert schedule, they were better aircraft and much more mission capable. And the pilots were now beginning to get more out of their training flights. Everyone knew that the ADC ORI team would soon be arriving to test the 95th FIS. There would be a lot at stake, and the ORI team would not be kindly.

I continued to attend the daily four o'clock maintenance meetings, and through those meetings I demonstrated to the maintenance personnel that I was their commander and not just the commander of the pilots. I praised them when they met their goals and I encouraged them when they worked hard and failed. I also was hard on any of them who did not show me that they were doing their best. I continued to get rid of several people who would not or could not cut it. I gave anyone a second chance, but failing that, they were gone. I also showed them that I had no favorites. When I found that several pilots refused to adhere to AF grooming standards, especially in the hair department, I required that the officers stand inspection with the enlisted men. When I had the First Sergeant take down the name of an officer for an unsatisfactory haircut during one of these inspections, this message was not lost on the enlisted men. After a few of these combined inspections, the grooming standards in the 95th improved immeasurably.

These things did not improve my popularity rating, especially with the pilots, but I was looking at the overall readiness of my squadron. In the summer of 1970, I began to see some forward movement, but I was not at all certain we were an operationally ready unit. I would soon find out.

Nuclear Matters

A quick way for a squadron commander to fail an ORI and get fired was to mishandle your nuclear responsibilities. With the 95th I owned two nuclear sites, one at Dover and one at our dispersal base in Atlantic City.

Once weekly we practiced breaking out the MB-1 Genie rockets from our storage site on the north side of the base and convoying them to our ramp where they loaded onto the aircraft. Strict security rules had to be in effect and the handling and loading operations had to be conducted according to very stringent safety regulations and procedures. Errors, even of a minor kind, were enough to call a halt to the loading and end up as a failure. All personnel who were involved with nuclear weapons, including of course the pilots, had to be entered into a Personnel Reliability Program (PRP). People had to be screened before being allowed into this program and their medical records had to reflect that these people were in the PRP. This was an administrative nightmare that also created many problems for commanders and their staffs. On my last ORI for ADC in 1963, as Team Captain, I had to flunk the 27th FIS at Loring for their mishandling of their nuclear loading. I knew that it meant the end of the career of the commander. He was let down by his armament people. But he was the CO and he was responsible. I remember calling Gen Elder at Colorado Springs and made a request for giving this guy a break. He told me to use my own discretion but be prepared to defend it when I got back. I did what I had to do.

As time passed on, I felt that we were getting closer to maybe sneaking by, but one area I worried about was our WSEMs. WSEM (Weapons System Evaluator Missile) were electronic devices that simulated the radar missile fired by the F-106. The WSEM was mounted on the rails and when the pilot fired against a target, all the electronic signals were transmitted to the WSEM and recorded. In the air, when the pilot squeezed the trigger, the armament doors opened, the rails extended, the signals went to the WSEM. Naturally the WSEM did not leave the rail as a real missile would. On the ground, these signals were registered and evaluated to determine if all the correct signals were transmitted from the fire control system to the missile and that under real conditions the missile would have been guided to the real target. During an ORI, to record a "kill" not only must the pilot have the dot centered on his radar scope, but the WSEM must record a "hit" too. While our MA-1 radar systems were beginning to improve considerably as a result of our aircraft improvement program, our WSEM kills still lagged.

In other matters, I was encouraged that many of the problems the 95th had encountered in its pre-deployment ORI had been substantially corrected. Certainly, maintenance morale was higher and we were beginning to meet our aircrew training requirements more closely. We were fulfilling our alert responsibilities, and we had not had any aircraft accidents or incidents. Gradually I was weeding aircrews and others who were detrimental to the program. Either they jumped on the program and joined in to making the 95th the best it could be, or they were gone.

Problem Children

About 17 percent of the 95th was African-American airmen. I had no problem with that and 95 percent were good and productive people. But often when there was a problem with a maintenance airman who was black, I heard from him that his supervisor, who was white, picked on him and was racist, etc. My answer to this complaint was to transfer these cases and put them under the supervision of MSgt. Wilson who was the Flight Chief Sgt. Wilson was black and so this move erased the racial motive. MSgt. Wilson was also one tough supervisor who brooked no nonsense on his flight line. Usually, after about a week or so under MSgt. Wilson, the problem child either corrected himself or was back in trouble. This time for good.

I had a pilot, Captain, who did not indicate to me that he was serious about his flying or about his additional duties. He had gotten into trouble in Okinawa during the Korean deployment by bouncing checks at the Naha AFB O club. Somehow, he managed to wheedle out of that problem, but I put him on notice. Shortly after, someone in the CAC (combat alert center), told me that he believed that this Captain had been falsifying his flying time in the logbook. A few weeks later, this pilot requested a weekend cross country flight. I approved it. I then advised the CAC sergeant (in confidence), to follow his flights as he flew from base to base, record his takeoff and landing times and give them to me after he returns to Dover. When I received this data and compared them to what the Captain had recorded in the aircraft logbook, it showed that he was falsifying the hours he had flown by an inordinate amount. When I confronted him with this information, I also gave him the choice of resigning from the Air Force or face a court martial. He chose the former and I was soon rid of him.

In 1970, the military did not permit homosexuals to serve. The reason was that it was in the best interests of good order and morale that homosexuals not be mixed in with men, the majority of whom are straight and who do not want anything to do with homosexuals nor appreciate their advances. I noticed how often the First Sergeant would report to me that "airman-so-and-so" was in the hospital with injuries from having been beaten up by someone unknown. When I would visit these patients, they would never say who it was that had beaten them. I finally got it out of the First Sergeant that these guys were probably homosexuals who had made advances on the wrong guy and had been beaten up in exchange. I then changed our policy that from then on, we would Article 15 the homosexuals and try to get them out as they presented only more problems that we didn't need. During my time in the 95th, we did get rid of a few more, which made me wonder how many of them were out there, and why they got beat up so much. The sorriest case I had was a young airman I had who after a bad altercation went AWOL. His parents called me from a small town in Pennsylvania and I felt badly for them as they sounded like fine people. A few weeks later he was picked up in a park molesting young children and was imprisoned. I had to call his parents and tell them. It was a terrible call.

The Air Force was cracking down on drunkenness and domestic abuse. Unfortunately, these two often went together. I had several cases where one of my sergeants would beat up his wife and end up in the pokey. On these occasions I always visited the family and tried to do whatever I could to help the injured wife. I first made sure that she received medical attention. I would then advise her to press charges against the husband, lest he do it again. In almost every case, the wife declined the next day to press charges and the husband went free. This did not prevent me from bringing the culprit in to my office with his supervisor and giving my best ass chewing along with threats to have him thrown out of the AF if he does it again. But I always thought the AF policy on this was wrongheaded.

These cases and many others were very time consuming, and many were not resolvable. I thought many times about my dreams of being a squadron commander, with nothing much else to do but lead a great flying organization of fighter pilots and sleek aircraft and having a great old time. Lots of flying and lots of beer calls. But I didn't know about all of this other stuff and this other stuff was taking all the fun out of it. I was meeting myself coming and going. My days were long and many of my nights were interrupted with calls on the red phone that were usually bad news. My little black book of things I must do was jammed with items that had been in there for weeks. I had to be very careful to keep myself focused on what were the priorities and to delegate to those I could trust the items I hoped they would take care of. I had to stay mentally sharp and physically fit. I also had to remember that I had a wife and a family who deserved my time as well. As Thomas Paine wrote: *"These are the times that try men's souls."*

As the summer wore on, we continued our aircraft improvement program. It was slow going. Some of the birds just wouldn't get better despite everything that maintenance tried, while other aircraft went in and out of the program in less than a few weeks. However, as more aircraft came out and returned to the daily flight schedule, we could see that the aircrew training program began to improve immeasurably and that resulted in a lot less pilot unhappiness. I also perceived better results in our 20th Air Division and NORAD exercises. The daily aircraft OR rate improved slowly but was still below ADC averages for F-106 units. I could now see that under Van Daele, things in the ops area were getting better every day leaving me to concentrate on maintenance and supply problems.

Every squadron in ADC watches the ORI schedule and tries to learn when they are next on the list. ORI's are "no-notice," i.e. the team will fly in unexpectedly. Still, every unit knew about when they were due as every squadron was to be evaluated once a year, and no less than once every 18 months. But the 95th FIS was a special case. It had been given an ORI in the fall of 1969, prior to its deployment to Korea. The results were unsatisfactory, but because they had to deploy, the ORI Team was instructed to give them a pass. This, the ORI Team members did not want to do and as it turned out, the 95th FIS embarrassed ADC in Korea by its poor performance. The ORI Team was anxious to return to the 95th as soon as possible, and this time their conclusions would not be tampered with. Some were out to right a wrong that they felt impinged on their reputation as a professional inspection agency. All of this was told to me and it did not make the challenge facing the 95th any easier. More than anything, I wanted to pass this test. I had never been a member of any Air Force organization that had failed a readiness inspection. I did not want this to occur to a unit of which I was the commander.

CHAPTER TWENTY-SIX: MR. BONES FACES HIS BIGGEST TEST

ORI Team Captain: "We're here to help you."
Unit Commander: "We're pleased to see you."

Two big lies heard at the start of every ORI.

In the middle of September, the ORI Team arrived at Dover. The Team Chief arrived by T-33, and the others by various AF aircraft and some even came in by commercial air. Gen. Robinson, the Team Chief, whom I knew well and respected, presented the letter of notification, which also contained instructions for the squadron. I immediately notified Gen. Gamble at 20th Air Division. All aircraft were grounded and maintenance stopped. We would fly with what we had ready to fly. Aircrews were given an examination which covered their knowledge of nuclear safety and procedures. The ORI aircrew examiners then began reviewing everything pertaining to our Combat Alert Center, small arms storage, handling of the aircrew cryptic code cards which all alert aircrews carried with them, and then tested our CAC enlisted personnel.

The next item was a breakout of our Genie's from the nuclear site and then loading them on 15 of our aircraft parked on the flight line. This is one task in which there can be no mistakes, even of the most minor category. God knows, we had been practicing this for weeks, but this was the big one. I watched carefully as the air police set up the security area and then the trailers carrying the nuclear rockets flowed out of the site, across the taxiways and onto our ramp. I then watched several of our aircraft loading teams do their work in a precise manner carefully using their checklists. Every act was closely scrutinized by members of the ORI Team. I thought my armament men were doing a good job.

It was a busy day and from what I saw and heard, I believed we had done okay so far. Reports back from my supervisors did not indicate any serious disqualifying problems in their areas. But tomorrow would be the crucial day. I went home tired and worried. Many thoughts raced through my head as I tried to be sure I had not forgotten anything that we had overlooked and the Team would look catch. I did not sleep well.

Early, the next morning began the flying phase. The squadron commander was always the first pilot to be scrambled. The ORI Team wanted to know whether the CO was a flying CO or a paperwork CO. The targets consisted of one bogey at low altitude (below 1,000 feet), and then a second high altitude target, above 45,000 feet. The target aircraft would be T-33s and B-57s. Both of these aircraft carried chaff dispensers and electronic radar jamming equipment. The intercept against the high altitude target would be head-on. In order to obtain a complete "kill" on a missile firing pass, both the MA-1 radar system must show a solid lock-on and firing signal with the dot centered, plus the ORI Team examiner would have to ensure that the WSEM received all required signals for guidance to the target. My sortie went well as I picked up the T-33 on my radar at about 10 miles. The T-Bird was flying an altitude of at 500 feet about 50 miles east of Cape May over a very rough and choppy Atlantic Ocean. I got an easy kill with the armament door opening. I then climbed to 40,000 feet and picked up the B-57 at about 20 miles. The B-57 tried to jam my radar but my anti-jam system defeated its attempts. I got my second MA on him using the Genie selection. After recovery I watched as the rest of our aircraft scrambled. I liked the way our flight line was operating and we had only one abort and one late takeoff out of 20 scrambles. From the feedback I was getting from Van Daele, our aircrew scores were good. We were running above 75 percent, with no aircrew errors. My biggest concern now was how we were doing with the WSEMs. This had been a major concern during our aircraft upgrade program and we had not come close to attaining the success rate I wanted. One thing was becoming clear as the exercise progressed, ORI Team members were showing signs of surprise that the 95th FIS at this point was not the same squadron they had evaluated less than a year ago.

At the end of the exercise, the ORI Team Chief asked to see me privately. Gen. Robinson told me that the squadron had put on a very good performance and clearly had made substantial improvement since prior to its Korean deployment. He even mentioned confidentially that many of his team personnel were hoping to flunk the 95th because of what had happened a year ago. Still we were not out of the woods yet. A few too many of our WSEMs had failed to record sufficient incoming signals from the MA-1 fire control system to determine a kill. His men were still working on trying to assess whether they could record a kill or not. He would let me know as soon as they made a decision.

That night, Robinson called me and told me that he would have to fail the squadron. There were two doubtful WSEMs that had insufficient signals to declare a kill. The Team gave us one but not the second. We failed by one kill. I made a faint attempt to change his mind, but I knew from experience that the decision was made and could not be altered. General Gamble and Col. Joe Joiner came in the next morning for the out briefing by the Team Chief. All in all, we did very well. Maintenance, operations, nuclear loading and handling, the CAC, and all other important areas were rated "satisfactory" with a few "outstanding's" thrown in. The squadron failed because we could not come up with a sufficient number of "kills" due to WSEM failures. We missed by one. However, Gen. Robinson went out of his way to remark how impressed he and his Team were with the improved overall performance of the 95th FIS. He saw it as a squadron on the move upward and he commended the new spirit that his Team observed. He also congratulated my supervisors. I ended the briefing by thanking the ORI Team for their professionalism and for their objectivity. I told them that when they returned, they would see an even better unit. When the Team departed, I spoke with Gen. Gamble. I was concerned about his reaction to our failure. But he was complimentary to me, saying: "If you are going to flunk an ORI, that's not the worst way to do it." I promised him we will get the job done next time.

That afternoon I addressed all of my squadron supervisors and aircrews. I thanked them for their efforts, and that while I was not happy about failing, I was pleased that so many of our sections were rated high in performance by the Team. I told them that we could expect the ORI Team to return within 60 to 90 days. On their return they will just test us on the flying exercise. So we will strive to improve in that area and ensure our WSEMs are up to speed. On that we will concentrate. Our goal to become the best F-106 squadron in ADC had not changed.

In the ensuing weeks we concentrated on our WSEM program and with that emphasis I began to see our success rates steadily improve. I also saw squadron esprit improve and relations between aircrews and maintenance become much better. I was also told that I would be receiving a new Chief of Maintenance soon, one with experience and he was a rated pilot as well.

In November, the ORI Team returned, and we conducted another flying exercise similar to the previous one we failed in September. Again, I lead the squadron off with two kills on the first sortie of the day and there followed an excellent day of shooting for Mr. Bones. This day there was no doubt about the outcome as both aircrew MA's and WSEM kills exceeded well above what we needed to pass. The 95th FIS was now "whole". We were now back in the fold and a truly operationally ready F-106 squadron fulfilling our air defense and NORAD missions. It had been a long journey since April and along the way many former 95th personnel had been sent off. Many had been reprimanded and told to shape up or ship out, and many new heads had arrived to inject new spirit, with more experience and dedication to the task. My popularity rating had sunk into the sewer with most of the older crowd who had survived the upheaval, but I had long ago accepted that as the price of success. What I wanted more than acceptance was mission accomplishment and a unit that could meet its flight responsibilities with safety and success. For the first time in a long time. I was beginning to feel that maybe I was seeing the light at the end of the tunnel.

One last item about the ORI: In both the first ORI in September, which we failed, and in the follow-up, which we passed, I lead the squadron by flying the first sortie. I had good luck both times with two successful MA's on each sortie, one on the low and one on the high target. This was mentioned by the ORI Team Chief in the out-briefing because ADC wanted their squadron commanders to be "flying commanders" and set the example for their aircrews. In both instances I scrambled in the aircraft assigned to me, aircraft number 72505, which had my name painted on the port side near the cockpit entrance, Col. Gene Cunneely, Commander. 95th FIS. A framed picture of that aircraft hangs in my office, where I am writing this memoir. It was a departing gift from the squadron and painted by one of my pilots, Captain Jensen. A gift so dear to me it graces the cover of this very book. My Crew Chief was a New York airman named Cook.

He was proud to crew the commander's aircraft and he worked hard to keep it in top condition and well-polished. I showed my confidence in his skills. When I arrived at the aircraft, I asked him if it was ready to fly. He always answered, "Yes Sir." I then told him "to button it up and let's go fly." He liked that. I had an aircraft with my name on it in the 80th FBS in Korea, the first fighter aircraft on which I ever saw my name emblazoned. I was so proud. I also had my name on an F-86D in the 317th FIS and on an F-106 in the 318th FIS at McChord. But I did not always fly those birds. But at the 95th I tried to have tail number 505 assigned to my sorties. I guess there was a little ego involved there.

Getting through the ORI was a rough ordeal, and, in retrospect, it almost did me in. Since the squadron returned from Korea, with the exception for my trip to Idaho to pick up the family, I had been working on the average six days a week, 12 hours a day. There seemed to be no end to the problems that kept popping up in all areas and many of them I had little ability to solve. They ranged from problems that directly affected mission readiness such as aircraft OR rates. Sometimes the number of aircraft we had able to fly and sit alert was pitifully low. This was, of course, reported to higher headquarters which often resulted in phone calls from 20th AD and ADC asking for explanations and just what I was doing to resolve the problem. Often the problems involved just personnel failures, airmen getting drunk and ending in the base jailhouse, or a wife beating, or airmen caught pot smoking in the barracks. Some of the problems involved officers who came in with domestic problems, or some who just refused to adhere to AF regulations pertaining to personal grooming. I had never dreamed that as a commander I would have to spend so much time on such matters that prevented me from putting my time toward more crucial subjects that truly were priority. Failure of leadership at the squadron's intermediate levels was the problem and I continued my crusade with ADC Personnel to replace the weak supervisors I had with stronger and more experienced people.

The ORI took its toll on me—especially on my emotions and psychological outlook. It didn't help that I was physically exhausted from the long days and nights, but when we finally got through it successfully, I gradually returned to a more normal composed attitude. We had that hurdle behind us at last. However, there was still a lot of work to be done. The 95th FIS was by no means near the level of overall operational readiness that I wanted. That I knew. So, it was back to work for me to continue to raise the squadron to where it should be. I was still committed to that goal as long as I was the commander.

One last thing, when we flunked the ORI by one faulty WSEM, as Squadron CO I might have reacted differently. I knew the ORI Team was out to get us. The Team was still smarting because their previous ORI finding of the 95th, i.e., that it was an unsatisfactory unit, was overruled by higher authority. And they wanted to set the record straight this time. I was certain that the ORI Team was determined to flunk Mr. Bones for sure. And so, they did. But it was by a whisker, and Col. Robinson had to accede that the 95th was a greatly improved outfit. In my response to Col. Robinson and the Team, I declined to whine or whimper over the results, knowing deep down from my experience, that the Team could have given us the benefit of the doubt on that last WSEM and given us a pass. Instead, I chose to complement the Team on its expertise and professionalism; after all I was a past member of that organization. And I invited them to return as soon as possible. More importantly, I wanted them to go back to ADC Headquarters and tell them, especially Gen. McGhee, that the 95th FIS was under new management and was no longer at the bottom of the pack.

Life after the ORI

Despite our finally successfully passing the ORI and wiping the books clean, I never felt completely comfortable with the 95th FIS. This squadron had been mortally wounded by terrible mismanagement and lack of disciple by its previous leaders. It would take many more months and more personnel changeovers to clean out the rot. Some of it would probably be there after I leave. Our aircraft were still below par and the operationally ready rate was unsatisfactory too much of the time. Some of it was beyond our control (parts), but a lot of it was not.

CHAPTER TWENTY-SEVEN: THE FAMILY ADJUSTS TO THE EAST COAST

Oh, East is East, and West is West,
And never the twain shall meet.

-Rudyard Kipling

Dover, Delaware

Our house on the base was a lot roomier than the one we had at McChord and we needed the extra room. The house was made of brick and was situated on a large corner lot in the center of the officer's housing area. A highway separated the base housing area and many of the base recreation facilities from the flight line. It was a two story house with three bedrooms upstairs and one downstairs. Young Gene had the downstairs bedroom which also had a bathroom next to it, so he had some privacy the others didn't. The house also had a nice dining room and an ample kitchen and living room area. The heat came up from the floors, a system I did not care for and I believed was inadequate for the cold winters we would experience in Dover.

That summer, Gene and Jimmy attended hockey school at Wilmington which was about an hour's drive away. They had to be driven each morning and picked up in the late afternoon. It was a tough grind for Alice, as I rarely could take the morning shift. But the boys were able to resume their hockey and they did really well showing those Delaware kids that they could play. They also got to see Bernie Parent, one of the great NHL goalies who came to several of the sessions to instruct. I wanted to have the boys continue their hockey, but the Wilmington rink was privately operated and was too costly for us to join. However, at the end of the school, I was approached by one of the members who asked if the boys would play with their youth teams. I explained that it would not be possible for us to join the club. To my pleasant surprise, he offered us a "scholarship," which would waive the initiation fees and allow us to join with only a small monthly charge. So, the family was able to join the rink club and the boys were able to continue their hockey. For the next two years Alice and I drove the boys back and forth to Wilmington so the boys could play hockey and the whole family could ice skate. That was a great break for them. This meant many early morning trips to the ice rink and late pickups after their practices and games. But Gene and Jimmy soon became leading players on their teams and favorites with their coaches. Hockey was definitely their sport and young Mike soon followed in their footsteps. Mary learned to skate and so did Katie, although Mary wanted to continue her competitive swimming which she started in Tacoma. We soon found a YWCA in Dover that had swimming teams and Mary joined that group. All of this, of course, kept Alice and me very busy, especially on weekends.

Living on the east coast also has some other advantages for us. Alice and I were now able to see more of our families than before. Alice's mom came down to stay with us as well as my dad. We were also able to visit them in New York and did so in the summer and on the holidays. The children loved visiting my dad in Patchogue who had a quaint home near the beach and a pool. From there I took the boys to see the Yankees play at Yankee Stadium. I wanted them to see the great Mickey Mantle play, the successor to my idol Joe DiMaggio. But Mantle went down swinging about three times that day and looked awful. We went to Fire Island where they had their first encounter swimming in the surf of the Atlantic Ocean. We also did some fishing. We took a cruise around Manhattan that included the Statue of Liberty. The kids finally met many of their cousins, aunts and uncles they had not met before for the first time. Our children also were able to visit Philadelphia with all of its historic places, and while living in Dover we spent many wonderful summer days at Rehoboth Beach in Delaware.

Dover AFB provided some excellent recreation for the families. The officer's club had a nice pool which was badly needed in the heat and humidity of the east coast in summer. It also had a golf course which the boys and I enjoyed. The kids were enrolled in an excellent Catholic school in Dover which provided them sound religious and academic training, and they made fast friends there.

Larry Van Daele, the son of my new ops officer, was the same age as Gene Jr., and they became close friends. (They still are to this day.) Though they had many different interests, one that they shared was a love of ice hockey. The Van Daele's had been stationed at Selfridge AFB near Detroit there they became huge Red Wing fans. About this time, the Philadelphia Flyers were beginning their franchise. The Spectrum where the Flyers would be playing was about an hour's drive from Dover. Yogi and I decided on a lark to buy four season tickets. For two seasons, the Van Daele's and Cunneely's drove from Dover to Philadelphia to root for the hapless Flyers. They lost most of their games, but they were an exciting team to watch and when they did win, it was so sweet. We had a great time taking our kids to these games; we sometimes took the wives. It was a treat I always looked forward to because it brought me away from the squadron for a few hours where I could leave my troubles behind. The Van Daele's provided me and my family much in the way of support and pressure relief for which I was always grateful.

I knew that my days at Dover would not be long. Squadron commanders usually left after about two years. In that time period they either were fired for not being up to the job, as was my predecessor, or promoted and then reassigned So. I expected to be leaving in the summer of 1972, if I lasted that long. This would be hard on my family as this would mean another move. The boys and Mary would then be in high school, and that is always a rough time of life for a teenager to have to leave his friends and move to a new area. I dreaded that thought. It would also be tough on Alice. I expected that my next move would be a remote assignment, that is, one without my family going with me. I expected it to be Vietnam as that war was still raging on with no end in sight although rumors of Peace Talks were in the air. The thoughts I kept mostly to myself but I knew that Alice was well aware of them. We both knew that there would be some very tough decisions ahead that we would have to make affecting the welfare of our family.

Tom Curry in the News

On a morning in May 1971, we received news that Alice's brother Tom, who was a New York City patrolman, had been shot and seriously wounded. He was in St. Luke's Hospital in Manhattan and was in critical condition. I immediately called Gen. Gamble and requested emergency leave, explaining to him what had happened. We arranged for the Van Daeles to take care of the kids while Alice, little Kate, and I set out for New York. Each half hour we listened to the national news. The lead story concerned the shooting of two NY police officer by members of the Black Panthers. The officers were guarding the home of the New York district attorney when they were ambushed. Each officer was shot several times, but both were still alive although in grave condition. I drove as quickly as I could up the New Jersey Turnpike and into Manhattan. Alice was almost frantic with anxiety, fearful that Tom would be dead before we would arrive. At the hospital, we found New York's Mayor, Robert Lindsay, the Police Commissioner, and many other high-ranking officials as well as journalists. Tom was still alive although he had been shot five times. I asked to see him, but Alice was too shaken to go in. When I went in, Tom was in an oxygen tent with many tubes going in and out of him. However, he looked at me and recognized me right away.

"Gene, Gene," he said. "What happened?"

"They shot you, Tom. Do you remember, they shot you?" I said.

Tom grabbed my hand and squeezed it. He groaned a little. Again, he asked, what happened. But he appeared strong to me for a guy with five bullet holes in him.

"You are going to make it Tom," I said, "You're tough."

When I came out, I told Alice, your brother should be dead, but he isn't. "You know, I would not bet against him."

The doctors performed an extraordinary job on my brother-in-law. One bullet they never were able to fully get out. He did recover, but he never worked again. And the damage those bullets did caused him much pain and suffering for the rest of his life. At the time he was shot, Mary Lou, his wife, had given birth to a new daughter only weeks before, their third child. From then onward, Mary Lou had the burden of taking care of her husband and raising three children. She showed tremendous courage, loyalty, dedication, and love in fulfilling those responsibilities. Tom, though slowed down and handicapped in many ways, did what he could as a father and husband. He was a warm and generous man, a strong family man and a devoted Catholic. His faith was what made it possible for him to survive and live the life he did. I never loved or respected any man more than I did my brother-in-law, Tom Curry. Whenever possible, we would visit Tom, and he and I would recall our early times together when I was courting his sister Alice Marie, and when we came together in Korea during the war.

When we returned to Dover, we kept in daily touch with Mary Lou. Though she was told about her son's predicament, Tom's mother Katherine did not visit him in the hospital as his condition was still grave and he looked bad and could not converse with her. You can bet that she kept her Rosary beads busy on behalf of her on and that certainly worked its magic. Tom eventually survived and after much rehab was able to return to his home. However, he was never able to resume a normal life. But he did everything within his power to do what he could to get back to where he was before the shooting. He marveled everyone by displaying courage and cheerfulness despite the pain and frustration be suffered. Alice went up to New York to visit him several times during our days in Dover, and that was such a blessing for her. I tried to see him as often as I could, but I still had much work ahead of me in the 95th. There were still many problems facing this ill-starred unit that I had not come close to solving, and the pressure on me to do so had not been lifted.

My Children Are Growing Up

In the middle of 1971, I was 42 years old and my five children included three teenagers. I began to realize how fast time was running by on me and how much time I was spending on Air Force matters compared to time spent with my children. It would not be very far ahead that some of them would be leaving the nest. The thought of that unsettled me. I prided myself on being organized and being a good manager of time. But the 95th's never ending problems just ate up my hours and my workdays were long and burdensome. I never dreamed that being a fighter squadron commander would be so grinding. Each day brought new problems that were unexpected. My phone calls from higher headquarters were endless. (Alice Marie despised the red phone in our house that sat next to the bed. Whenever it rang, she jumped and exclaimed, "Now what?"). She was right, it was never good news. Whenever I could I spent times watching the boys play hockey or baseball or Mary swim, played ball with Mike, and games with Katie. Having Yogi Van Daele on board was a great help and getting a new chief of maintenance also relieved me of my close oversight of the aircraft maintenance problems that plagued the 95th. Although we had passed our ORI, like all other squadrons, we were constantly being inspected by various teams from Air Division, NORAD, and even USAF. Every so often Gen. McGhee, the ADC Commander would drop into a squadron without prior warning and scramble the alert aircraft to test their operational capability. And General Mike Price, the ADC Director of Ops, enjoyed dropping in on fighter squadrons on weekends to inspect things. It seemed to never end. The most difficult thing to find was time to spend with family and that became a rising priority with me. I wanted to succeed as a commander and I wanted to bring the 95th FIS up to the highest standards of combat readiness, but I did not want to surrender my family and their security at the same time. This was a hard choice.

Although this was a difficult period in my life, I kept reminding myself how fortunate I was and how well God had treated me in spite of my inadequacies. First of all, I had a happy marriage. Alice Marie, who took a huge gamble in leaving her job and her home to become the wife of a military man, was a loving and supportive partner. She had given me five wonderful children, each of whom I loved and cherished dearly. Alice Marie, not surprising, was a dedicated mother who placed the health, welfare, education, and security of her offspring first and foremost. She was also a loving wife who was always there to provide me solace, advice, and inspiration, whatever it was that I needed to persevere in my duties. No matter how bad things were at the squadron, I knew that when I walked in the door of my home, life got instantly better. When she put her arms around my neck and gave me a kiss, my troubles faded away. Alice was also a great "mother hen." Although her husband's popularity rating was pretty low among the officer corps, Alice's was very high with the wives. That was because Alice was kind and open to everyone, regardless of rank. She was adept at listening and, only when asked, providing some sound advice. Alice never indulged in gossip nor spoke ill of anyone. And that did not go unnoticed. Looking back, I know now how much my wife contributed to whatever success I attained in my career.

Teenagers and the Future

Gene Jr. was a handsome boy, well-built and athletic. He was a good student and had many interests outside of sports. Jimmy was noted for being always in good humor. He too was an excellent athlete and a good student. While Gene was brown haired, Jimmy was a brunette as was his sister Mary. Mary was a beautiful girl. She enjoyed swimming and won many ribbons in competition during her teen years. Little Kate was talkative and loved animals. She always had her own pets in addition to our family dog, Brandy, whom we brought with us from McChord. Young Mike trailed his older brothers around and tried to compete with them unsuccessfully at his age. They teased him unmercifully, which caused him to cry for help from me, but I gave him little consolation. I felt that some adversity early on would build character. Maybe I was wrong, but as Mike grew older and bigger, he more than held his own.

I began to feel badly about how much time I had spent away from my children as they were growing up. When we were in Colorado Springs, I was on TDY about one-third of the time. As operations officer of the 318th FIS, I worked long days, sat alert, had TDY's in Alaska and at Walla Walla, plus six months in Korea. Now as a commander of an ailing squadron, I was putting in long days and nights, plus alert duty both at Dover and at our DOB in Atlantic City. I was also required to travel to overnight meetings at Division Headquarters in Virginia and to ADC in Colorado Springs for squadron commander's meetings. All of these commitments meant time away from home which was precious to me. And time was getting short. My tour of duty at Dover would end in 1972, and my next assignment would undoubtedly mean more family separation. But I had little control over these possibilities. I had volunteered to take on this job and I would see it through to the best of my ability. We had passed our first big test, the ORI. We had weeded out some of the deadwood people who were hurting the unit and were getting in new and more promising replacements. Our senior enlisted personnel in maintenance were getting better and a new Chief of Maintenance was inbound. Morale was improved, although still below average for a fighter squadron. We still had serious supply shortages, and our aircraft OR rate lagged below the command standard. But we were working on it. My little problem notebook was still filled with items that never seemed to decrease. While my own personal attitude was still positive, I sensed that this assignment would be the major test of my military career. Everything I had ever learned about leadership and management was being put to the test on a daily basis. To stay on an even keel, I tried to attend Mass whenever I could during the week. Asking help from the Lord was not something I shied away from at all. Many of the problems that came before me defied my experience and, at first blush, appeared to have no answer. Yet I discovered that by turning many of these problems over to Divine assistance not only provided me an inner peacefulness, but I found that somehow, most of these worst problems were then resolved in a manner I would never have imagined. Stepping up my spiritual life, plus the security and support I received at home from Alice Marie and my children, kept me going with confidence. What will be, will be!

MY STORY:
PHOTO ALBUM

My grandparents James (holding the baby) and Mary on his left. True Irish immigrants in America. (New York City, c. 1890)

The Cunneely's and the Reed's enjoying a cool day at the Rockaway Bungalow: Uncle Jim Reed, Aunt Anna, Aunt Lillian, Grandma Mary, (unknown), Aunt May and Dad.

My father, Joseph. He had but only one year of schooling, yet he was amazingly sharp in math and a brilliant bridge player! (c. 1920's)

My father Joseph with Grandma. (Rockaway, New York)

My parents, Joe Sr. and June Cunneely. (1920's)

Alice's brother, Thomas Curry and her mother, Kathleen Curry. (New York City)

My grandparents James and Mary Cunneely (1920's)

My grandparents James and Mary Cunneely (1920's)

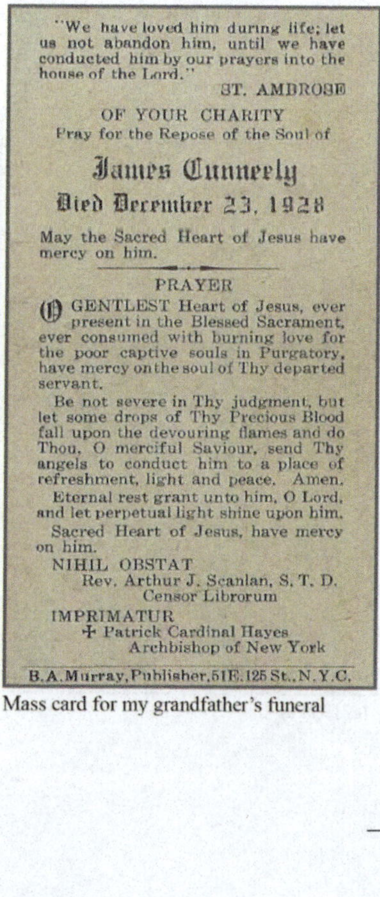

"We have loved him during life; let us not abandon him, until we have conducted him by our prayers into the house of the Lord." ST. AMBROSE

OF YOUR CHARITY
Pray for the Repose of the Soul of

James Cunneely

Died December 23, 1928

May the Sacred Heart of Jesus have mercy on him.

PRAYER

O GENTLEST Heart of Jesus, ever present in the Blessed Sacrament, ever consumed with burning love for the poor captive souls in Purgatory, have mercy onthe soul of Thy departed servant.

Be not severe in Thy judgment, but let some drops of Thy Precious Blood fall upon the devouring flames and do Thou, O merciful Saviour, send Thy angels to conduct him to a place of refreshment, light and peace. Amen.

Eternal rest grant unto him, O Lord, and let perpetual light shine upon him.

Sacred Heart of Jesus, have mercy on him.

NIHIL OBSTAT
Rev. Arthur J. Scanlan, S. T. D.
Censor Librorum

IMPRIMATUR
✠ Patrick Cardinal Hayes
Archbishop of New York

B.A.Murray, Publisher, 51E. 125 St., N.Y.C.

Mass card for my grandfather's funeral

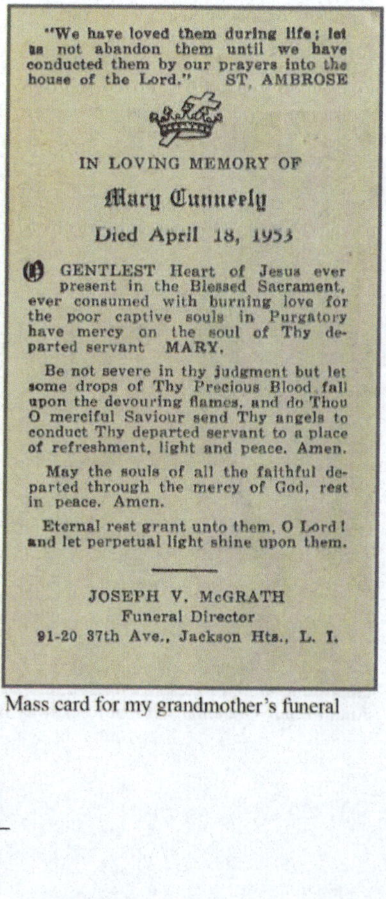

"We have loved them during life; let us not abandon them until we have conducted them by our prayers into the house of the Lord." ST. AMBROSE

IN LOVING MEMORY OF

Mary Cunneely

Died April 18, 1953

O GENTLEST Heart of Jesus ever present in the Blessed Sacrament, ever consumed with burning love for the poor captive souls in Purgatory have mercy on the soul of Thy departed servant MARY.

Be not severe in thy judgment but let some drops of Thy Precious Blood fall upon the devouring flames, and do Thou O merciful Saviour send Thy angels to conduct Thy departed servant to a place of refreshment, light and peace. Amen.

May the souls of all the faithful departed through the mercy of God, rest in peace. Amen.

Eternal rest grant unto them, O Lord! and let perpetual light shine upon them.

JOSEPH V. McGRATH
Funeral Director
91-20 37th Ave., Jackson Hts., L. I.

Mass card for my grandmother's funeral

Family at Rockaway Beach, New York.
(c. 1930's)

Me, as a young man in New York City.
(c. 1930's)

Aunt Lillian, Grandma and I, at Rockaway Bungalow. (c. 1930's)

Aunt Lilian, me, Grandma Mary and my Dad. (New York City, 1943)

Chaminade Flyers #8 (Second row, second in) Team Captain in 1947

Taking my cuts at Chaminade High School. Probably a double. (c. 1947)

A happy photo of my brother Joseph, finishing his tour in the U.S. Navy. He joined the Navy at 17 years of age. He then saw a great deal of combat, especially in the Battle of Leyte Gulf.

Joe Jr. with Joe Sr., our father, at Great Neck Train Station. (1945)

Joe Jr. with Grandma and Aunt May.

Joe Jr. on leave from war.

The family at Great Neck: Joe Sr., Joe Jr., Patricia, myself, May, Grandma Mary (taking a nap), Aunt Lillian and Aunt Anna. (1948)

My graduation portrait:
Fordham University Class of 1951

Alice Marie Curry
Cathedral High School, Class of 1946

My brother's wedding to Patricia Halloran. I am the best man. (Flushing, New York, 1947)

Alice and myself at Rockaway Beach. (New York, 1951)

Alice Marie Curry. (New York City, 1951)

Flight School: Class of 1952 (Bainbridge,GA)
I am the furthest right of the back row.

T-6: My first aircraft to solo. (Bainbridge Air Base, GA, 1952)

Flight school for T-33 jets: Pinecastle AFB (Orlando, FL, 1952)

T-33: Pinecastle Air Force Base.
(Orlando, FL, 1952)

Officially a fighter pilot:
Lt. Gene Cunneely.

Lt. Cunneely (Orlando, FL, 1952)

My F-86 Sabre: *Flying Tigress.* (Korea, 1953)

'Nothing I Could Do':

Pilot Leaves Disabled Jet, Leaps To Safety

"Too bad we had to lose a good ariplane, but there was nothing I could do."

That was the comment Friday of Fist Lt. Eugene P. Cunneely, 26, of Great Neck, N. Y., after he had parachuted to safety from a disabled F-86D jet interceptor.

Cunneely, in his first parachute jump, landed several miles inland from the Pacific Ocean and Pacific Beach. He was uninjured.

'TOO MUCH ANXIETY'

Cunneely said he bailed out at 8,000 feet after his all-weather fighter got into trouble at 24,000 feet. Cunneely also said that although he suffered "too much anxiety" before the parachute opened, he enjoyed his long ride to earth under clear skies.

"I could see for miles," he said.

Upon landing in a swampy clearing, Cunneely continued, he set off a flare. His jump was seen by the pilot of another jet fighter and also by Frank Milward, a Ground Observer Corps spotter on duty at Aloha.

SAVED BY 'COPTER

A 43d Air Rescue Squadron helicopter piloted by Capt. Richard C. Kirkland flew to the scene from McChord Air Force Base and picked up Cunneely an hour after his jump.

Air Force officials have not released any information regarding the cause of the crash. Cunneely, who has been stationed at McChord for two years, said the plane may have fallen into the sea.

Nothing I Could Do
News article from local Tacoma paper about my Friday the 13th bailout. (January 13, 1956)

Alice Marie Cunneely, on her wedding day, September 17, 1955 (Tyndall Air Force Base, FL)

Aunt Anna, Patricia (my sister-in-law), Dad, and Aunt Lillian. (1955)

Curry family wedding, Alice is second from right. My father in-law and mother in-law are standing to the left of the bride.(1960's)

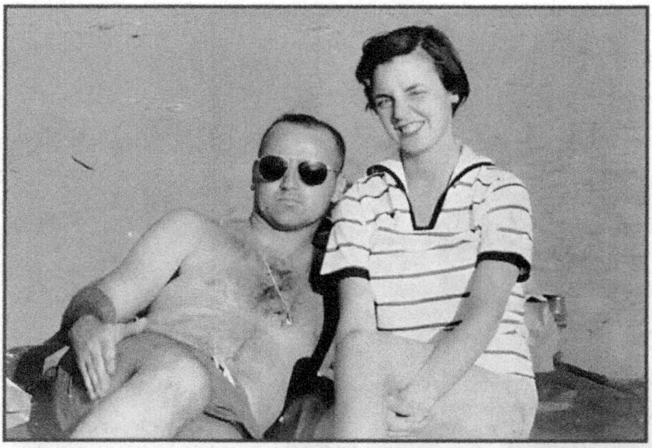

Alice and I, catching rays at a Washington state beach. (1955)

Alice has one in the tub. (Tacoma, WA, 1958)

Gene Jr., Dad and James (Tacoma, WA, 1959)

James, myself and Gene Jr., at Camp Hale, CO (1959)

Our first three children, Gene Jr., James and Mary (Colorado Springs, CO, 1959)

Aunt Anna, Alice, the kids and Dad (Colorado Springs, CO, 1959)

Grandma Curry with her grandkids:
Jim, Gene Jr., and Mary; (Colorado)

Our family at North Bay, Ontario, Canada;
Dad, Alice and the kids-No Kate yet (1964)

Family at North Bay, Canada, hockey game... I dropped the first puck! (1964)

Family portrait in North Bay, Canada (1965)
Top: Alice and I
Bottom: Gene Jr., Mike, Kate, Mary, James

Family portrait in North Bay, Canada (1966)
Top: Alice and I
Bottom: Gene Jr., Mike, Kate, Mary, James

My brother-in-law, Patrolman Thomas Curry on duty in New York City. (1960's)

Grandma Curry, my mother-in-law.

Tom Curry and his kids: Kathleen, Mary Elizabeth and Thomas Jr. (Trout Lake, NewYork, 1960's)

Lt. Col. Cunneely about to take an
F-106 to the skies (1968)

General Harry Thyng:
A great pilot and American hero.

Harold "Yogi" Van Daele pilots his "Six".
71st FIS

Colonel Chuck Yeager:
A great pilot and American warrior.

"Plan your flight; fly your plan!"
318th FIS, McChord AFB, 1968

Touchdown! F106 from the 318th.

Cabin on Lake Pend Oreille. (1970)

Mary, Katie, and Mike at Lake Pend Oreille (1970)

Fun times at Lake Pend Oreille with Alice, Deedee Grashio, Dad and Sam Grashio (Summer of 1970)

Gene Jr., Larry Van Daele, and Jim at Lake Pend Oreille, building the dock! (1973)

Alice and I vacationing in Mexico.

Katie with Uncle Joe [my brother, Joe Jr.] (1980)

Dad, myself and my brother, Joe Jr. Spokane, WA (1986)

Family Portrait: Gene Jr., Mary, Dad, Mike, Katie, Jim, Alice and I. (1979)

Mary Graduates from Eastern Washington University, Class of 1982

My beautiful daughters: Katie and Mary

Alice and our girls:
Katie and Mary

"Cunneely & Sons" My boys: Gene Jr., James and Michael

Our first grandchild, Gene III, and Alice taking a swim in Pearl City, Hawaii, 1987

The McClary's: Connor, Casey and Brendan (c.1998)

The McClary's: Casey, Brendan and Connor (c. 1999)

Mike and Amie's Wedding:
May 9th, 1998
Left to right: Katie, Gene III, Gene Jr., myself, Amie, Mike, Connor, Alice, Casey, Brendan, Jim and Mary.

McGee, Mary Lou and family. (New York, 2000)

Father Thomas with his sister's Kathleen and Mary Elizabeth, (New York)

Mary Lou, Tom Sr., Mary Elizabeth, Tom, Thomas, Kathleen, Jens, Mary and Kids. (New York, 2000)

My son Jim and his family.
Clockwise: Collin, Jim, Rachel and Gail
Reno, NV (2004)

Mike, Katie, Thomas Curry, Alice, myself, Gene Jr., Mary and Jim. (Spokane, WA, 2005)

Jim, Collin, and Mary.
(Seattle, WA; c. 2005)

Katie with her sons:
Connor, Casey and
Brendan McClary

Our dear friends:
Nancy and Bill Curry

Myself, my daughter Katie and her son, Brendan. (2013)

Gail and Jim with Collin at his 8th grade graduation. (June 2014)

So very proud of my first grandchild, Gene Paul Kainoa Cunneely III, for his service in the United States Coast Guard.

Alice celebrating her birthday (August 18) at the lake.

Alice and I at the cabin.
(Summer 2015)

Proud Grandparents
with ALL of our
grandkids at the lake.
(Summer 2015)

Alice's 87th birthday party at the lake cabin. Clockwise: Grace, Rachel, Alice, Michael and Collin

Alice and me, celebrating our 60th Anniversary; we got married on September 17th, 1955! (2015)

Kanani graduates (with two degrees!) from Eastern Washington University, Class of 2016

Grandchildren with Alice and Gene, just missing Collin and Rachel. (Summer of 2016)

Gracie's graduation from Gonzaga Prep, Class of 2018. Mike, Gracie, Amie and Michael. (2018)

Getting ready for the big day! Kanani and Micah's Wedding Shower BBQ at the cabin. Gene Jr., Kathy, Micah, Kanani, Gene III and Mallory (July 2018)

"Please welcome the Duncan's!"
Kanani and Micah's Wedding at Schweitzer Mountain Resort
July 7th, 2018

Michael, Grace, Brendan, Kaylee and Casey at Green Bluff. (2019)

Michael on the mound. (2021)

Three generations of Gene Cunneely.

All nine Cunneely grandchildren at the lake cabin. (Summer of 2021)

The Cunneely family gathered at Forty One South on Lake Pend Oreille, to celebrate the marriage of my grandson, Gene III to his wife, Mallory. (Sandpoint, ID, August 2021)

From bottom left, clockwise: Jim, Gail, Collin, Rachel, Mallory, Gene III, Connor, Gio, Casey, Michael, Adele, Josh, Grace, Mike, Amie, Kaylee, Brendan, Micah, Kanani, Kathy, Gene Jr, and me.

Meeting my first great-grandchild!
Micah, Kanani and baby Ellis Beckett Mahealani Duncan, born on October 1st, 2021.

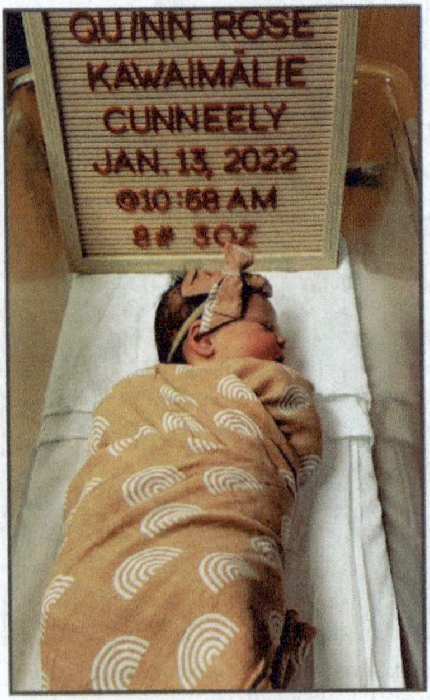

Quinn Rose Kawaimālie Cunneely, born January 13th, 2022 (Fairfax, VA)

Gene III with his wife, Mallory, and daughter, Quinn. (Virginia, April 2022)

With my first great-grandchild:
Ellis Duncan (Summer 2022)

With my second great-grandchild:
Quinn Cunneely (Summer 2022)

Collin Cunneely's Graduation from University of Washington, Class of 2022. Collin, Rachel and Gail.

Amie and "Little" Mike at his graduation from Gonzaga Prep, Class of 2022.

The lake cabin at night (Summer of 2022)

My great-grandchildren: Quinn (5 months), left; Ellis (8 months), right. (Spokane, WA, Summer 2022)

Celebrating the Fourth of July, 2022 at the lake with my first grandson, Gene III, his wife, Mallory, and their daughter, Quinn.

My granddaughter, Rachel with my great-granddaughter, Ellis at the lake. (Summer 2022)

The McClary Boys: Connor, Brendan and Casey. (Summer 2022)

Gene Jr.'s growing family, left to right: Micah, baby Ellis, Kanani, Gene Jr., Kathy, Gene III, baby Quinn and Mallory. (Summer 2022)

Easter Sunday breakfast with my three grandsons, the McClary Boys, at my favorite post-Sunday Mass breakfast spot in Spokane: Perkins!
Left to right: Brendan, myself, Connor and Casey. (April 2023)

Watching my grandson, "Little Mike", pitch his famous fastball at his college, Spokane Falls, game. (May 2023)

My *third* great-granddaughter: Charlotte Avery Anuhea Duncan! Born June 15th, 2023.

CHAPTER TWENTY-EIGHT: PROMOTION

'Tis a lesson you should heed, try, try again:
If at first you don't succeed, try, try again;
Then your courage should appear,
for if you will persevere,
You will conquer, never fear;
Try, try again.

Spiritual Week in the 95th FIS

In October 1972, our squadron, in cooperation with the Dover AFB Chaplain Corps sponsored a Spiritual Week. The purpose was to promote attendance at base religious services and more awareness of the many things that the Chaplains do in support of the military and their families. At this point in time, the Vietnam War was still raging and was a huge political problem with a large portion of the country's citizens opposed to it. This opposition to the war spilled over to anti-military attitudes that caused morale problems to both active military and veterans of the war. All the services faced retention and recruitment problems and most military personnel chose not to wear their uniforms off base. Support for the war was divided even among the military, and I had to admit, I privately questioned our strategic decisions not to employ the full military power we possessed to destroy the enemy. MacArthur's words, "*In war, there is no substitute for victory*," were seemingly ignored again at the cost of thousands of American lives. Many of my friends were now in North Vietnam prison camps not knowing when or if they would be freed.

During the 95th's Spiritual Week, the Chaplains visited the squadron, mingled with our airmen in the squadron areas, ate with them, and sat in on the aircrew briefings. On the Sunday ending the week, I addressed the congregations at both the Catholic Masses and the Protestant services. With the above problems in mind, this is what I said:

I wish to convey my thanks to the Chaplain Corps of Dover AFB for their assistance and attention to the 95th FIS during this most fruitful week. Not only did their presence in our midst, throughout our work areas, add a significant spiritual uplift, but their warmth and affability gave us new dimensions on our daily duties. We again thank them and extend an open invitation to them to return at all times.

I wish to commend the many men of the 95th and their families for joining me here at worship here this morning. We of the 95th, who are united by our military mission, also recognize that we are bound together in our dependence on Almighty God. I can say this to you men who fly airplanes, and I include those who fly and launch our nation's spacecraft, are men who have always openly confessed their faith and ultimate dependence on God. It has been said that there are no atheists in foxholes. I add that neither are there any in the cockpits of our F-106s.

We in the U.S. military forces today are, for the first time in our history, the targets for attack by many of our fellow citizens. Those who attack us include writers, Churchmen, Congressmen, entertainers, professors, and students. They perceive the man in uniform as a representative of an establishment dedicated to self-perpetuation by the use of fear. They not only denounce the size and costs of our nation's military, but further charge that the military is a tool of a government bent on the use of those forces to propagate its international policy. And as such, the man who chooses a military career is seen as a man who must subscribe to killing and destruction as an integral and necessary part of our national policy. Thus, the very morality of our profession is called into question.

What a paradox this is to us who wear the uniform and have done so proudly and voluntarily believing that service to one's country is second only to service to one's God; and that the two services, especially in our republic, which was founded by those who fought mightily for their religious and political freedoms are the highest and most honorable goals for any citizen. "Pro Deo et patria," for God and country, are these no longer the inspirational goals of the Christian patriot? Was it immoral for us to vow obedience to our freely elected leaders? Are we immoral when we train our fighter aircrews to be brave, aggressive, and skillful?

Is it a sin against God's commandment "Thou shall not kill?" to maintain our weapons ready at all times for attack?

Are our military chiefs murderers?

And what are we to think about our comrades in arms now in prison camps in North Vietnam, downed fighting against tyranny?

Have we permitted our dedication to the service of our country blindfold us from a higher order of values that demand a greater moral commitment?

We pray together this morning to seek His answers to complex questions such as these which deeply trouble us. We pray for the spiritual guidance we need to sustain us in overcoming the routine problems of accomplishing our mission, while knowing that to what we are committed is right and just in His eyes.

We declare to Him, and to our fellow citizens, that we will willingly use our armed force only to defend our beloved homeland, its freedoms and our loved ones.

We pray for these things together this morning united, not only by our military bonds, but united also in our aspirations for wisdom and peace.

Amen.

Troubling Times in Our Country

I believe that my talk surprised men. But it also echoed what many of the listeners also had in their minds. The Vietnam years were trying times to be in the military, trying and confusing. Not only did we witness anti-war and anti-military revolts, but there also came forth the counter revolutionary culture change that included increasing drug use among the young, the free sexual movement, and the racial civil rights actions, not all of which were non-violent. The shooting of my brother-in-law Tom was a result of the racial movement going on in that era that turned violent and antiestablishment. These problems had its effect on the military as well. Drug use in the military became more prevalent with pot smoking leading the way. The thought of someone working on an F-106 under the influence of marijuana was frightening. In 1972, racial problems broke out on some military bases; the most noteworthy was at Travis AFB, California. I had a large percentage of minority airmen in the 95th and I called a meeting of the Senior NCOs to ask their cooperation to catch any problems early before they got out of hand. I told them that I personally was the squadron Equal Opportunity Officer, so come to me.

A few weeks later, a problem did occur on the base. Two airmen held up in a barrack began harassing black airmen walking by outside on their way to the Airmen's Club. These two were using language intended to cause trouble calling the airmen forbidden names such as niggers, etc. My First Sgt. called me as a group of 95th airmen (blacks) were in an adjoining barracks planning to attack the perpetrators. I went in there and pleaded with them not to do so and to allow base authorities to handle the situation. I wanted to keep 95th personnel out of this problem. However, the Wing CO was slow in responding. I told him that he had better send his Air Police in there soon and make an arrest or he would have a full scale race riot on his base. For some reason, he did not want to do it. But when the problem arose again for a second day, he finally took action. It was just in time as again for a second night, I was in the barracks holding off a very aroused bunch of 95th black airmen who were losing faith in the system. However, when the two rabble rousers were arrested things calmed down. It was a close call. In the long run, I believe my action in being with my black airmen siding with them and counseling them to not engage in any violence that would hurt their career, demonstrated that I was a fair and racially neutral commander. I received good feedback from my senior black NCO's.

Doubts about the New Commander

The Vietnam War was a problem in other ways. In the squadron, we had pilots who had flown in Vietnam, and those who had not. As time went on, the number of war returnees grew in number. When I took command of the 95th, I realized that I was not welcomed and I knew that from what I was about to do to the squadron, and the changes I was going to inflict, I would become further disliked. One of the things that some of the pilots resented, especially the Vietnam returnees, was that I had not served in Vietnam. Now, I had flown combat in Korea long before any of these pilots were commissioned, and I had been the first F-106 aircrew to lead a squadron deployment on an overseas deployment. But not having served in Vietnam was the big thing with these people. I had to admit to myself, that I wish that I had completed Vietnam before taking command, but that did not happen and I could not change that. I was still qualified to command this hapless group and make it better, and I could still fly the F-l06 as good as or better than any one of them. I scheduled myself to lead a night four aircraft training mission. The mission called for us to take off in formation and fly a navigation sortie joining up with a KC-135 tanker over Arkansas. I briefed the mission. Following our refueling, we would be returning to our training area for intercept work against a B-57 target ship equipped with chaff and electronic counter measure equipment. After intercept training, we would rejoin and recover at Dover in formation. If Dover closes in weatherwise, we will recover at our DOB Atlantic City. The weather would be questionable with heavy cloud cover in the area of the tanker and at refueling altitude. The takeoff went on schedule with a good element join up. We climbed up and headed south and soon entered clouds in which we stayed all the way until we reached the area where we met the KC-135. The flight south was bumpy and I had to request an altitude change from TC to find a more suitable flight level for formation flying. I finally made radio contact with the tanker and soon picked him up on my radar. Still in close formation and still in clouds all the way, I eased our flight into the tanker's six o'clock and began our join up I now had a Judy (radar lock on) with the tanker at about a quarter mile. When our flight all had visual contact with the KC-135, my wingman and the element separated. We then began our refueling, I was going on the boom first. Each aircraft flying in close formation with the tanker on a dark night above Arkansas, took its turn coming up behind the large imposing KC. The voice of the Boomer showed an experienced guy was in charge there and each of the Sixes had no trouble filling up. When number four had its fill, I thanked the tanker and we broke off and headed back north to our training area over the deep dark Atlantic. After checking in with our 20th Air Division controller, our four F-106s made several intercepts on the B-57. Following that we rejoined into four ship formation and returned to base coming back over land at Cape May, New Jersey, descending into Dover. Breaking into two elements, we landed in formation. We had just completed an all-weather, air defense training mission that included night formation takeoff and landing, weather intercept of a KC-135 tanker and night aerial refueling, a rejoining and return navigation to perform night intercepts on an electronic jamming B-57 target aircraft. Then our four ships rejoined as a flight and returned to base for a night formation recovery and landing. After our maintenance debriefings, I met again with the aircrews in the ops room. There we went over the flight from start to finish. We learned what we could from everything that took place so that next time we would do it even better. It was now about midnight. The pilots were tired and ready to head home. But I could tell that they were pleased because they had just completed a complicated and difficult mission which tested their skills. This was how it was supposed to be done in a professional way and safely. And their new squadron commander led them. I sensed that that was the end of any doubts any of them may have had about their new CO's skills as an airborne leader.

New Chief of Maintenance

Nothing that happened to me as Commander of the 95th was more welcome than the arrival of a Chief of Maintenance with some maturity and experience. Lt. Col. Jim MacDougald was the man I was waiting for and his arrival provided leadership in the one area where the squadron was still weak. While many of maintenance's problems were not within its control, such as the shortage of critical parts, still that side of the house was in need of strong supervision by someone who had years of firsthand experience in the business. MacDougald had that, and he also was a pilot who would be flying the product of his unit's labors. I was truly happy to get him and again I owed a big thanks to my friend at ADC, Col. Lou LaSalle, who was looking out for me. I now had two strong leaders, Yogi Van Daele, running ops, and Jim MacDougald, overseeing aircraft maintenance. I liked both of these men, and I respected their judgment. I told them what I wanted and what I expected of them. My door was always open for them to give me their ideas and recommendations, even when they were adverse to my own. But once I made the decision, I expected them to conform. They never let me down. I have always believed that the secret of successful management was finding the right people, put them to work, tell them what you want done, and then leave it up to them to get the job done. I now had the right people. All I had to do was watch over them and keep the ship on the right course. It had been a long tough period for me trying to be both chief of maintenance as well as commander. But now, I had a full team on board. Patience and perseverance had won out. I felt that there would be better days ahead for the 95th FIS.

Women in the Air Force

My good friend Col. Lou LaSalle, Chief of Personnel at ADC was after me to accept some females in the 95th FIS. The AF was now pushing women and there was no doubt, with the shortages of manpower caused by Vietnam, WAF could fill in lots of gaps where needed. But I was adamant that I had enough personnel problems as is with the men and that I did not want any females to make matters worse. LaSalle did not like it, but he left me alone on that subject. But I got blindsided by my own Executive Officer.

Lt. Col. Mancus was a good man and an effective administrator. He knew the regs and kept the paperwork flowing. However, when I returned from a weeklong TDY to ADC, I found three WAF working in the Orderly Room. Mancus had asked for some temporary help and ADC sent us three WAF from Tyndall AFB on a TDY basis to pick up the slack. We were short of people in the administrative section, but Mancus knew I did not want any females. I really chewed him out on this one. He assured me that they would be no trouble. But he was wrong. The three young women were attractive and they were also flirtatious. They soon found their way to the Alert Hangar where the pilots on duty there were just delighted to show off their loaded F-106s and charm the gals with their tales of dating do. The following weekend, I learned that some of my Lieutenants had hosted the gals on a water skiing trip to a nearby lake. Now I had to forcefully remind my officers about the restrictions of fraternizing with enlisted personnel, male or female. Next, two of the gals began to work in the NCO club at night as bar waitresses. I had no power to stop that, but I had bad vibes about it. I was correct. One of the girls gave the come-on to a black NCO, (she was Caucasian). Some white NCOs did not appreciate that and a fight broke out which brought in the Air Police. I was called out in the middle of the night to reclaim my TDY WAF's. Finally, two days later I was visited by Captain Logan, Commander of the WAF Squadron on Dover. She had her First Sergeant with her and their complaint was that my three female airmen wore skirts that were not in accordance with regulations; the skirts were too short. She thought they were giving her WAF Squadron a bad name and hurting their reputation. That did it. I called in my three WAFs and told them their TDY was over. I sent them back to Tyndall. They cried, but my heart was hardened. I was trying to run a fighter squadron and chasing these three around was taking up too much of my valuable time. No more women in the 95th as long as I was CO.

It's Not All About Flying

A commander of 700 people runs into a lot of stories, some happy, many not. Some of the problems you can fix, some you cannot. People are often their own worst enemies. They make bad choices for whatever reason and then don't know how to get out of it. In my 21 months as commander I dealt with divorces, separations, wife beatings, drug abuse, drunkenness, homosexual transgressions, desertions, obesity, cowardice, gambling addictions, and lying. Managers in corporate situations have similar problems. But in the military, the closeness has more impact on mission readiness which might result in loss of life and failure to successfully defend our country. The 95th FIS's mission included defending the nation's capital, Washington, D.C., and several major military installations and cities from aerial attack. What could be more important? Good morale and esprit were essential. And that was my responsibility.

The Promotion Board Meets

In early 1971, the promotion board met to consider selection of eligible

Lieutenant Colonels to the grade of Colonel. This would be my first year of eligibility. I knew that this would be it for me because if I were not selected this time, while filling the position of a squadron commander, my chances of selection would decrease severely. My Evaluation Report, written by General Gamble in March 1971 was an outstanding one, and I felt good about my chances. However, a day before the list was released, I and other commanders, were meeting with General Gamble and his staff at Fort Lee, where his Division Headquarters were located. That night, the General had us to his home for cocktails and a buffet. It was a pleasant evening, but the General said nothing to me about the promotion list although I knew that he had already seen it. I went home a little concerned. But the next day, I received word from him that I had been promoted to full Colonel. My first call was to Alice Marie. No one deserved this bit of good news more than she, because no one had sacrificed more during our married years in the military to make it possible. When we were married, I was a First Lieutenant earning about $400 a monthly, including flight pay. Thanks to Alice, we made that salary work for us. And through the years, Alice was always frugal, rarely spending money on things we could do without. We never went into debt other than for our mortgage. And she always made it a point to put some dollars away for the future. I never had to worry about the budget, even when the children came along, one after the other. As I achieved field grade status, (Major) things got a little better financially and now, making full Colonel would really be helpful.

This promotion was hard earned, I had to admit. And it would mean more salary and benefits that would be helpful to my family. But it would also mean that I would be receiving an assignment soon to God knows where - probably Vietnam. That would mean a family separation of at least a year and maybe more. So that was the downside of the promotion upside. Alice and I knew that this would be coming and we knew we had to face it and do the best we could with it. In the meantime, we intended to enjoy our promotion and wait for what was in store for us and then make the best decision we could for our family. Decision time was coming and soon. God give us wisdom!

Report I received in March 1971 from General Jack Gamble, Commander, 20th Air Division, ADC. This OER covered the period 29 March 1970 to 28 March 1971, my first year as Commander 95th FIS, Dover AFB.

Lt Col Cunneely's performance during this period has been superior. He has clearly demonstrated his ability in his most responsible and demanding job. When he took over as commander, he had many serious problems to contend with mainly in the maintenance and supply area. He also faced a problem between operations and maintenance which caused a lack of confidence on the part of each party in the other. He rapidly but methodically identified each problem area and planned a corrective action. He made personnel changes where necessary and lent strong personal leadership to improving morale and management. He accepted advice from higher headquarters in the technical areas of supply and maintenance, the total result being a steady increase in U.E. flying, improved OR rate, and reduced NORS rate. The quality of his aircraft improved dramatically. As the quality of squadron performance improved, the support of division/region operations improved. He is very active in base activities and has worked hard to establish good relations with his host base. He has demonstrated at every step his qualification for promotion to temporary colonel and his potential for greater responsibility.

STRENGTHS: Outstanding ability as a leader. Calm and effective performance under pressure.

SUGGESTED ASSIGNMENTS: Professional schooling followed by a responsible command or staff assignment. This recommendation would enhance then utilize his high potential.

I put on my eagles on July 26, 1971. It was a good feeling and an experience I never imagined would happen back at Fordham University in June of 1951, when I was sworn in as a Second Lieutenant. I recall that Alice Marie was there with me on that day. When we married in 1955, I was a First Lieutenant. My monthly income was about $400. Now as a Colonel, it would be over $1,900. With five kids, we would need all of it.

Life is like a road with many twists and turns. You are never sure of where it will lead you. But, if you have your goals in mind, and you have an idea of where you want to get to, with hard work, sacrifice, and a little bit of good fortune you will get there. In my case, I had all of the above, plus I had a marvelous helpmate in Alice Marie. She kept me going through the toughest times with her warm and supportive disposition and encouragement. I also had my children whom I enjoyed dearly. They always made my home a place of peace and joy. They made all of the challenges Alice and I faced more than worthwhile.

CHAPTER TWENTY-NINE: I COMPLETE MY DUTIES AS 95TH FIS COMMANDER

The best laid schemes of mice and men
go aft astray and leave us naught
but pain and sorrow for promised joy.

-Robert Burns, from the poem "To a Mouse"

Inspections, Inspections, Inspections

Though I had been promoted to Colonel, my responsibilities as squadron commander did not abate. While I expected to be receiving orders to a new assignment after the first of the year (1973), there were still many problems facing the squadron and lots of improvement to be put in place. And I intended to continue working on those problems and putting in those improvements until the day I was relieved. Thankfully, I had now a much better team to help with those problems and I set about working them hard to bring the 95th up above average in every category in which fighter squadrons were measured. And squadrons were being measured all the time, every day and every month. If Squadron CO's only had to worry about their annual ORI, that would be reasonable. But that was not their only concern. Every Headquarters had its own inspection team up to the Pentagon, and somehow, they all found time to drop in on the fighter squadrons and inspect them for something. In addition, the ADC Commander, General McGhee would pop in unexpectedly in his Lear Jet and inspect the alert hangar and then have you scramble the alert aircraft to see if they were truly combat ready. General Price, the ADC Deputy for Ops liked to take some squadron's F-106 and drop in on you on a weekend and inspect your Combat Alert Center, your alert hangar and your nuclear storage facility. Then there was NORAD with its large scale exercises with SAC bombers as the faker (enemy) forces which usually lasted 48 hours. I even had a group of career happy Pentagon whiz kids drop in and check the efficiency of our "tail number scheduling." They spent hours looking over our daily flight schedules to see if the aircraft scheduled for those flights actually flew those sorties. Tail number scheduling was the latest buzz word at the Pentagon. It was a moderate success in SAC and in Commands that flew big planes once every three or four days on long duration sorties. It was a total failure in fighter squadrons which flew aircraft in big bunches on short sorties on which the planes were subjected to great airframe stress. (Lots of G-forces) The whiz kids wrote me a really bad report on my tail number scheduling success rate. None of them had ever been closely associated with fighter operations.

I also had a team come in to inspect my Nuclear Storage Site. One of them climbed the fence and when he reached the barbed wire angling off at the top, he pronounced it as unsatisfactory because a penetrator could somehow extract the metal framing that held it in place. I was speechless at that finding. Another got on the ground with a ruler and found the grass within the storage area was a half inch too high. Fortunately for me, cutting the grass was the responsibility of the Base Commander. From my many years as an inspector from higher headquarters, I would never have believed that some people would be doing the things these folks did. They clearly had no conception of what the primary problems squadron commanders and their NCOs faced in meeting mission requirements on a daily basis. Their visits with their ensuing written reports, which had to be answered up through the chain, only burdened front line personnel more than necessary and ate up valuable time that should have spent on more vital problems.

Problems in the Nuclear Site

One thing that Alice Marie truly hated was the Red Phone. There was one located downstairs and one in our bedroom. When it rang, she would jump. Instinctively, she just knew that it meant trouble in the 95th and I might have to leave. I can't tell you how often this happened just at dinner time or, even worse, in the middle of the night. It happened one early morning, about two AM. A call from the CAC; there had been a shooting at the storage site. My heart just sank. How could such a thing have happened? Any problems in the nuclear area meant special reports to Air Force, the DOD and the Atomic Energy Commission, immediately. I couldn't believe it. On arriving at the site, I learned that two air policemen, apparently bored or looking to inject a little excitement into their long night of duty, decided to engage in a quick draw contest. One of them was a winner. The other was hit in the shoulder, but not mortally wounded. On the floor of the facility was a chalk drawing of where the wounded policeman fell with lots of blood still pooled about. I asked the Air Police sergeant who had been notified besides me. He replied, "no one." I told him I would notify the Wing Commander and he should call the Base Commander immediately. When I got the General on the phone and relayed to him what had taken place, his reaction was volcanic. He understood immediately that the Air Police and the security of the storage site came under his jurisdiction and not under the 95th FIS. When he arrived at the site, he was totally beside himself. He continued walking up and down, back and forth, wringing his hands and asking out loud, "How could this have happened?" I felt badly for him. I think he saw his career coming to a swift end. There are some things that commanders have little to no control over. At any given time, one of your people might do something completely absurd or unthinking that could have dire results and could end your career. It's happened to the best commanders and they paid a price for someone else's misdeeds. Such is life.

Orders Come

As the fall of 1972 moved on, I was as busy as ever, but now I was concentrating on the finer things and tweaking away at lingering problems in personnel, supply shortages and budget. The Atlantic City DOB was running smoothly under Major Johnson and I rarely had to worry about it. I would drop in there once a month, perhaps spend the night on alert and return the next day. I now had some more time with the kids and used it to watch Gene, Jimmy and Mike play hockey at the Wilmington ice rink. Gene and Jimmy also were now attending Holy Cross High School in Dover. This was a fine Parish school and Alice and I were pleased with their schooling there. Mary was swimming with the Y teams and doing well winning her share of ribbons. The boys and I also enjoyed playing the base golf course with the Van Daele's, taking on the Van Daele's in racquetball and, of course, driving up to Philly to watch the struggling Flyers hockey team. All in all, life was not too bad; however, I knew that coming next summer, a big change was in the works.

I had been in touch with Lou La Salle about my next assignment. Lou knew all the guys at Colonel's Assignment Branch in the Pentagon and I asked him to nose around. I expected that I would be headed to Vietnam and I hoped for a good flying job such as F-4s, F-105s, or even A- l0s. Lou told me he would see what was up for me. But in December, I received another assignment shock. I received word that I had been selected for an assignment to the US Military Training Mission in Riyadh, Saudi Arabia reporting in June 1973. This would be an unaccompanied assignment. I was aghast and completely disappointed. I called Lou. He told me I could talk to the Colonel Assignment Branch guys, but warned me that from what he had heard, I had an uphill battle. I then drove over to the Pentagon and spent several hours with the Colonels' Assignment guys. I have to say that they were more than hospitable and sympathetic. They spent a lot of time with me and we even had lunch together. I pleaded for an assignment to Vietnam, preferably a flying job. Their response was twofold. One, they did not have a great need for full colonels in Nam. Colonel spots over there were not that plentiful. Second, the Army General who headed the US Military Training Mission in Saudi O.P. Smith, had fired three USAF Colonels in the past year. Colonels Branch had sent him ten Personnel Folders of AF Colonels for this position, and mine was the only one he accepted. Why, they did not know. But that was the way it was and they were tired of Brig. General O.P. Smith, and since he wanted me, that's whom he was going to get. All of this did not make me feel one bit happier. The thought of a year's separation from family was not something I was looking forward to but doing it in Saudi made it even worse. I drove back from the Pentagon, a very disheartened man. Alice and I now had to make some big decisions and some delicate plans for our family's future.

With this on my mind, there was still a squadron to command and I promised myself that I, would complete the job I was set to do until I turned the command over to the next guy. The problems of command did not let up and every day brought with it new and unexpected hurdles to overcome. These I shared with my fellow commanders, especially with my old buddy Doug Brenner who was now commanding the 49th FIS at

Griffiss AFB, New York. Doug and I swapped tales with each over the phone on a regular basis, and occasionally he and I would recover from one of our training flights at the others base and visit. It always helped our morale to learn how the other guy was coping with many of the same problems you had. Both of us spoke longingly about our days with the 318th FIS and how we missed being in an outfit that was so well stocked with superb personnel and excellent leadership. Another co-commander friend of mind was Mike Compton who had the 27th FIS at Loring AFB, Maine. Mike and I were in the 80th FB in Korea and also were at North Bay together in the 60's. I was surprised to learn that Mike had been given command of the 27th FIS. Although he was a bright and effective staff officer with fine communication skills, I thought he was a weak pilot. I remember he had several problems in Canada when the RCAF aircrew evaluation officers flunked him on his instrument flight tests. Compton got into serious trouble when he was alleged to have submitted some TDY per diem vouchers that did not pass the smell test. He came by Dover one day to tell me about it and I could only commiserate with him. I had never known Mike to be anything but an honest broker and I could not believe he would cheat to pick up a few extra bucks. However, to my surprise, Compton was relieved of his command of the 27th and shortly after retired from active duty.

I began to work hard on two personnel situations. First, I wanted to do everything I could to keep Van Daele on active duty. Yogi was a Reserve officer and the Air Force was starting to involuntarily retire Reserve officers with 20 years of service unless they were full Colonels. I wrote a strong appeal on his behalf and received both the endorsement of General Gamble and General McGhee, the ADC Commander, to obtain a waiver for Van Daele. Second, I began a campaign to have Jim MacDougald, my Chief of Maintenance, replace me as Squadron Commander. I was very impressed with MacDougald's leadership abilities and I thought it would be a valuable thing to have a man with lots of maintenance experience become commander of a squadron that had so many past difficulties in that area. I soon received support from General Gamble on that score also. Most of the personnel problems I found in the 95th when I look it over had been resolved. Fortunately, a good many of the problem children were shipped out. Those who stayed changed their attitude, at least on the outside. I no longer had to conduct unit inspections on grooming and uniforms. Of course there were always a few pot heads who had to be dealt with and a wife beater here and there. I had one airman report in to the 95th caught trying to rob a bank in a small town west of Dover, but the FBI took care of that matter. As in every organization in the world, large or small, the people you have make it happen, one way or another, for good or for bad. Just make sure you have the right people and take care of them.

The Last Inspection

Early in 1973, rumors began circulating that another visit by the ADC/ORI Team would be coming soon to the 95th FIS. I wasn't worried. The reason I still wasn't worried was not because I was leaving the squadron anyway; I wasn't worried because I knew that the 95th would pass this time easily. I did not intend to leave a squadron that had just failed an ORI and I let that be known to my senior officers and NCOs. So we began to work hard at getting our aircrews and aircraft up to snuff on all the things that we knew we would again be tested on. But then something intervened that really got me worked up.

I mentioned earlier that every headquarters came up with its own evaluation branch which commanders would send down to test units under them to prepare them for the ADC/ORI. And so did General Gamble have an evaluation branch headed by a young Captain, for whom I did not have too much respect. Now these intermediate evaluation teams were in a tough position. If they evaluated a squadron and rated them "Satisfactory" and then the same squadron flunked the ADC/ORI, the Division Commander would likely fire his own evaluators. So the best avenue for the Division evaluators was not to give the squadron a high rating. And this was what the 20th Air Division Evaluation Team did. They came in and wrote a very nasty unsatisfactory report on the 95th. When I went over the report with Van Daele and McDougald and my NCOs, I knew right away what had occurred. And I told that to General Gamble. The General was not happy with my response. And I was not happy with his belief that his evaluation team's report that stated my squadron would probably fail its ORI. I told the General that his team did not know what they were talking about and that they were just covering their a--. General Gamble ended the conversation by telling me that he was considering firing me. That certainly was within his authority, but I did not believe that he truly intended to take that action. I meant to have him eat his words.

Winding Down

After my conversation with General Gamble, I called together my staff and told them that although I would be turning over command of the 95th FIS soon, I did not intend to leave under a cloud. All of us had worked hard to bring this outfit back from the bottom where it had languished as the worst in ADC-a real loser. I was not going to leave it with it back in that same situation. My reputation was on the line, and so was theirs. I promised General Gamble that the 95th was an organization that his Division could be proud of and I meant to show him I was right. We all agreed to make it happen.

The holidays of 1971 were great. Alice and I made it a point to visit with both her and my family in New York. While we had been at Dover, her mom had visited several times and Alice had gone up to New York to see her brothers, her sister and her nephews and nieces. Katie and Mary had accompanied her and she saw some Broadway shows as well. We spent some time that summer at my dad's place at E. Patchogue and saw a lot of my brother and his wife Pat. We also toured the historic places of Philadelphia where our nation's history began. Soon I had to break the news to our relatives that I would be leaving the following summer for Saudi. Right after the first of the year, the Air Force advised that I would be required to attend a series of preparatory schools and conferences for my new assignment. This necessitated my surrendering command of the squadron in February 1972. Beginning in March, I would begin a series of TDY's to different seminars and visits to organizations preparing me for my assignment to Saudi. I was surprised to see that I would be on the road for the best part of four months being educated about the diplomatic and military aspects of my responsibilities as part of the Air Force's military training mission in Saudi Arabia. I began to realize that my next assignment would be taking me into a strange new world with which I had no familiarity. In the interim, I needed to wind down my responsibilities to the 95th FIS, turn over command, decide on my family's future and work out my itinerary. Things were beginning to move at a fast clip and Alice and I had to decide what to do with my family when I was overseas. Life was not getting simpler…

Decision Time

Alice Marie and I looked at many options. These included: she and the kids staying in the Dover area for the year that I would be in Saudi. We would have to rent a house for that year. The advantage here was the kids could stay in the schools they were in for at least one additional school year before having to make a move. A second choice was to move the family now to a place they would enjoy better than Dover. One option we discussed was a move to Spokane where the family could use our cabin on Lake Pend Oreille which they loved so much. This would get them out of the hot and humid Dover climate. However, this would mean an early move across the country before I left for overseas, and in the end, it would result in two moves for the family within about 15 months. Alice and I had long discussions on this matter before we decided to move the family to Spokane. It was a very difficult decision and one that we have second guessed many times over the years. But once we decided, that was it; we put our decision into motion. Other than being near her family, Alice did not particularly care for Delaware. She did not like the climate and it was a hard two years on the both of us. There was never a warm feeling of friendship and loyalty within the organization. Because we had replaced a commander who was well liked, and because we were outsiders sent into to clean up a mess, we engendered little affection from those who were there when we arrived. Additionally, they knew that I had effected the dispatch of many of their compatriots out of the squadron and brought in handpicked replacements. These included all new supervisory people, a new operations officer, a new chief of maintenance, a new electronics officer, a new first sergeant, and many new NCOs. The 95th FIS was pretty much a new squadron from the one I had taken over and a far more effective fighting unit. They knew it, but they still resented it, and they still resented me.

And so, Alice and I decided to cut our ties to Dover for good. In February 1972, I turned command of the 95th FIS over to Lt. Col. Jim MacDougald, my Chief of Maintenance. (I was pleased that ADC had accepted my recommendation about Jim.) The squadron gave us a fine going away party at the officer's club and many nice things were said. The aircrews presented us with a painting of my F-106 climbing out over the Atlantic on a dawn mission. In my parting address, I told the officers and their spouses that I understood what a difficult 21 months it had been for them. I realized that I had made many changes they did not welcome and with which they did not agree. I may not have been 100 percent right in all of my decisions, but I wanted them to know that every decision I made, I made believing it was in the best interest of the 95th FIS and the accomplishment of our air defense mission. I then told them this. Though not all of you will ever think highly of what I did as your commander, you can be very proud that the 95th Fighter Squadron is now perceived as one of the best in the business, and all of you are part of it. Alice and I then gave the audience an Irish blessing.

The next day, I handed command over to Jim MacDougald. The squadron flew a 16 ship formation over the base in my honor. The door had now closed on another chapter of my professional career, by far the most difficult chapter to date. But I was proud of what I had accomplished. There was one thing I was frustrated about and that was that the ADC ORI Team had not come a second time before I turned over command. I was still a little burned about Gamble's threat to fire me believing that the 95th was in danger of flunking its next ORI. I wanted to show him how badly wrong he was. That would have to wait.

February 1972

At this point in my life and career, I took stock of where I was and where I was headed. I was nearing the age of 43, with a wonderful family of five beautiful children. But I now began to consider whether I had let my career concerns taken me away too much from my fatherly priorities. Every chance I had, I dedicated to being with them and those were very important and precious times. All of my children were lovable and unique and deserving of my attention and affection. Seeing to their needs and ensuring that they were being met required that their mother and I coordinated closely and unselfishly. School, sports, homework, and social events kept us on the go. Dinner hour was a special time for our family as well as Sunday Mass and breakfast afterward. I especially enjoyed watching the boys at hockey and baseball and Mary in her swimming. With the Van Daeles we looked forward to the Flyer's hockey games in Philadelphia and we also took in the Phillies baseball games. We golfed the Dover AFB course and played racquetball. We took in the wonderful Delaware beaches and drove up to New York to visit my dad and brother as well as Alice's family. So, we tried to use our time in the East as best as we could as we knew it would come to an end soon. And I knew that I would soon be leaving my family behind for at least a year's separation at a critical time in their lives. And it worried me.

Following is the OER I received in February 1972 written by Brig. Gen. Jack Gamble, Commander 20th Air Division and endorsed by Lt. Gen. Thomas McGhee, Commander Air Defense Command.

> FACTS AND SPECIFIC ACHIEVEMENTS: Colonel Cunneely's performance was characterized by hard work, long hours, and a keen interest in his people as individuals. During this period, the squadron maintained its excellent flight safety program. They developed an outstanding air refueling program with an Air National Guard KC-97 unit which provided them considerably more than the normal refueling sorties. Refueling was combined with other phases of training so that maximum individual training in a variety of areas was accomplished on each mission. He and his people always aggressively supported our air defense exercises and gained the respect and admiration of our weapons controllers. His emphasis on fire control system maintenance over a sustained period achieved excellent results which are currently being reflected. He has maintained a reliable alert force with a high degree of success on active air missions. He has greatly improved harmony in his organization through innovations such as his monthly NCO-officer luncheon which is used as a "commander's mailbox" is one means he has used to broaden his open door policy and to maintain close contact with his people. Colonel Cunneely has on several occasions made presentations to civic groups on the air defense missions of his unit. He actively supports the equal opportunity program.
> SUGGESTED ASSIGNMENTS: Assign to operational staff position. Colonel Cunneely is an excellent pilot and has an extensive flying background which would qualify him for this kind of staff duty.

At this date I was a Command Pilot with approximately 4,650 hours flying time and 42 years of age.

PART V: FAMILY SEPARATION, I SEE A VASTLY DIFFERENT PART OF THE WORLD

CHAPTER THIRTY: JUST WHEN YOU THINK THINGS CAN'T GET WORSE... THEY DO!

Life is a struggle, accept it.

-Mother Teresa of Calcutta

By handing over my duties as commander of the 95th FIS after 21 very trying months, you would think I would be relieved and somewhat pleased. My accomplishments had been amply rewarded. I now wore an eagle on my uniform and the daily, even hourly pressures of command responsibilities had been lifted from my shoulders. I could now sleep through the night without fear that the red phone would awaken me with some terrible news of an accident or incident or a crime involving members or property of the 95th FIS. That was now the concern of Jim McDougald. But Alice and I had much ahead of us that we knew were going to be unpleasant for us and our children. And there was no way to get around it. We had to make the best decisions we could and face the results of those decisions with a stout heart and perseverance.

My first order of business was to report to Williams AFB, Chandler Arizona for checkout in the F-5 fighter. The USAF had sold the Saudi's a squadron of F-5s with more to follow. The Air Force wanted me to be familiar with this aircraft as I would have a detachment under my supervision at Dhahran where the RSAF had their F-5 stationed. I flew by commercial flight into Sky Harbor Airport, Phoenix and was transported by staff car to Williams AFB which was located a few miles southeast of Phoenix. In the following two weeks I was once again a student pilot learning to fly a new aircraft. But the F-5 was an easy checkout and a joy to fly. Over my two weeks I had 10 flights which included some air to ground strafing, skip bombing, dive bombing and air to air gunnery. These were skills I hadn't utilized since my Korean War days in the F-86. I surprised myself and my instructors with my excellent scores. I felt very good about myself, but mostly this could be attributed to the superior characteristics of the F- 5 which provided an exceptionally good platform for launching ordinance on target. I truly enjoyed the F-5. It was an honest aircraft, dependable and easy to handle in formation. You could take the F-5 and point it straight up until the airspeed went to zero without worry of a spin. The nose would just drop, airspeed would build up and you would be flying again.

My time at Williams was enjoyable due to the flying. We did most of our flying south of the city over the desert. The weather was always sunny and clear, no worries about fog, low visibility approaches, and radar controlled landings. From Williams, I flew commercially into Los Angeles. I was met by an old friend from my ADC/TACEVAL days, Col. Norv Heath. Norv was retired and was now working for Northrop Aviation, the builder of the F-5. He was the head of the Tac Eval Team in 1958 when I joined the group in Colorado Springs. In the next few days, Norv squired me around the Northrop facility where I saw the F-5 being built on the assembly line. I also got to meet the heads of the company and the chief test pilots. I was very well treated by all of them. It became obvious to me that in my new position as chief advisor to the Commander of the RSAF, I could have some influence on their choices of any future weapons systems they might buy. At least that is how the Northrop people saw it. As I soon found out before my departure to Saudi, I was greeted and hosted by many more defense system manufacturers who wanted to talk to me about what their company had been offering the Saudi's in the hopes I would support their product. I was being introduced into a world that I had not before known about. This was the world of FMS... foreign military sales. The amount of dollars the USA brings in through selling weapons systems to the Armies, Navies and Air Forces of other countries is enormous. These sales must first be approved by the State Department, the Intelligence Community, the Congress, the Executive Branch and, of course, the Pentagon. When approved by all of the above, the sale then becomes the business of the individual service to put into effect the contract and oversee the delivery of the weapons with the training and support that goes with it. The US Military Training Missions then get into the act. And that was my next assignment.

Col Heath and I had a great reunion. He invited me to dinner served by his wife Billie at his beautiful home in Pacific Palisades overlooking the Pacific Ocean. We talked over the great times we had in Colorado Springs and the wonderful collection of guys and gals we had on that Team. I told him how much I learned in that job and how it had helped me in my career, especially in my last job. The trip to Northrop was very helpful to me and I left feeling that I could speak with some firm knowledge about the F-5 when I arrived in Saudi.

The Royal Saudi Air Force F-5 Tiger

More Preparation... More TDY

My preparation for my new job was just beginning. In the next three months I was rarely at home in Dover. The USAF in cooperation with State Department, the CIA, the FBI, the DIA and several other agencies laid out a program of indoctrination and education for me that was designed to tell me all I needed to know and more about why we were in Saudi and why we were determined to maintain cordial relations with this nation that supplied us with that precious commodity which fueled the American industrial economy, the world's leader. After my three weeks that included my visit to Randolph AFB, Williams for the F-5 checkout and then to Los Angeles to meet the Northrop people, my next TDY was to the Foreign Affairs Executive Seminar held at the Foreign Affairs Institute in Arlington, Virginia. This seminar was conducted by the State Department and lasted three weeks. To attend, I rented a car at Dover on Sunday afternoon, and drove to Arlington, a drive of about two and a half hours. There I turned the car in and checked into a hotel. During the week, I walked to and from the classes and ate most of my meals at the hotel. Friday, after the last session, I would rent a car and drive back to Dover to spend the weekend with the family. I repeated this routine for the three weeks of the seminar. The seminar itself was truly interesting. A list of distinguished speakers were on tap every session and they covered a multitude of topics that overlapped foreign affairs, the Vietnam situation, economic difficulties, the future for America, oil, and so forth. My classmates were from different arenas. Some were military, but most were in other government positions some diplomatic and some would not disclose. The discussions were lively and not all of the lecturers found universal agreement from the student body. We also visited the Congress. We got to meet Senator Buckley, (R) NY, and the brother of William F. Buckley. Many of the speakers were impressive and introduced us into areas that I had not known much about prior. Examples were Bill Small, head of CBS news, Richard Scammon, Director of the Election Research Center, the head of the RAND Corporation, the Dep. U.S. Rep. to the UN, Dolf Droge, a top White House Advisor to Pres. Nixon, the head of the USIA, and many more top university professors who spoke about various problem areas facing the country. All in all, this was a most exciting and educational seminar for me. From the world of fighter piloting, I was catapulted into a world which dealt with problems that spanned the globe. These problems were complex and required interaction between agencies within our government and sometimes in cooperation with foreign bureaus. My mind was beginning to turn away from the 95th FIS and ADC and was starting to hone in on how I would face dealing with situations that would have some international implications.

In April, I attended the USAF Special Operations School Military Assistance Advisory Course at Eglin AFB, Florida. This was a one week course and to be truthful I don't remember much about it. It was pretty much boiler plate stuff on how to get along with your advisee (in my case the Saudis), and to be careful about their customs and knowing the rules governing foreign military sales. Following this, I flew to Wright-Patterson AFB Dayton, Ohio for a few days more of training for MAAG/Mission Personnel on foreign Military Sales. With all that behind me, I now headed back to Washington for interviews with the CIA, the FBI, NSA, the Saudi Arabian Ambassador to the USA, a meeting with the USAF General in charge of FMS, and several luncheon meetings with various lobbyists and sales reps from Lockheed, Westinghouse, and Northrop. My meeting with the General in charge of FMS sticks in my memory. He was a three star, but I don't recall his name. He made a big point that I would not permit the defense contractors "buy" any favors from me and that I should not take any gifts from them. I should pay for all of my meals. I took this to heart. When the Lockheed and then Westinghouse guy insisted on taking me to lunch, I went but absolutely paid my own part of the check. And these were not cheap restaurants either. I remembered this when about a year later, this same General along with a few others, was caught accepting favors from a big defense contractor and spending weekends at the contractor's Maryland hunting lodge where there was much partying and even a few ladies in attendance. I was not happy about this story.

William B. Robinson

It was clear that things were not right in Saudi and they had undoubtedly honed in that the culprit was Army Gen. O.P. Smith, my new boss. The man I was taken to meet that impressed me the most was Under Secretary of the Air Force for International Affairs, William Robinson. I was told that he wanted to see me personally and privately and discuss with me my assignment. Those who took me to see the Under Secretary referred to him as Willy. I was escorted to his impressive office where he came from around his desk, shook my hand and greeted me warmly as if we were old friends. He asked me how much I knew about the man I would be working for next in Saudi. I told him I knew little, but what I had heard was most unusual. The Secretary then launched into an extended description of Army General O.P. Smith, my new boss. Smith had fired the last three Air Force Colonels he had under him and went through more than ten Colonels resumes before selecting mine. Why mine, he had no idea. That's why when I asked to decline this assignment; Colonel Branch said no, you have to go. Smith was a real problem for the USAF. He was an Army guy but most of the ongoing FMS business in Saudi was Air Force stuff. Further, he was constantly barging into Air Force business with no expertise. He intercepted all incoming and outgoing messages from the states into Saudi and in many cases pocketed those without letting them go through to their destination. He was too much in bed with the Saudis and was suspected of doing some inside dealing with them that were against the law. Morale in USMTM was low and due in large part to Smith's overbearing policies and bizarre behavior. Robinson then asked me to do something I never thought I would ever be requested to do in my military career. He asked me to keep him informed about Smith through back channel communications (U.S. Mail) and to come to the Pentagon and brief him personally about Smith at my six month R&R leave. Also to let him know if things became unbearable for me by Smith's idiosyncrasies. He told me that the SECDEF was aware of his instruction to me. I came out of this meeting stunned. I had never before been subjected to such a situation where I was instructed to actually observe and evaluate my superior officer and report his failings to higher-ups behind his back. This defied all the known rules of officer loyalty that I had learned and adhered to during my career. However, the Secretary was clearly an upstanding man and he made a strong case that this was a situation that called for a break-with protocol. I trusted him.

Every Time We Say Goodbye... I Die a Little

The days were now slipping by and my departure date for overseas began drawing closer. I tried to spend as much time at home with the family as I could. In March, I drove the boys to a hockey game in Wilmington one night. On the way home we entered a deep fog that slowed us to a crawl as visibility was reduced to a few feet. At a stop sign half way home, a drunken driver rammed into the rear of our Mercury Marquis Station Wagon and totaled it. No one was seriously hurt. State Police drove us home and I had to buy a new Mercury station wagon for the family in time for our leaving Dover. Because I had to leave the USA before the trial came up, the guilty driver was let off.

I spent lots of time watching Gene Jr. and Jim play hockey and baseball, also Mike. Mary had her swimming meets and brought home her share of Blue Ribbons. I got in some F-l06 flights with the 95th courtesy of Col. Van Daele, slowly but surely the days went by and in May I drove to Spokane and rented a house for my family for 1972-1973. I stayed with the Grashio's and fortunately found a nice house in the St. Thomas More Parish which would meet our needs. While there I registered Gene and Jim in Gonzaga Prep and Michael in St. Thomas More. With Sam's help I hired a carpenter, Sam Bossio to add a room on to our lake cabin that June. With all this accomplished, a 95th FIS F-106 picked me up at Geiger and flew me back to Dover AFB. I still had my doubts whether Alice Marie and I had chosen the right course or not.

But now the die was cast. There was no turning back. The transportation office had arranged for Alice, the kids, and Brandy the dog to depart by AMTRAK from Philadelphia to Chicago and from there westward to Sandpoint, Idaho. We were busy cleaning the house and preparing for the trip. There was so much to do and so much on our mind that the full impact of it all was barely grasped. The children were being ripped out of a very secure situation and transported thousands of miles away, separated from friends and schools and all the other things that were important to them. On top of that, their father would not be there to help. The whole thing would be dumped on their mother to solve and sort out. Alice and I would not only be separated, we would be unable to communicate for days, even weeks. Alice would be on her own to make every decision. This was every military family's worst nightmare. Following four years at a very busy and pressure filled job as Operations Officer of the 318th FIS, that included a six month tour of duty in Korea, and then two years as a CO of the struggling, debilitated 95th FIS at Dover, I would now be faced with leaving my family for over a year to be 13,000 miles away from the ones I loved and cared for most dearly. This was a most difficult situation to face and to adjust to. Mostly, I felt badly for my wife Alice Marie, who again would have to take over and be both Mom and Dad to our brood which included three teenagers. While I was confident in her, I was also concerned for the pressure this was putting on her. Military life brought with it many sacrifices and none were greater than repeated family separations. This one would be the toughest yet for ours. Dad drove down from New York to be with us the last few days and say goodbye. He was a huge help. Our children loved him dearly. On the day Alice and the kids left, we drove them to the train station at Paoli. It was an emotional goodbye. It was difficult to find a bright side to speak about. Young Jim was especially down and in tears. So were my daughters. When the train left the station headed west, I could hardly control my inner emotions. I truly felt that I had left my family down in some terrible cruel way to fend for themselves. I drove back to Dover nearly in tears with Dad. The next day he dropped me off at the Philadelphia Airport from where I left to Charleston, S.C. and from there to Rhein-Main, Germany on my first leg to Saudi Arabia. It would be the beginning of the strangest assignment of my career. An assignment that would call me upon me to react in ways I would never have dreamed of, in situations I could never have imagined—except when people who should know better, perform in ways outside of what been trained and are expected to.

From Germany, I was flown to Torrejon, Madrid, Spain, where I spent the next few days sightseeing and awaiting MAC transport to Dhahran. In Madrid, I hooked up with some USMTM people who provided some advance insights on the problems facing me in Saudi. Finally, I left Madrid on a C-141 and flew into Tehran, Iran and then on into Dhahran. This was a part of the world that was completely new to me. Most of what I flew over was desert and arid, mostly uninhabited and desolate. It was hard to believe that these countries contained wealth that the rest of the world envied and craved. Oil was a major reason that I and many others like me were coming to these area to serve and assist these nations in their desire to improve their military posture so that they could ward off potential enemies and defend their assets. The countries had what we needed, oil, and we had what they needed, technology to procure the oil and military arms and know-how to defend their sovereignty. It was a simple trade off.

My Tour in Saudi Begins

My arrival in Dhahran Air Base was in midafternoon on a 120 degree day. By the time I retrieved by luggage from the rear of the C-141, I was soaking wet. Dhahran, being near the Persian Gulf (or as the Saudis liked to call it, the Arabian Gulf) always had high humidity to go along with its high temps.) Dhahran was a RSAF fighter base on which it operated an F-86E squadron, two F-5 squadrons and a squadron of RAF Tornado fighter aircraft. Also on the base was the King Faisal Academy which trained the RSAF's enlisted people. The base also had a basic pilot training program operated by a British civilian contractor. All this, plus the commercial aircraft terminal with its incoming and outgoing flights by SAUDIA Airlines and other international carriers, made Dhahran Airport a terribly busy one. As I would have small detachment here under my Supervision, plus USMTM Headquarters, I would be spending a lot of my tour at this place.

After checking into the VOQ and then into USMTM Headquarters with my official orders, I then reported into to meet my new boss the infamous Gen. O.P. Smith. I have to admit I was anxious to meet this man who had created such a stir all the way up to the office of the Secretary of Defense. His reputation for being difficult to work for, utilizing questionable personal strategies, and for interfering into areas where his expertise was clearly lacking, had been causing serious morale problems as well as delaying progress on important FMS projects. Smith was also suspected of pushing his own career interests ahead of the interests of USMTM by cozying up to the Saudis royals on many issues that were outside his area of responsibility.

Smith greeted me warmly. He was an Army officer of southern stock. He was of a slim frame, well-built and in very good physical shape. His uniform displayed a raft of combat ribbons that reflected an excellent Vietnam service record. Smith told me how pleased he was to have me on board and how much he needed me to take charge of the Air Force detachments and shape them up. He repeated this several times and emphasized how much work there was to be done. He needed help badly and he had not been receiving any from either the Air Force or the Army types in the country. He went on at length how much of the load he had to personally carry and how little he could trust anyone. He openly admitted that he had to review every communication that came in and went out of the country to ensure that his people were not going behind his back and disobeying his wishes. He did this by getting out of bed at five in the morning and going to the communications shack and looking at all of the dispatches. He admitted to pocketing many of them. Smith's paranoia was evident. At the same time, he continued to tell me that he would be depending on me to make things better and to support him. With a heavy conscience, I assured him I would do my best. Smith also told me that he wanted me to take over command of the living quarters at Riyadh. This job had always been filled by the ranking Army officer, but this time I would be the senior officer as the new Army commander was still a Lt. Col. The majority of the Riyadh detachments composed of both Air Force and Army personnel, mostly Army by about three to one. I knew immediately that this would pose a problem and so did Smith. But he didn't care. I believe he wanted to inflict this indignity on the Army guys at Riyadh and he didn't care if they liked it or not. Later in my tour, this problem erupted and the Army guys again came out badly. Smith invited me to his home for dinner that evening which I accepted naturally. He had a nice comfortable home on base and after a few cocktails, I enjoyed a fine home cooked meal. Mrs. Smith was a genteel southern lady and I liked her very much. I also got to meet some of the other USMTM officers stationed at Dhahran and who lived on base. All in all, I felt I had made a good start with the General.

I was now anxious to travel to Riyadh, the capital of Saudi Arabia, where I would be permanently stationed. Riyadh was about 250 miles to the west of Dhahran. Living there, I would not have daily contact with General Smith. In fact, we could only talk to each other via, short wave radio a very awkward device where a communications airman sat next to me and threw a switch when I said "Over." That meant Gen. Smith would speak next. The airman, of course, was privy to what the General and I said to each other. During the course of my year there, he overheard some really thrilling exchanges. My estimate is that the communications airman probably heard General Smith fire me at least six times. He also heard these firings executed verbally with some pretty salty language. There were two ways to travel around Saudi. One, I could hop on an RSAF C-130. These flew daily on a circuit around the Kingdom. In fact, any Saudi citizen was permitted to travel on these flights. It was common to see Arab men, women and children, with their packings, plus goats, etc. filing up the rear ramp of these cargo aircraft and going off from coast to coast. The aroma of these aircraft cargo spaces was staggering. The second and preferred way was to fly Saudia, the Kingdom's national airline. This was what I usually did. When I did, I normally flew in uniform and I was ordered to fly in first class. When I arrived in Riyadh, I was met by Lt. Col. Corzine, who was the acting chief until my arrival. Our compound was only about a mile from the airport and I found my apartment to be quite comfortable. It had two rooms plus a bath. It had air conditioning, which was a must of course. I also had a car, a 1960's Chevy that did have AC. The compound resembled a modern motel of two story apartments. It also housed a dining hall, a bar, a movie theater where we had nightly films, some meeting rooms, a MARS station, a communications station and another secret area which housed the CIA personnel. It had an area of three or four acres and was located conveniently to the airport and downtown Riyadh. Maintenance of the compound was contracted out but had to be closely monitored. Water was trucked in, as it was to every other building in the city. We received our food flown in by Air Force aircraft and it was good old American repast. We had an Army mess sergeant and our dining hall was operated in an excellent fashion. We were served at table by mostly Sudanese waiters and the prices were absurdly low. In summary, our living conditions were better than satisfactory. We were stationed in a country that was as diverse from our own as it could be in culture and climate. We would all have to put in our year, away from family, and doing a job which certainly was far from what would we prefer to be doing. But this was the military, and you do what your orders tell you to do and you do it the best you can.

I Think of Home

By the time I arrived in Riyadh, it had been almost two weeks since I last saw my wife and children. I thought of them constantly and worried how their rail trip across the country worked out and how they were doing at the lake cabin. Were they as unhappy as I was about this turn of events? How was Alice Marie doing trying to keep everything going with five unhappy kids, a dog and God knows what they found at the cabin? Thank goodness, Sam and Dee Grashio were there to help out. And what about the home I rented for them in Spokane which they had not even seen yet. I missed them so much and I worried about them and here I was 13,000 miles away and no way to talk to them. I did the next best thing. I sat down and wrote the first of many letters I would write from Arabia to Idaho. I had a job to do here and I found myself in a very strange assignment with many complex sidelines that had to be dealt with. It would be a long year I felt, and there was no use whining about it. In previous tough situations I found myself in, I focused in on what I had to do and then did my best. Despite my worries about my family, I had the greatest confidence in Alice Marie. She was a fortress of common sense and always made the right decisions. She was a tireless homemaker and placed the welfare of her children first and foremost. If anyone would come through this year in good style, it would be my wife. And that made me feel as good as I could under the circumstances. So, I decided to get on with it and not spend too much of my time thinking about home. I decided to keep busy as I could at my job, and that would make the days fly by faster.

I had written some postcards and a few letters as I had traveled from Dover to Saudi which I hoped the family had picked up at the Sandpoint post office. When I finally received Alice's first letter from

Sandpoint, I was relieved to learn that my family had enjoyed their rail trip across the country and arrived safe and sound in Idaho in the wee hours of the morning. There they were met by Sam Grashio and brought to the old cabin on the lake. (Thank God for Sam and Dee Grashio.) Our station wagon was there, and the cabin even had some food in it to take care of the family until Alice could get into town and stock up. Naturally, they were tired and a little down from the long trip. Still, they were at home at the lake and pretty soon they were enjoying swimming and diving off the dock into the cool but pristine blue green waters of Lake Pend Oreille. This was a letter I wanted badly to read and at least I now knew that despite the great distance between us, my family was safe and at home in their second home at Jack's Cabin and would be enjoying the wondrous beauty of northern Idaho for the next few months. I was thankful and relieved. Now I could get down and work at the job I was sent to do in this crazy place.

USMTM

The United States Military Training Mission (USMTM), in Saudi was headed up by General Smith who was stationed on Dhahran Air Base, which was on the east coast of the Kingdom. I and my staff were head-quartered in Riyadh, the country's capital that was about 250 miles west of Dhahran. Riyadh was a busy growing city that housed King Faisel's palace as well as the palatial homes of his ministers, most of whom were his sons or close relatives. Also in Riyadh were the nation's important ministries including Defense, the Treasury, Education, Petroleum, the National Guard, etc. To make it more difficult for the country team to coordinate matters, the American Embassy was located in Jeddah. This most cosmo-politan city in the Kingdom was located on the Red Sea, the west coast of Saudi Arabia. The Saudis wanted this separation of American units for their own suspicious reasons, but the effects made it very hard for us to communicate with each other and ensure we were all on the same page. As there was neither phone service across the country nor mail service that was secure and efficient, communications were slow and unreliable. The USAF portion of USMTM that I supervised consisted of four detachments. At Dhahran I had a small detachment consisting of an F-5 pilot who flew with the Saudi squadron and a few maintenance personnel. The training of the F-5 unit was being run under an FMS contract to Northrop Aircraft.

This contract not only included flight training to aircrews and ground training of maintenance personnel; it also included English language training to the Saudi airmen so that they could read the tech orders. This contract was under the scrutiny of an Air Force team from Wright-Patterson AFB, whose job it was to ensure that these contracts, which were government to government, were being fulfilled to the letter.

At Taif, a small air base in the mountainous southwest of Saudi, I had a detachment of aircrews and maintenance people who worked with two RSAF Helicopter units. It was always pleasurable to visit Taif, as the weather there was usually cooler, and one could enjoy the outdoors during the daytime. The King and many of the wealthy Saudis had vacation homes in this area where they went in the summer to escape the blistering heat. I also enjoyed flying the helicopters, with instructor pilot, of course. In Jeddah, I had a detachment with the Saudi C-130 units. These aircrews and maintenance people were active flying the C-130s all over the Kingdom and the Middle East. These units had a good safety record and I had very few problems with this detachment. During my year in Saudi, I enjoyed my trips to Jeddah. Our compound in Jeddah was right on the Red Sea. There were many good places to eat in Jeddah as it was not as restricted as the other cities in the Kingdom. There were many more foreigners in Jeddah, and it was the major port of entry for business people, tourists and the Hajjis, those Muslims who came each year to perform their pilgrimage to Mecca. During my year, I tried to visit each of my detachments on a regular basis, at least once every 60 days. But, because I most often was called to Dhahran to see General Smith, I was at that Detachment the most frequently. At Riyadh, I had the most people under my control, approximately 12. All but one were officers and they represented various areas of expertise. I had air defense ground control officers, aircraft maintenance, helicopter and armament people, aircrews both fighter and multi-engine and an administrative sergeant. I also had assigned to me a civilian interpreter and political advisor. His name was Darwish I. Darwish. He was a Palestinian and he literally took care of my interface with the Saudis that required matters of protocol. He also took care of my travel arrangements. He was at my beck and call 24 hours a day.

With regard to Riyadh, General Smith gave me a problem immediately after my arrival when he told me that I would be the Commander of the Riyadh compound, a job that had always been held by the Army

Colonel. I was not about to say no to the General, but I hoped that this would not result in any problem down the road. My hopes were dashed when the new Army Colonel arrived. First, he was not a Colonel, he was a Lt. Col. He was on the Colonel's list but had not yet pinned on his eagles. Thus, he would not be the senior officer on the base. But shortly after he arrived, he came to see me. His name was Sam Vincent. I greeted him warmly in my room and offered him a drink. However, after some small talk, he came to the point. He questioned me about my being the Compound Commander, a job traditionally held by the Army ranking officer. I answered him candidly. I did not ask for the job. If asked I would probably decline it as my other duties gave me more than enough to keep me plenty busy. The Air Force side of the house, with its large sale of F-5s and some other contracts in the works were going to keep me on the road. However, General Smith made a point of it and made it clear that he wanted me to command the compound, and I was not about to debate the issue. I told Sam, as far as I am concerned you are the second in command here, and when I am not here, you are in charge, and I will back you. Just keep me informed of what you are doing. Vincent left. I could see he was not happy about what I had proposed. I hoped that this would be the end of it. Alas, it wouldn't be.

A Call from Dhahran

Shortly after, I received a call from Dhahran. There had been an F-86 accident resulting in the death of a Saudi pilot. I was asked to come over and help in the investigation. I had flown the F-86 in Korea in 1953, 19 years ago. The Saudi pilot crashed shortly after takeoff. He was number four in a four-ship flight. The suspicion was engine failure. By the time I arrived, the wreckage was being removed from the field just north of the base. Not only that, but the pilot's corpse also had been taken out into the desert and buried without autopsy or even a close visual investigation. (This is Arab custom.) When I was brought up to date by the Board, they had pretty much decided it was engine failure. The pilot had made an Emergency call, but they were unable to explain why he had failed to eject. "Did he blow the canopy?" I asked. They did not know. "Did you find the canopy?" Again, they were unsure. I suggested that we find the canopy and, if we can, recover the pilot's helmet. These could tell us if he attempted to eject. The

Board, composed of Saudi officers and a few Americans seemed hesitate. I said well, let's get some airmen and go out and search now. I got up, put on my boots and my flying suit. A few of the Yanks followed and finally the Saudis got up. When we drove out to the crash site a large group of Saudi airmen showed up. The Saudi officer briefed them of what we were looking for. We then spread out and walked, 10 yards apart and searched the crash site. Sure enough, within a half an hour, we found the canopy and the pilot's helmet. I remembered in the F-86, when you ejected, you first blew the canopy. When you did this, you had to duck your head, because when the canopy left the rails, it dipped in and, unless the pilot ducked, it could possibly strike him on the head and prevent him from ejecting. The pilot's helmet confirmed that that is what happened. It had a deep gouge in it. This unfortunate aircrew would have made it out alive except for the canopy knocking him unconscious. The canopy left and he stayed with the aircraft. Thanks to our being able to find the helmet we found the answer to this fatal accident. The Saudis were impressed with my contribution. However, this only caused me more headaches. A month later, I was asked to participate in another fatal aircraft accident investigation that was one of the worst experiences I can remember because the fate of the pilot was at stake.

General Hashim

As the principal advisor to the Commanding General of the RSAF, I actually worked for General Hashim. He received me in his office shortly after I reported into my headquarters in the Defense Building near downtown Riyadh. He was already a legendary figure and actually he was Egyptian not a Saudi. He was tall and handsome and very active. He has claimed his fame years earlier when the Saudis and the Yemenis got into a little dispute over some border issue. Hashim flew some combat sorties in the F-86 and did some damage to the enemy before the affair was settled. The General was aggressive and stated up front his wishes to upgrade his air force with the most modern equipment he could get, including the F-15. This was even before his squadrons had not even mastered the F-5 and the new base at Taif where what would be the equivalent of a Saudi Air Force Academy would be built was still far from completed. But this was Hashim and his aggressive attitude spilled over into personal life. He took for his second wife a Lebanese woman whose net worth was estimated at $50 million dollars. Shortly after I was in Saudi, Hashim flew his RSAF Lear Jet to Jeddah on some personal business. I hitched a ride to visit my Jeddah detachment. Both the first and second Mrs. Hashim, and a few of his children also came along. It was an interesting experience for me. The newest Mrs. Hashim, who was younger and much better looking than the first, sat up in the better seating section where I also sat. The older Mrs. Hashim and the children were in the rear. The newer Mrs. and I conversed at length. She was a modern Arab woman, well educated in western schools, and spoke English extremely well. However, when we were served luncheon by the steward, he took my order first and served me first. I noticed it and so did Mrs. Hashim. She did not like it, but she knew that in Saudi Arabia that is the way things were done.

Hashim made piles of money from the graft he received from the award of sales to foreign companies for contacts and for weapons, etc. This was a way of life in Saudi. It was called Baksheesh, under the table payments. Hashim built a home in Riyadh, not far from our compound that was truly magnificent. It was contained within a surrounding wall a block square. When you entered through the large gate doors, you found yourself in a courtyard lined with palm trees with several fountains on which colored lighting was aimed. On a terraced landing above the courtyard there was a beautiful swimming pool and two bathhouses, one for the ladies and one for the men. The home itself was a large structure with many rooms all heavily carpeted with the finest rugs from all parts of the world. Each room had sculptures and paintings of the most expensive taste. The house also contained a billiard room, a large barroom, a theatre, several dining facilities, and Lord knows what else there was that I did not get to see. The night I was there, Hashim had a large dinner party at which there many dignitaries from many companies, including diplomats, military and contractors. After a long cocktail hour at which only the most expensive liquors were served, (this in Saudi where all alcohol was banned), we adjourned to the outdoor courtyard for an Arabian dinner of lamb. Here you reclined and ate with your hands (your right hand mainly). It was an exquisite evening in an exquisite setting under a clear, star filled desert sky and a crescent moon. I was truly impressed. I had never seen such a place as this, not even in the movies.

But Hashim's dream house did not last long. Prince Sultan, the Saudi Minister of Defense, son of the King, took for himself a 16-year-old bride. He decided that he would like to give his bride a worthy wedding gift, such as a new home. He too was impressed with Hashim's villa. The story goes that he asked Hashim how much he had paid for his home. Hashim, who was only a Lt. Gen., had to be careful in his answer. Clearly, the actual cost of his home far exceeded what his salary would have permitted him to build. Hashim told the Prince that he had built it for one million dollars. The Prince told Hashim, "fine," he would buy it from him for that price and a check would arrive the next day. And so, the General moved out, and the Prince and his bride moved in. Shortly after, Hashim retired, and a new General took over as Commanding Officer of the RSAF. His replacement was General Asaad Al-Zuhair, who was totally a different person from the man he replaced. A small man, a devout Muslim, and a complete gentleman, Al-Zuhair was a pleasure to work with. We had a good relationship for the rest of my tour.

Baksheesh

I next became involved in contract negotiations between the RSAF and the British on a huge multimillion dollar contract. British Aircraft Corporation (BAC) had bid on a contract to operate several of the Saudi airports across the Kingdom and I was asked to sit in as an advisor to the RSAF. These negotiation meetings took place at Dhahran, and they were interminable. They lasted for hours and the debates between the British and the Arabs over every detail in the contract drove me to distraction. After a while I realized there was more going on than I knew about. It, of course, was the money not the money called for in the contract but the money that would be paid under the table to the assorted Saudi higher ups. In Saudi and the Middle East, it is called Baksheesh. In the USA, we call it graft. It is a way of doing business in the Arab countries, and if a company wishes to do business there, it has to pay up. The US Congress has prohibited American companies from doing it, but many did anyway in order to compete. Sitting in on these seemingly endless hours of negotiations between the Saudi and the British gave me a front row seat at how this under the table operation took place. To these people, there was nothing unethical taking place; it was just business as usual. In a nation as oil rich as Saudi Arabia, the dollar amounts were enormous.

Arabia

Although I had been told much about Saudi before my arrival, and I had read scads about its history and culture, nothing quite prepared me for what I experienced when I got there. First, the climate was hot, very hot. Arriving in June, the temps daily exceeded 100-degrees, often hitting as high as 130. On the coastal cities of Dhahran and Jeddah, the humidity caused you to perspire to the point that I had to change uniforms at least once and sometimes twice daily. The heat resulted in a shortened workday. Our office hours were 0800 to 1300. At 1300, the Saudis went home, changed into their cool robes, and relaxed through the hot part of the day. The souks (shops) closed and did not open until sunset when they again opened and did the majority of their business. Then the Saudis enjoyed coming out and enjoying mingling, eating, and snorting on their water pipes. The Saudis loved the desert. Families that could do so, would drive out to the desert to picnic and frolic similar to the way Americans would head for the beach. The geography of the country was amazing, mostly desert similar to the southwest USA. But across the south there was also a vast breath-taking stretch of land consisting of mountainous and dunes, the Rub al Khal endless changing shapes. This area was named the Empty Quarter. It was bleak, uninhabited and yet, hypnotically beautiful. I was always amazed, as I flew across the deserts of Saudi Arabia to look down and see caravans of nomads winding their way across its empty plains. Where were they headed? Why were they out there in such a place so far from any civilization? How could they survive the dust storms? How did they navigate to wherever they were going? In the southwest, the terrain was mountainous and contained much forest filled countryside. The west coast along the Red Sea was loved and the sunsets there were some of the most glorious I had ever witnessed. The Red Sea afforded some of the best snorkeling and scuba diving experiences in the world.

But the Kingdom was dominated by the wealthy royal family and its multitude of princes and their families, all descendants of the promiscuous King Saud. They were shrewd in sharing their wealth to ensure that the tribes were taken care of, and that no Saudi citizen went without the basics. Most of the low paying jobs were performed by laborers imported from Africa. In addition to the regular army, the King had his National Guard whose mission it was to defend the royal family and the Saudi government. There was very little public opposition to the government, as that would be quickly quelched. Sharia Law, an Islamist government, was in effect, and all the laws were based on the Koran. Friday was their day of worship, so our work week was that Thursday was our Saturday (a day off) and Friday was our Sunday. In addition to Friday being their day of worship when the Arabs attended the mosque for prayer, it was also the day when civil punishments were administered. These were performed in the public square in downtown Riyadh. Decrees were read describing the crimes and the punishment. The prisoner was led out and the punishment administered. They had their right hand cut off with a sword. A rapist or murder was decapitated by sword also. Large crowds attended these executions although they were bloody affairs and carried out unemotionally. I attended once and I did not find the spectacle satisfying. While I admired how swiftly the Saudi system brought criminal to justice, tried them and then administered punishment as compared to our system which often takes years, still I wondered if the poor creatures had received due process.

During my year in Saudi, I met and dealt with many Saudis, mostly military, but also some civilians. I liked most of them and I found that most of them liked Americans. Their biggest complaint was America's support for Israel. This was a subject that they usually brought up. It was never resolved. I enjoyed visiting the souks and while I did not buy much, I enjoyed their system of bargaining on prices. I got used to the call to prayer from the minarets, and I was impressed with the dedication of the Saudis to their faith. At noon, work stopped in the Ministry of Defense, and the Saudi officers and airmen hastened to the prayer room that faced westward toward the city of Mecca. There they knelt on their prayer rugs and prayed to Allah. Several times I was in conference with Gen. Zuhair when the noon prayer call came. He excused himself and I left his office as he unfurled his prayer rug and went down on his knees. I would never see an American three-star general do that. During Ramadan, the month of fasting, we were especially courteous to our Saudi compatriots. As soon as the sun broke the horizon, signaled by the firing of a cannon, they were bound to fast until sunset. This included fasting from food, tea, coffee and sex. This left many of them slightly irritable to deal with. I always refused offerings of tea or coffee in their presence during Ramadan. My Arabic was not great and limited to a few conversational phrases. But I used them as frequently as I could, and the Saudis appreciated my efforts. My assistant Darwish was my helper in that effort. He also wrote my holiday cards and letters for me in Arabic.

Foreign Friends in Saudi

Our compound in Riyadh was open to many of foreign friends in Riyadh who were non-Arab. They were permitted to dine in our mess hall at supper and to attend our movies. Most of these people were associated with the various companies that were doing work for the Saudi government. They included Germans, Swedes, British, South Africans, et al. There were also American contractors living in Riyadh from Lockheed, Westinghouse, and other firms. Many had their wives and families with them. Being able to eat some good old American food and see some American films helped their morale a little and we enjoyed their company. The mess made a few dollars too. The water system that supplied the city of Riyadh had been operated for years by two German brothers by the name of Schubert. Unbelievably, beneath the desert floor south of the city was an underground river. The Schubert's operated large turbines that brought this water out and then shipped it to Riyadh where it was trucked to the various compounds, reservoirs, holding tanks, and water towers to be used by the pop ulation. It was something of a tradition that the Schubert's invited the Military Commander of the Riyadh compound to have the keys of their villa. I was given those keys. It was about a 30-minute drive south of town across the desert, but the Schubert's villa was an oasis. It contained a swimming pool, a tennis court, a sauna, lots of shady areas to rest and relax, and always good company. Whenever I was in Riyadh on the weekend (Thursday/Friday), I headed to the Schubert's. There I played tennis, swam, and enjoyed the company of my newfound German and British friends. There was always some sadiki (homemade alcoholic brew) on hand. I also brought some beer with me which the German's loved. I also brought food from the mess including steaks, sausages, and potatoes. These weekends at the Schubert's gave me my only escape from General Smith who made it a point to call me daily and make my life as miserable as he could.

I Begin to Learn the Job

Since our workdays were short, ending at 1300, that left lots of time to kill. Keeping up morale was difficult. Many of my people played tennis; some were off to the various corporate compounds that had swimming pools where they could keep cool. Others had different hobbies with which they tried to while away the off hours, watching the calendar that recorded the waning days of their year-long tour. Supervising my airmen and the compound was not time consuming and I left much of the day-to-day stuff up to Col. Vincent's discretion. My major concerns were that none of our men violate the laws against approaching Arab women. Secondly, they would not drink off the compound. General Smith allowed the USMTM personnel to receive a ration of liquor flown in from Bahrain on his C-54 and distributed to each detachment. The Arabs did not inspect this aircraft because it had diplomatic status. Therefore, each of us had our share of alcoholic beverages for personal use and with which to entertain our non-Arab guests and stock the compound bar for unit parties. Our use of liquor in Saudi was a delicate problem that had to be carefully handled. While the Saudis undoubtedly knew we had alcohol, they did not bother us about it as long as we controlled it and did not permit its use by Arab citizens. This was one area that I continually stressed repeatedly in my Commander's Call. Under no circumstances was anyone permitted to leave the compound under the influence of alcohol or provide any liquor to an Arab civilian. The gravest consequences would result. Fortunately, during my tour, we did not have problems in that area.

And so, my days dragged on. I was kept busy at my advisory duties and in my visits to the detachments. I began to better know and trust the men working under me and recognize their capabilities and knowledge. And they started to understand what I expected of them. The Foreign Military Sales (FMS) business was complicated. It involved the DOD, the defense contractors. the State Department, the Air Material Command (who oversaw the contracts), the Saudi Ministry of Defense, the RSAF, and USMTM, who provided advice and assistance to the RSAF. Also involved were the various Saudi princes, member of the Royal Family who were imbedded in the RSAF squadrons and actually wielded the decision authority, although they might only carry the grade of lieutenant or captain. So, it took some time for me to understand how to get things done and what were the lines of authority and decision making. And decision making in Saudi was very slow. It seemed every decision, even the very smallest, took forever to be finalized. I discovered that General Al Zuhair although he had three stars, did not make many decisions on his own. Almost everything was decided by the Minister of Defense, Prince Sultan or his brother, Prince Turki. In the F-5 Squadron at Dhahran, Sultan's son, Bandar, was a Lieutenant fighter pilot, who actually controlled the organization. (He later became the Saudi Ambassador to the US.) And so, you were never sure where the final decision was coming from, but it was probably validated by the member of the Royal Family.

In other matters, coordination usually had to be affected with the American embassy in Jeddah and, of course, General Smith made it his business to see everything that came into and left the country. Smith would get up early in the morning and go to the communications shack at Dhahran. There he would confiscate all of the TWXs (telegrams) that came in overnight and read them no matter to whom they were addressed. Some of them he confiscated. Nothing got by him. He also saw everything that was being sent out of country. That is what Secretary Robinson instructed me to communicate with him by mail. Smith's paranoia was rampant. He would tell me in private how he was aware other people were out to get him. He did not name these other people. He showed me a safe that stood behind his desk in which he kept TWXs that he had intercepted and told me how he was going to retaliate against these enemies. He would rant about these faceless enemies of his for hours. He would become red in the face and his language would be sprinkled with expletives of the finest order. I did not argue, nor did I disagree. In fact, I displayed empathy. I was not about to make my long tour in Saudi Arabia any more miserable than it was going to be. Usually after one of these long rants, Smith would invite me to dinner. I happily accepted.

CHAPTER THIRTY-ONE: ARABIAN NIGHTS ... AND DAYS

is-salaam 'alaykum... Peace be upon you.
wa-'alaykum is-salaam... And upon you be peace.

-Universal Arab greetings

What surprised me the most was how busy I became fairly soon after I got into the job. Yes, there were lots of hours when I could not get anything done because the Saudis just did not work in the afternoons. And when they did work, their pace of decision making was awfully slow. Still, each month I visited the detachments which were spread out from coast to coast and took about eight to ten days to complete. Meetings with the various contractors took much of my time and I tried to avoid them. I really could not help them much because I left it up to the RSAF to come to me and request information or advice about aircraft or weapons systems they were interested in. I did not see myself as a salesman for American war products. When I was asked, I gave my best advice without regard to the American contractors who were constantly coming to see me. Later in my tour, I became embroiled in a bitter dispute between the RSAF and the Lockheed Corporation that ended up in the Pentagon and the State Department. It almost cost me my job, but I figured that if they fired me from Saudi Arabia, they could not send me anywhere that was worse. A good many of my travels were back and forth to USMTM Headquarters at Dhahran at the beckoning of General Smith. He would frequently go into a tirade over something, get out the hot wave radio, fire me, and then order me to "get my Air Force a-- over to his office ASAP." I would then get my man Darwish to get me a reservation on the next SAUDIA flight to Dhahran, pack my overnight bag and go. This occurred so many times during my tour that it became a standing joke in the Riyadh compound. The radio man who sat next to me working the switches on the short-wave radio would turn his head and try not to display his embarrassment (or laugh) as he overheard the General rip into the Colonel in language usually reserved for sailors of the lower grades. (My estimate is that Smith fired me at least ten times during my tour.) However, when I arrived at his office following this urgent, threatening, summons, his greetings were opposite to what was expected. Smith would say something like, "Gene, damn I'm glad you're here. I got a mess on my hands, and you are the only one I can rely on to fix it," or something to that effect. Usually, Smith would just be upset over some little thing related to his paranoia or confused over some Air Force matter he did not understand. He just needed someone to vent his anger to and then everything would be all right again. He was a complex man, very bright, but insecure. The question of firing me would not be mentioned and many times I would end up at his house for drinks and dinner. I learned that I could best handle General Smith by never showing that I feared him or was intimidated by his threats. I answered his questions candidly and when I gave him my advice, I spoke with authority and confidence. When he gave me an order, I carried it out, without question. Early in my tour I began to feel that while Smith was a wild man and incompetent to be running USMTM, he was not going to make my life any more miserable than it was already being 13,000 miles away from Alice Marie and my five children.

An Aircraft Accident That Becomes a Murder

There always seemed to happen events that took me away from my primary duties of supervising my detachments and providing sound military advice to the RSAF Commander and his staff. About a month into my tour, I was called upon to sit on a board of officers in what turned out to be a very sad and tragic case. A few weeks before my arrival in Saudi, a RSAF Lt. Col., who was the CO of a Tornado fighter squadron at Khamis Mushayt Air Base in the southwest part of the Kingdom, was involved in an incident that turned out to be a national tragedy. On what should have been a routine solo test flight the aircraft returned with a slight dent in the right wing that also had some blood stains mixed in. The pilot stated that he had experienced a bird strike on his return to base. Nothing further was thought of this incident until about 15 days later, a man, apparently a nomad, reported to the local police that his wife was killed a few weeks earlier when a jet aircraft buzzed their desert encampment several times and on the last pass struck his wife in the head while she was scurrying to safety carrying their infant child. The husband buried his wife's body in the desert. When confronted with this accusation, the Colonel vehemently denied it. But after an investigation was conducted, which included a visit to the scene of the accident it became clear that the husband's version was the truth. The pilot was tried and convicted of murder. In Saudi Arabia, the husband could demand an eye for an eye; that is, he could demand that the pilot be given the death sentence. And the husband did just that. However, the Colonel appealed for his life to the King on the basis that he was doing his duty in practicing low level combat tactics, something that all RSAF fighter aircrews must be skilled in and that his hitting the woman was accidental. And so, a board was formed composed of "experienced" fighter pilots, including me, to determine whether the Colonel had a point and therefore his life should be spared. This was just an awful situation to be in and I wanted no part of deciding whether this man should live or die. I was sent to Saudi Arabia to assist the RSAF in their mission of defending the Kingdom and that I intended to do to the best of my ability. But playing judge and jury on a capital case was something I had not bargained for.

For over a month, each night, five nights a week, our Board met in Riyadh, took testimony and reviewed the trial testimony. It was an extremely slow process. Witnesses gave their testimony in Arabic which had to be translated to English for the Board. Questions by the Board members also had to be translated for the witnesses and so forth. The translation requirements, including the recording of the proceedings into written form, slowed progress and delayed completion. At one point, the Board was flown to Khamis Mushayt and then proceeded by helicopter to the site of the accident miles out in the desert. How the chopper pilot was able to find the site, I'll never know. There we saw the place where the wife, a small woman, in her terror, picked up her child and ran to a small rise where she thought she could be safe from the aircraft that was coming down on her. But she did not get there before the wing of the plane struck her in the forehead, slicing off the top of her skull. A large rock was placed where a few parts of her head still remained on the hot desert floor. I remember when one of our members lifted the rock, he was greeted by a large desert scorpion guarding bits of the dead woman's brains. This scene had a very serious effect on all of us.

After listening to the pilot's defense of his actions, the Board was not impressed that he had any plausible reason to be harassing an innocent nomad family far out in the desert, putting his aircraft and himself in danger flying at such low altitudes, and then aiming his ship at a woman carrying a child, killing her for no good reason. Then, of course, he lied about it, something no officer, much less a commanding officer, should ever do. Our Board could find no reason to exonerate him and so we reported. It was a most depressing experience, and I was glad to have it over with. The report was forwarded, and we were never given any feedback officially. However, the rumor was that the King stopped the execution of the Colonel and sent him instead to life imprisonment. Imprisonment in a Saudi jail is no piece of cake I am told.

Ramadan

Fasting is the fourth pillar of the Islam faith and takes place in the ninth month of the Muslim year. In 1972, Ramadan came in October. My man Darwish prepared me for it by warning me about the way I should treat my Saudi colleagues and the phrases I should use. Saudis must fast from food and drink and sex from the moment before dawn when a white thread can be distinguished from a black until the setting of the sun. To perceive this white thread, a few soldiers were stationed on top of a small hill not far from our compound in Riyadh. Beside them was a small cannon. When they first saw the white thread of dawn, they fired off the cannon. In October, this was about 0600 and it brought everyone a foot out of the bed. The Saudis could become a little irritable during Ramadan, not all of them, and we were briefed to decline offers of tea or coffee during our meetings as they could not join us. Everyone was happy to see the end of the fasting which was the great feast, Id al-Fitr. On this day we exchanged cards, much like Christmas cards, and I visited several of my RSAF colleagues, including General Zuhair at his residence. I enjoyed these visits. I met the children of my RSAF friends (not the wives or the grown daughters). I saw their homes and we had good conversations that excluded business and politics (Israel). I found that they were normal people just like us in America with the same kind of concerns. Although I was homesick and my thoughts were always wandering back to how Alice Marie and my children were doing, I did enjoy the work I was doing in Saudi and how much I was learning about this important part of the world. I realized that the Middle East is an area that most Americans do not have sufficient knowledge about, and that included me prior to this assignment. Yet the Middle East is a part of the world that because of the West's need for energy coupled with the growth and volatility of Islam, will have a major impact in the future. It's an area that we should pay more attention to and soon.

Christmas Break

Although time seemed to pass much too slowly for me in Saudi, all of sudden I realized that I was approaching my six-month break I would be heading home for a leave in December that would include a stop at the Pentagon to brief Secretary Robinson on how things were going with General Smith. I really began to count the days and I made my travel plans. Thanks to my British friends, Len and Marie Taylor, I arranged to meet them in London where I would have a stopover on my way to Washington, D.C. Len was a former RAF pilot who was now a flight instructor for the budding RSAF pilots. We enjoyed tennis at the Shubert's watering hole. We became great buddies, and they often joined me at the compound for dinner and drinks and the cinema. As it turned out we did meet at Heathrow and after a whirlwind tour of Piccadilly and Soho and a few other London highlights they set me up for the night in a wonderful British country inn. They picked me up the next morning and after a great English breakfast, they delivered me back to Heathrow for my flight to Washington, D.C.

My briefing at the Pentagon went well. I told Robinson that I was getting along with Smith although he had fired me about five times already but had rehired me each time. Smith was definitely a loose cannon and is capable of doing damage, but so far, he had not stepped over the line. He was listening to me on Air Force matters, but he was too cozy with the Saudis. He was piloting the Air Force C-54, something he probably should not be doing, but I could not stop him and no one in Saudi could either. Robinson seemed pleased with my report and told me I was doing a good job handling Smith. He also asked me about my next assignment. I told him that I would like to return to a flying unit if possible. He asked me about coming to the Pentagon to work for him. I was flattered but noncommittal on that. Secretly, I had negative feelings about coming to the Pentagon. After six months into my Saudi Arabian tour, I began to give a lot of thought to my future career options. I had just completed some challenging assignments: four years as operations officer at the 318th FIS that included a six month TDY of South Korea plus hundreds of days and night on alert and TDY's to Alaska, Walla Walla or somewhere else, 21 months as Commander of the 95th FIS, a most difficult assignment dedicated to bringing back to life a unit that was in most dire straits, and a year totally apart from my family in Saudi, leaving Alice Marie to cope with handling all the problems of three teenagers in new schools, in a new house and taking care of every last detail, 24 hours, seven days a week. I began to conclude it was time I spent more time at home with my wife and children before it was too late. I wanted my next assignment to provide a better home life in which I would spend more time with my children whom I felt I had neglected for purposes of my career. I was a full Colonel now and I had worked hard to earn that grade. But what good was that if I lost being close to my children when they needed me.

And then, fate seemed to have stepped in. While still in the Pentagon after seeing Secretary Robinson, I ran into an old friend, General Donavon Smith. I knew Smith from my days at ADC on the TACEVAL Team. He and I struck up a close relationship as I had helped him in some small matters dealing with his command. Smith asked me about my next assignment, and I told him about my thoughts on the matter. Smith was now the Commander of Alaskan Air Command. He asked me how I would like to come to Alaska and work for him in his operations section. My heart jumped at that suggestion, and I told him immediately I was interested. I told him I wanted to talk it over with my wife first and then I would let him know as soon as possible. I always like my TDY's in Alaska and I just felt that the kids would love it too. I was not so sure about Alice Marie. Getting a job in ops would give me a chance at flying fighters again and that was a plus. That night, I headed out of Washington, D.C. for Spokane and home. It had been six months since I had last seen my wife and five children. I had stomach butterflies. I couldn't wait to see my family again. I hope they hadn't forgotten me.

CHAPTER THIRTY-TWO: TOGETHER AGAIN

Absence makes the heart grow fonder.

-Davidson

Due to weather problems, I had to spend the night in Seattle and did not arrive in Spokane until the following morning. In Seattle I received what was not good news. My mother-in-law was in the hospital in New York in critical condition and Alice had flown that morning out to be with her. When I arrived the next morning to a snowy Spokane, Sam Grashio met me and drove me to the house that I had rented for the family. The kids were all at school and Alice called with the news that her mother had passed away. This was a great sadness to our family as my mother-in-law was a fine woman of the old school, totally dedicated to her family and the Catholic faith. Her husband was initially a good man, but early on succumbed to the temptations of the drink. His alcoholic failings prevented him from meeting his responsibilities of fatherhood and bread winning. Early on he lost the respect of his wife and children. In later life, his wife left him. Only after he gave up drinking through AA was, he able to regain respectability in his later years living with his son Jack. But it was too late to resume his marriage. My children loved their grandmother who visited us many times in Colorado, Canada and Dover. She reminded me in many ways of my own grandmother who had raised me from childhood to manhood.

Naturally, my wife was despondent at the loss of her mother and would be staying in New York until after the funeral. Coming at the time of my arrival back in Spokane after six months made things difficult, but there was nothing we could do to change that. I reassured her that I could manage things until she returned, although I wasn't positive about that as I didn't know where everything was and what the schedule was for the kids. I soon learned. Gene and Jim were attending Gonzaga Prep. They caught a bus to the school at a bus stop which was not far from the house. Michael walked to St. Thomas More School, which was just down the street, and Kathleen attended kindergarten at Evergreen School. To get there she walked out through the back yard fence gate and across the field to the school. However, Mary was a freshman at Marycliff High School which was on the south hill and a half hour drive from our house which was on the north side. Just getting all the kids up, dressed, fed and on their way was a complicated task that took lots of preplanning. I again realized what Alice had to deal with daily without help. But the kids were very helpful and for the most part things went smoothly. Getting dinner ready every night and lunches put together for their school days kept me busy. Not to mention, making beds, washing dishes, even some vacuuming carpets, etc. What a difference my schedule became from what I was doing in Saudi Arabia.

Still, I was at home again with my family. It was great. I could not help but notice that even in six months, the children had grown in size and, three being teenagers, changed in other ways too. I again realized how important it was for me to spend as much time with them as I could. Gene and Jim had a big adjustment to make from Dover where they had made many friends and were very comfortable. While the Prep was a good school, they had few friends there and they knew they would be leaving Spokane in a year. Fortunately, they were playing hockey which they truly enjoyed, and I enjoyed watching them play. Over the years they had become very good players in this tough sport. Michael, at St. Thomas More School, was doing fairly well, but was not the best student and needed parental support. Katie was a good student in kindergarten, and this forecasted her future as a fast learner throughout her school years. Mary had the toughest problems adjusting. She found the girls at Marycliff to be a bit snobbish and cliquish. This was hurtful and caused her quite a bit of unhappiness. Like the boys, she did not know any of the girls and they did not accept her quickly.

Alice and I talked regularly on the phone, and I was exceedingly happy when she arrived back home in Spokane after her mother's funeral. We looked forward to a happy family Christmas together before I had to return to the Middle East to finish my tour. We had a great family time putting up a tree and decorating it and the house with Christmas lights. I especially enjoyed Alice's home cooking again after months of mess hall food. The days flew by, and I hated to think of having to say goodbye again and return halfway around the world for another five months away from home. But that is the lot of military life. Those who serve understand that this is par for the course, and we must accept it without whining. We only ask that others (civilians) understand and render support wherever they can. During our years together in the service I never heard Alice Marie complain. She was the ideal military wife and I attribute to her whatever success I attained professionally. Despite the frustrations or exasperations I encountered on the job, I was always able to leave them at the door of my home. When I entered that door and was greeted by my lovely wife, I could put away the cares and headaches of the job and enjoy the peace and warmth of my home. What more could any man ask for, I ask you?

During my leave, I brought up my next assignment with Alice. I mentioned that Secretary Robinson wanted me to come to the Pentagon, but I was not enthusiastic about that. Alice agreed. Washington, D.C. did not appeal to her either. Then I asked her how she felt about Alaska. To my surprise, she did not give an immediate negative reply, as I expected she might. I told her that General Smith offered me a job and I told him I would give him an answer soon. Alice asked me what I wanted. I told her I was inclined to accept Smith's offer. Alice said, “Then, go ahead and take it. Alaska should be a new experience.” And so, the next day I called General Smith and told him I am available for the job. He was delighted and said he would call Colonel Branch to make it happen. That was a decision that I looked back at many times over the years.

During my leave I was able to watch Gene and Jim play hockey several times. I enjoyed Katie perform on stage in a Christmas play, conversed with Michael's teacher, spent some time with Mary, who was a cheerleader for the Prep, and took the kids skiing on Mt. Spokane. My dad came out to visit for Christmas, and the kids always enjoyed having him because of his great sense of humor and generosity. It was a great vacation for me to be home for the holidays and to be with the ones I loved the most dearly and cared for the most deeply. Again, sitting around the dinner table and just talking about what was going on in their lives was important to me and to them. Sundays, it was morning Mass and a breakfast at a favorite restaurant, a longtime family tradition. Christmas morning, the kids were up early to open their gifts, of which there were many, thanks to their mother's diligent shopping and their grandfather's generous habits. I was so happy being home, but in the back of my mind, I was counting the days when I had to begin the long journey halfway around the world to start the last half of my 12 month tour. Little did I realize what awaited me in the coming five months in the way of frustration, pressure, and strife that brought me to the edge of my being fired from my job. If I knew what was forthcoming, I might have considered desertion. However, my time at home was great for me. It reaffirmed my decision to find an assignment where I would be able to spend more time at home with my children especially during their teen years, and I was determined to meet that responsibility. I was hoping that General Smith would come through for me on that Alaska assignment. I would have to await word on that.

Time came for me to leave on my long journey back to Riyadh. It was a long, long, lonely journey. However, I did feel refreshed having spent some wonderful days with my family. After a half year's separation, I found them all healthy and doing reasonably well following a heart wrenching separation from friends, their schools, and familiar surroundings moving to totally new situations. My terrific wife, Alice Marie, performed a great job in leading them through this transition, making all the tough decisions, driving them everywhere, setting them up in their different schools and activities, and, at the same time, moving them into a new house. Alice Marie was the epitome of the military wife. I was not only proud of her accomplishments; I was in awe of her as well. While I was terrible unhappy about having to return to Saudi for another five months, at least I was reassured that things at home were in good shape and my family was safe and sound and doing well.

On my return to Riyadh, I found a letter from General Smith, welcoming me to Alaskan Air Command. I would be assigned to command Operations at Elmendorf AFB, Anchorage. Smith indicated that I would be contacted by DCS/Personnel regarding my assignment and other matters of interest within the next few days. (This never happened.) I was very pleased to receive this news and I fired off a letter to Alice and the family advising them of my safe arrival back in Saudi and the news of our new assignment. I wondered how living in Alaska would be received by the kids. I was uncertain that all of them would be pleased about living in snow country. However, having spent lots of time in Alaska myself, I was fairly certain that many of the kids will get to really enjoy what this great state had to offer in the way of scenery and outdoor activities. Also, I found the Alaskan people to be friendly, energetic and independent, fun to be around. So, I was fairly confident that once they arrived and settled in, they would eventually enjoy living in Alaska. I truly hoped I was right about that. At least, in this assignment, I would be home a lot more than I had been than in my three last assignments, and I was looking forward to that.

Back in Riyadh, nothing much had happened since I had left. General Smith was glad to see me back and I stopped in to see him on my return at Dhahran. As usual, he complained that his enemies were conspiring against him. He had intercepted several TWXs that were defying his explicit instructions. He was really down on the Army contingent and also the Air Force detachment that was in country overseeing the Lockheed Contract on the F-5 sales. This group was out of Air Material Command, Wright-Patterson and did not report to USMTM. I told him I would look into that and see what I could do about it. Smith also had received complaints about the booze distribution to the detachments. Smith had used the C-54 to pick up liquor for USMTM personnel in Bahrain and then dropped the orders off at each detachment for each person who had put in an order. This, of course, was illegal in Saudi, but the C-54 had diplomatic privileges and was not subject to customs search. Smith was hopping mad about these complaints, telling me how was putting himself out doing this for the troops and taking a risk doing it. I agreed totally and advised him to cut the whole booze program out completely. I said, "You have been too good to the troops doing this for them. The hell with them. We can live without liquor while we are here cut it out, General." Smith was flabbergasted at this response. He agreed with me. "You're right," he told me, "that's exactly what I am going to do. They can all go to hell as far as I'm concerned. Cunneely, you hit the nail on the head, I'm cutting them off." He loved that idea, and he spent the next hour chuckling over how the troops would react when he broke the news to them that the liquor rations were cut out for good. But I knew he would never do it. He would threaten them, and they would shut up with their complaints. Then Smith would relent, and the booze runs would go on as usual. That's the way General Smith liked to operate. I knew that even as I encouraged him to do it. It made him feel good and it made me feel good too. Let him get after the Army guys, and while he was doing that he was off my back, at least for a little while. Alas,

I was totally back. I had five months to go, and I would do my best to help the Saudi Air Force and the USA accomplish the mission I was sent half-way around the world to do. My visit home refreshed me greatly and my morale was as high as it could possibly be under the circumstances. I tried to make every day count and each new experience a worthwhile one. In many ways this was an intriguing assignment, and I was involved in many issues of major national and international significance. So, I vowed to do whatever I could putting my experience and intellect to best use.

CHAPTER THIRTY-THREE: THE FINISH OF A LONG HARD ASSIGNMENT

To thine own self be true, and it must follow,
as the night the day,
Thou canst not then be false to any man.

-Shakespeare

In the election of 1972, Richard Nixon was reelected for his second term as President, but he was under fire for failing to end the Vietnam War during his first term, although he had drawn down the size of our troops deployed in that country by nearly 500,000. Henry Kissinger had been involved in the Paris Peace talks for many months, but the North Vietnamese kept stalling and the killing went on. And the anti-war protests continued. But in January 1973, a final agreement was reached after Nixon resumed serious bombing over North Vietnam. In March 1973, our POWs were released and were flown home from Hanoi. Included in that group of great heroes was my close friend Colonel Dave Winn. He had survived over four and a half years in captivity. My prayers for him and the others had been answered. It was wonderful news for our country to have that war ended even though we were not victorious.

Back in the land of never-ending sun and sand, I felt that with only five months to go in my tour I should have an easy time finishing up my work there. But as usual I was very wrong. My last months in Saudi turned out to be the most eventful and controversial. Thankfully, the weather had eased, and the days were comparatively mild. Temperatures were mostly in the high 70s and 80s, cooling in the evenings into the 50s. This certainly was better than the summer temps we had when we often saw the mercury hit 120 to 130 degrees. The sandstorms also became less frequent, and we even had a cloudy day here and there and a shower or two. On the climate front, things were definitely better.

My first problem arose in a debate over whether the US should sell F-15s to the Saudis. I was asked to write a comment memo on this, and I did stating that I saw no reason why we should not do so. To my surprise, my position was hotly contested by the Embassy personnel and the Defense Department. I am sure the hidden issue here was Israel who opposed arming the RSAF with a state-of-the-art fighter. My position was that even with the F-15, the RSAF would pose no problem to the Israeli Air Force would the Saudis ever decide to engage them. I also stated that the RSAF could not maintain the F-15 without foreign government aid anyhow. My position still raised a lot of ruckus. I was certainly overruled, although the RSAF eventually were sold F-15s.

Lockheed Aircraft vs. Col. Cunneely

The wildest incident in my tour in Saudi came about unexpectedly. One of my staff Majors, an engineering type, uncovered a serious glitch in a contract the RSAF had with Lockheed. The Saudis had ordered four RC-130s (refueling tankers) for their fighter aircraft. Their specifications included that these tankers would have both a high pressure and low-pressure refueling system to accommodate the different types of RSAF aircraft. But in reviewing the plans for the first RC-130 to be soon coming off the production line in Georgia, my Major discovered that the aircraft had only one type of refueling system, not two as the RSAF had ordered. This we confirmed. When I broke this news to General Zuhair he immediately called in the local Lockheed representative stationed in Riyadh. I was present at the meeting. The rep was on the defensive and assured the General that I and my people were wrong. He said he would check it out and return the next day to brief the General on the facts. But he failed to neither return the next day as promised nor even call. The General was furious. He asked me to verify our meeting with the Lockheed Rep in writing, which I did. He then reported the affair to Prince Sultan, the Saudi Defense Minister. I could see then all hell was about to break loose. I had already advised General Smith the previous day by short wave, and I again updated him. I advised him to notify Ambassador Thatcher and the Defense Department as I was certain that this issue was going to get worse. And it did, but not in the way I thought. I believed that Lockheed Aircraft was in deep trouble. I was certain that somehow, they had screwed up the contract and had failed to meet the specs concerning the dual pressure systems. As the first aircraft was about to come off the production line, fixing the problem was going to delay delivery and cost a lot of money. And who was going to pay the costs?

But I was wrong about who was in trouble. Lockheed and the Defense Department concluded the trouble started with the Chief Advisor to the RSAF, one Col. Cunneely. How they came to that conclusion I couldn't fathom, but General Smith was on the short wave to me every other hour demanding to know whether or not I knew what I was talking about, how G.D. sure I was about the specs, and whether I'd be there when the Lockheed Rep said he would come back to see General Zuhair the following day. I was then told a team of Defense Department officials and Lockheed bigwigs were on their way to Riyadh. They would be seeing me, General Zuhair and Prince Sultan. General Smith told me "Good luck, you're on your own on this one. These guys are out for your butt."

When the team arrived, I met them at the airport and took care of their needs. They totaled eight in number, four from the Pentagon and four from Lockheed. We met in my office in the RSAF Building. While I did not know any of them, l could see that at least one of the Air Force guys appeared to be sympathetic. I believe he was from Secretary Robinson's office. After some sparring, I laid out the case. I first reminded them that my job was to give my best advice to the RSAF, and that retaining a solid relationship between our two nations was important. I then reviewed the contract pointing out the specs were quite clear in calling for two systems. The information I had was that the aircraft presently on the line in Georgia did not meet those specs. I asked the Lockheed personnel to confirm or deny that. The Lockheed guys flubbed around on the answer to that question stating that they were still confirming the status. I said, "If you cannot tell me the status then I will conclude that the aircraft does not meet the specs. Your Tech Rep here in Riyadh also did not answer that question and then insulted the Commanding Officer of the RSAF by breaking his appointment to return with the information. I felt that that was bad performance by your company." This last statement was not well received by the Lockheed personnel, but they had to agree. They then told me that this individual had been fired. When I concluded my presentation, I could see that the Air Force people were relieved. I think they had already thrown me under the bus, but now were feeling better that their guy in Riyadh (me) maybe wasn't so dumb after all. They then left my office to meet with the

Saudis and to try to straighten out this awful mess, which they eventually did at great cost to Lockheed.

When the meeting ended, I congratulated my staff for their work. I then went back to the Compound and called General Smith and gave him the results of the meeting. I told him that if they wanted to fire me, that was okay with me. But I did my job and what I told them was absolutely right, and I expected the Pentagon to back me up. And they did. And that was the end of it. Smith was tickled pink. I never heard it mentioned again by USAF personnel except from Secretary Robinson when I gave him my final out briefing. He told me I did the right thing. I also received plenty of thanks from the Saudis. They always enjoyed it when the Americans screwed things up.

The Oil Embargo and Scoop Jackson Visits

In 1973, again due to financial unrest in the Middle East and dissatisfaction with the failure of the United States to persuade Israel to evacuate the territories it conquered in the 1967 war, the OPEC nations, led by Egypt instituted another oil embargo against the West. Heavy diplomatic efforts by Kissinger failed from a government weakened by Vietnam and Watergate. Word came to us that Senator Henry Jackson of Washington was on a U.S. Navy aircraft carrier in the Mediterranean Sea and that a meeting was being set up for the Senator with the Saudi Arabian oil minister Sheikh Ahmed Zaki Yamani, who was the leader of the OPEC organization. The meeting was being arranged by our CIA people in Riyadh and would be held at a secret residence in the city. Our job was to meet the Senator and host him and his party while he was in town. However, a major problem cropped up at the last minute. One of the Senator's close aides, Richard Perle, was Jewish and the Saudis did not permit any Jews to enter the Kingdom. In that case, the Senator said, "No Perle, no Jackson." After much back-and-forth secret diplomacy, the Saudis relented and permitted Perle to accompany Jackson as long as he was kept in tow and his presence in the country was kept secret. Jackson and a few aides, including Perle, arrived in Riyadh, and we took care of them. We hosted them for a luncheon in our mess hall along with the Ambassador and it was a successful day. The Senator was well known as a supporter of the military and as I had much active experience in the State of Washington at McChord AFB, we had many mutual acquaintances, including Col. Sam Grashio. The meeting between the Senator and Sheikh Yamani came off successfully, although I did not see where it did any good in offsetting the OPEC nation's implementation of the 1973 oil embargo that sent gas prices sky high. The highlight of the evening affair, which I attended, was the long-heated debate between the Saudis and Perle over the Israeli situation. This was the first time I had heard of or seen Richard Perle, and I had to say I was impressed by his brilliance and coolness under fire. He was more than adequate in handling the several Saudis who came at him from many directions. Perle would later become an assistant Secretary for National Security. I was also pleased to meet and host Scoop Jackson, a great friend of the military and a very personable politician.

USMTM Inspects the Inspectors

General Smith did not like the Air Material Command guys who oversaw the Northrop contract at Dhahran, the F-5 program. This was a big contract that included not just aircrew training, but aircraft maintenance, plus the training of maintenance personnel, English language education, classroom instruction, et al. It was a very large lucrative foreign military sale and employed hundreds of Americans, most of whom were former USAF military people. The few times that I flew with the Dhahran F-5 squadron, I would fly with some of these former USAF pilots who were now making big money in Saudi Arabia. One pilot I flew with was a retired Brig. General, who was probably doubling his Air Force pay. It aggravated Smith that this outfit operated outside of his control. I asked my staff to research this area, and lo and behold, they came up with something. In an obscure paragraph in a regulation governing the responsibilities of the Military Training Mission was a clause calling for us to ensure that the AMC Team was fulfilling its responsibilities under the Foreign Military Sales provisions. I passed this along to Smith and asked him for the go ahead to conduct this inspection. Smith could hardly restrain himself. I formed a team with my staff people and worked up inspection check lists. I then sent a notification letter to the AMC Team Chief that we would be inspecting his unit to determine if they had been fulfilling their responsibilities as outlined in that obscure regulation. This Colonel had a triple fit when he received that letter and appealed it all the way to Dayton. But he lost that appeal because the regulation clearly stated his unit was subject to oversight. So, our small team flew over to Dhahran and we gave them an inspection. And we found some gigs, not very important ones, but defects nevertheless, that had to be reported and corrected. The Colonel was outraged, and General Smith was delighted. Since General Smith was delighted, so was I. As General Smith was prone to firing me about once every couple of weeks for some annoyance or other, any time he was happy, my life was easier.

General Smith: The Air Force Pilot

Smith was certainly the most difficult boss I ever had in my career. And that was due to his unpredictability. He could change his disposition in just a few seconds, and you had to be prepared for it. He was hard to read, but gradually I became better at it than my predecessors. I never allowed his fiery outbursts bother me, and I never showed him that his outbursts bothered me. When he shouted at me and told me I was incompetent and that he was going to fire me, I said nothing in return, but at the first opportunity I got the conversation back on track to the issue we were discussing. I always showed respect for his grade and position and always, except in very rare instances, agreed with him initially on a subject, only later attempting to get him to change his mind. Whenever he came to Riyadh, he would arrive in the USMTM C-54. This was an old WW II prop aircraft that was used to transport personnel and goods (booze) around the country. Smith would fly the aircraft himself in the left seat with our USMTM pilot in the right seat. This, of course, was illegal, but who was going to tell the General he couldn't do it. Whenever USAF visitors from the Pentagon showed up and saw Smith flying the C-54 they were horrified and asked me why I let him do it. I told them that I was not about to tell the General he could not fly the USMTM C-54. Since they were in the Pentagon where they wrote all the rules, I said. "You guys tell him he can't do it." They never did. Whenever Smith flew to Riyadh, he would call me in advance and tell me he was coming and don't bother to meet him. But I knew better, I always met him. He liked to be met and greeted like a General, and so I always did so. Now the Army Colonel, Sam Vincent rarely came out to greet Smith. And Smith would always ask me where's Vincent. I would say that Sam is busy and could not make it. But I knew that Smith really did not appreciate *not* being met by his Colonels. He loved to taxi his C-54 in and wave from the cockpit and receive a salute from me standing on the tarmac. Why not! I had enough problems without aggravating this crazy General and in the long run my attitude paid results.

The Army Revolts

For reasons I will never understand, the Army guys who lived in the Riyadh compound resented that an Air Force officer was in command. This happened despite my decision to turn the day-to-day oversight of the compound over to Sam Vincent the Senior Army officer. I did this because I really did not want to be saddled with this administrative headache, plus I was traveling much of the time visiting my detachments throughout the Kingdom. If only I asked Vincent to keep me informed on what he was doing and on any policy decisions he felt, I should be aware of. I felt that this was a reasonable compromise as Vincent was not that busy and really wanted to command the compound. However, I did override a few of his decisions, none of any great significance, and that seemed to get the Army guys upset. When, our meal prices were raised, not by me, but by higher headquarters, the deportment of the Army officers in the dining hall became rude and offensive to the waiters who, of course, were totally uninvolved in the price decision. It became so bad that I intervened and dressed down two Army field grade officers in the mess for their discourtesy. This was observed by a large number of those present consisting of officers and enlisted men from both services. I let it be known that I would not allow such conduct to continue. The officers I chewed out were caught by surprise and clearly never expected to be openly embarrassed as they were, especially by an Air Force Colonel. At least I reminded everybody that while I let Colonel Vincent run things in my absence, I was still the senior officer, and I expected all officers under my control to meet high standards in their conduct. Reports of this incident soon spread and became the talk of USMTM. I didn't mind that at all because the conduct of the Army personnel noticeably changed for the better from that point forward. However, a great amount of subdued bitterness remained.

A few weeks after the incident in the mess hall, Sam Vincent knocked on the door of my apartment. He and his number two asked to see me about a proposal they had worked up concerning the administration of the Riyadh Compound. They presented it to me, and I read it. After I read the proposal, Vincent asked me if I would endorse it. The crux of the proposal was that Sam Vincent would become the commander of the compound. As senior officer, I would become someone more like a Fleet Admiral on a US Navy ship. The ship would be under the command of a naval officer of a lower rank than the Admiral. In our case, the compound would be the ship and I would be the Fleet Admiral. Vincent would be the commander of the ship. My reaction to the proposal was negative. I told Vincent that I had no problem with his commanding the compound as I had no great desire for that added responsibility. But General Smith specifically directed me to be the Riyadh Compound Commander, and I was not about to go against his desires in that decision. I also advised Vincent that it would be extremely foolish and risky on his part to attempt to reverse his decision without providing a substantive rationale for doing so. But Vincent was stubborn and determined to go forward to Smith his proposal to change the command structure. A few days later he forwarded his proposal to me for endorsement. I sent the proposal on to Smith recommending that the proposal not be implemented as I felt that the Riyadh Compound had been operating satisfactorily under my command and that there was no precedent for the structure that Col. Vincent was proposing that violated the principles of rank. Prior to forwarding the paperwork to General Smith, I again tried to persuade Vincent to withdraw it; but he was determined, and the proposal went to Dhahran.

It didn't take long for the bomb to explode. It was only a few days later that Smith called on the short wave. His call sign was "Apache." The radio enlisted man chased Vincent and me down and told us that Apache was on the radio. Apache was hopping mad and wanted to talk to both of us immediately. I knew right away what this was about. On the radio, Apache was near out of control. He told both of us off in clear terms, threatening to fire us both for incompetence and stupidity, and for arguing about silly nonsensical things of no consequence instead of performing the mission we were sent over here to do. He told us to get our as-es over to his office ASAP and he would straighten this out in a GD hurry. Over and out! I looked over at Vincent and I could say "Thanks Sam, this is just what we needed."

Our meeting with Smith was an awful experience especially for Col. Vincent. Never had I seen an Army General chew the butt of an Army

Colonel like I witnessed that day. At the end, I began to feel sorry for him. It was a pitiful and braising verbal beating. If Vincent had any expectations of leaving Saudi Arabia with an excellent Evaluation Report or a promise of a future star on his shoulder, it ended that day. I will never understand why he dreamed up his proposal that had no chance of succeeding and which made no sense. But he did and he paid dearly for it. I could only surmise that his dislike for working under the supervision of an Air Force officer plus the embarrassing butt chewing I dropped on his officers, caused him to try.

From that day on, our relations were cold. I took more personal control of the Compound and made all of the decisions leaving Vincent pretty much out of the picture. The Army cadre gave me no trouble and things went smoothly. The Air Force people were happy but did not gloat and that was that. We then got back to the business we were sent to do.

Christianity in Arabia

The Kingdom did not permit freedom of religion within its borders. USMTM did not have any chaplains and so we had no religious services. Unless you were a Muslim, you had no means to attend public worship. I truly missed Mass and receiving the Eucharist. However, about once every quarter a Catholic priest did enter the country. He was Italian and came in with a faked passport. We met him and put him up in the Compound and sent word out to the Riyadh community that he was with us and the schedule for Confession and Mass during his stay. We always had great turnouts as so many Catholics and many Christian non-Catholics were eager to attend. I was certain that the Saudis knew about this priest but decided not to take action as long as we did not permit any Arabs attend and held the services within the Compound. Mass was held in our theater and was a great morale booster.

Middle East Travels

In addition to my monthly visits to the detachments under my supervision, I also made a few trips to Iran, Pakistan, Bahrain, Egypt, and Ethiopia. My trips to Tehran were especially interesting. Iran in 1973 was under the Shah and was a thriving country relatively free and prosperous. The Shah was intent on building a large military that would be the most potent in the Middle East. Iran was feared by its neighbors and disliked because it had loosened the strict Islamic rules for the Iranian people. Iran was buying huge amounts of military equipment from the USA and its FMS sales were in the billions. Tehran was a beautiful city close to the snowcapped mountains bordering the Caspian Sea, and I enjoyed my days there shopping and meeting the friendly Iranian people. It seemed that most of them whom I met had relatives living in the US. However, only six years later, the Shah was overthrown, and Iran was back under the Mullahs as a strict Islamic nation. In 1979, the Iranians took over the American Embassy in Tehran and took over 170 American hostages. Our nations have not been on good terms since. In Karachi I bought some camel saddles which we still have. It was a dirty, uncomfortable place. In Bahrain, the US had a small navy base. Life there was much freer than in Saudi with some nightclubs and good restaurants. Addis Ababa was a lovely area but with extreme poverty. Later, I was able to spend some time in Cairo and see the great sphinxes. But usually, my trips were short and blended with business. The Middle East is not my favorite area of the world for vacationing although it contains much important history. It is also an area that has had much impact on our own country because of oil and because the earth is becoming a smaller place than it once was due to globalization. We could no longer just ignore what is happening in these countries. And this came home to us with great force on September 11, 2001.

Disturbing News and other Frustrations

As my tour of duty in Saudi began to wind down, I started to look forward to returning home and beginning my next assignment. I then received some bad news about my good friend and mentor General Donavon Smith. Smith had recruited me to come to Alaskan Air Command, but I got word that he was diagnosed with a brain tumor and was forced to retire. This worried me a little as Smith wanted me to come to AAC and take over a position in Operations. The new Commander was Charles Carson. I knew Carson when I was at McChord. He took command of the 318th FIS when I was at the 325th Fighter Group in 1957. He had come from the Pentagon and did not impress people with his knowledge of fighter operations. He was considered a weak CO. However, he was now a Major General, a grade he apparently attained from his assignments in the Pentagon as an aide to the Chairman of the Joint Chief of Staff. I was not sure that this was good or bad news for me, but I was disturbed because I would not be working for Donavon Smith, who I knew to be a real excellent General. (Smith died within a year of his retirement.)

In May, I had to go on TDY to Wiesbaden, Germany for my annual medical exam. I flew on commercial air through Greece to Germany, spending a little time in Athens. It was a great respite and I had time to do a little sightseeing in Germany as well, taking a boat ride down the river Rhine. Although raised in an Irish family with an Irish name, I cannot ignore that half of me is undoubtedly German. In May, the weather was fair and mild, and it was great to be away from the sand and the heat and back to the western world again.

On my return to Saudi, one of my last exasperating tasks was given to me by the Embassy heads. The Saudis, for reasons only they could explain, were refusing to pay some of their FMS bills. One of the largest was one they owed to Litton Corporation which had sold them millions of dollars' worth of navigational equipment for their aircraft. They were now way behind in their payments, and I was tasked to go the Saudi Ministry of the Treasury to persuade them to pay up. Why this assignment fell to me, I will never understand, but it did. So, on this hot June morning I showed up for my appointment with this Saudi bureaucrat and five hours later and ten cups of shahi (tea) I still had gotten nowhere. It was the most brutal five hours I can ever remember putting in for any cause. We went around and around, coming back to the same place we began minutes ago. At one point, the Minister told me that the Saudis would not pay any late charges on the bill because that would be interest and the Koran forbade Muslims from charging or paying interest. At another point, he told me that Litton was a rich corporation, had plenty of money and could wait for their payments to arrive when the Saudi's were able to pay them. I came away empty handed as well as totally frustrated. I really began to count the days to my rotation.

Saudi Problems

Running the Detachments became routine as I had good people, and they knew what I expected from them, and they did their job. My biggest problems were given to me from the outside like the Litton payments. Another was a case of the crazy Prince, a second-generation nephew of Faisal who thought he wanted to be an air traffic controller. He bugged everybody at the Embassy and at USMTM and they finally sent him to me. I interviewed him and it soon became clear that this guy only wanted to get to Florida, USA. He had done his homework and found out that the USAF air traffic controller school was at Tyndall AFB, Panama City, Florida. He even knew when the next classes began, and he wanted me to set him up in the class and arrange everything for him. In his interview with me, he did all the talking and just told me what he wanted and what he did not want. This spoiled brat would not last two weeks in any AF class and I told that to the Embassy people. I told them I would TWX the AF to arrange it, but the State Department would have to authorize it and somebody would have to pay for it, not the Air Force. That was the last I heard from the Prince. I heard later that Prince Sultan dropped the curtain on that one.

A strict rule in Saudi was no intermingling with Saudi women. Any transgressions would meet with serious punishments. While fortunately none of my military were involved during my stay, we had some civilian contract personnel who broke that rule. In these cases, we were able to get the rule breakers back from the police and out of the country within 24 hours. Our failure to do so could have meant very bad problems for these very foolhardy guys. Another serious problem I faced was when one of our Officers approached a Defense Contractor salesman with an offer to improperly persuade the Saudis to purchase the Contractor's line of equipment in exchange for a job with that firm. The Contractor Salesman reported the incident to me. I confronted the officer who denied the accusation. However, I reported the incident to Headquarters, and I relieved him from duty. He was later reassigned.

General Smith and the Pentagon

As my time to rotate came closer, the problem of General Smith began to get worse. General Smith's three year's tour as Chief of USMTM was also ending shortly after mine that summer. Word came to me that Secretary Robinson wanted to see me at the Pentagon when I returned from Saudi regarding Smith. On one of his visits to Riyadh, Smith told me confidentially that he wanted to extend his tour of duty in Saudi and that he had asked Prince Sultan, the Minister of Defense, to support his request. He also showed me a sword that the Saudis had given him as a present. It was a magnificent sword, certainly a very valuable item, and I congratulated him on his earning it. However, I also knew that it was illegal for him to accept such a gift. I now became very worried about why Robinson wanted to see me. I was certain it had to do with Smith. Actually, Smith had not bothered me very much in the past few months and as far as the Air Force units were concerned, he kept mostly out of our business. In my opinion, Smith was a screwball and a dangerous officer to hold the position he was in, but I had found the secret of how to handle him and things were going fine for me and my detachments. I was much afraid that Smith would discover that I would be checking in at the Pentagon on my return to the States. Smith loved his job and the power he held. He had gotten close to the Saudi hierarchy and did a lot of favors for them including distributing goodies such as movies and booze. No doubt they wanted him to stay in his job rather having to deal with a new chief. I decided to play my cards very carefully in my last few weeks in Saudi.

Nearing the End

The days were growing very warm again in the Arabian spring and I was longing to see home again and spend some time on Lake Pend Oreille before heading north. Being away so long from family was wearing and having to spend so much time writing letters to my wife and then waiting for replies was frustrating. Alice was so good at writing, but her letters would arrive in group of two or three at a time with two to three weeks in between. I would read them in chronological order so they would make more sense. In the last few months of this long tour of separation, I devoted lots of hours reflecting about how much of my married life I had spent away from home and away from my wife and children. Of course, I knew that when I chose the military as my life's vocation, that disadvantage would certainly be a part of it. And I would never openly complain, and neither did Alice Marie. But now she and I were in our forties and four of our five children were teenagers. My decision not to take on a tour at the Pentagon, I felt, had been a wise one. In Alaska, I will be able to give more time to my kids in the years when they would need me the most. If I failed them then, my other successes would not be as meaningful. With General Smith gone, I had no idea what job I would have at AAC. It wouldn't matter. Job one for me will be my family this time. That was a promise I made to myself.

I did feel some emotions when it came time to leave Saudi. General Zuhair was especially grateful for my work on behalf of the RSAF. I came to like him very much and respect his integrity and dedication to his country and his religion. As was the tradition, he presented me with a going away gift, the traditional male clothing of Arabia. During my year in Saudi Arabia, I made many friends, and I came to enjoy their company and respect their views. We disagreed on my country's support for Israel of course, but on many other matters, the Saudi's felt that the Americans were their friends and would come to their aid should any other nation threaten them. This came to pass in 1990, when Iraq, under Saddam Hussein, invaded Kuwait and approached the borders of Saudi Arabia. The USA did come in to ward off that threat in Operation Desert Storm, soundly defeating the Iraqis in short order.

I also said farewell to my foreign friends who had been good to me, especially the Schubert's whose hospitality knew no bounds. Whenever I was able to escape from General Smith's fangs, I headed out to their home south of Riyadh in the desert Al Ha'ir, where there was shade, a pool, fine food, tennis and interesting company to while away the hours. As a goodbye thank you, I brought out from our mess hall some steaks, hot dogs and lots of German beer for a farewell get together under the desert stars. My Air Force guys also threw a bash for me in our compound bar. I received the usual going away gift of a brass tray, inscribed with a map of Saudi Arabia with the Muslim profession of faith on top. The party was lively, and I received a pretty good ribbing for the various problems I went through with General Smith and the Army. I was also congratulated for having lasted longer than the previous AF Colonels in the same position. All in all, I did feel that I had done the best I could for the Saudis in the year I had with them. They wanted me to extend my tour, but certainly without family that was not considered.

I also was invited to the home of Darwish, my interpreter and advisor. He was a dear man, a little on the chubby side and always in good humor. He was a huge help to me during my tour in so many ways. I relied on him to translate the Arab messages and letters I received and transmitted. He made all of my travel arrangements, many at the last minute due to General Smith's wild, urgent demands for my immediate appearance in Dhahran. Darwish was a Palestinian whose family had been booted out of their home by the UN when it was given to the Israelis. It was a home his family had occupied for decades and then, out of nowhere, they were forced out, almost overnight. It was a bitter experience. Dining with his family was a pleasure. I met his wife, who cooked the meal, and his children. They were such good people, generous and religious. They offered a different perspective on this very complex situation that is the Middle East.

Other Good Middle East Memories

Each March, the King holds an open house called the majlis at which he personally greets the public. I was able to meet King Faisel at this occasion in 1973. One greeted the King by kissing his shoulder. Faisel was in his seventies. He was a homely man and I never saw him smile. He appeared to be a man who did not particularly enjoy his job. However, he stepped into the Kingship when it was necessary and stabilized the country. He did many good things among them opening up educational opportunities for women and using oil dollars to modernize his nation. He was a strong anti-communist and was a friend of the US. Unfortunately, two years later, in March 1975, one of his nephews shot and killed him at the majlis. Not surprisingly, the prince lost his head (literally) for this rash act.

Another person I met during my tour was General Andrew Goodpaster. He was the CINCEUR at the time, i.e., the Commander in Chief of European Forces (NATO). Goodpaster was flying throughout the Middle East and while visiting Saudi, I flew on the same plane with him from Dhahran to Riyadh. He had a fabulous career in WW II, serving under Eisenhower, and also served in the White House when Ike was President. I was able to enjoy a great conversation with him on our flight and found him to be a humble man with a great sense of humor. I could readily appreciate why this man was so sought after for positions of the highest responsibility. Even after his retirement, he was called to take on special assignments of heavy international importance.

I Get My Orders

I packed all of my gear and shipped it out, made my flight reservations, and got ready to head home. First, I had to stop and check out with General Smith at USMTM Headquarters in Dhahran. This I feared. I had a premonition that Smith might have learned that I would be stopping by the Pentagon on my way home. He always seemed to know everything because he intercepted all the incoming and outgoing messages. I was afraid that he would ask me about that, and I did not know how to answer him if he asked. But I was miraculously saved. My flight was to depart Riyadh at three in the afternoon, but as often happens, there came in a fierce shamal (sandstorm) that closed the airport and delayed my flight for hours. In fact, it was not until 7:30PM that we were able to get airborne. That flight on Saudia Airlines was one of the scariest I have ever been on. We made it into Dhahran but all the way down on final approach that airplane was all over the sky. I was very afraid that we would not make it to the runway. I was never confident in the Saudia pilots even though none were Saudis. But we did make it. It was too late to see Smith, so I checked into the BOQ for the night. My flight the next morning departed at 0700. I never called him. I left a message saying that I was sorry I missed seeing him. I wished him the best and thanked him for his many kindnesses to me. And that was that. I was relieved that things turned out the way it did. As much as I despised the Arabian sandstorms, that was one I thanked Allah for.

My Loyalty is Put to a Test

I flew to Washington, D.C. with stops in Greece and London. As anxious as I was to get home, I had one last hurdle to overcome. Secretary Robinson was insistent that I see him as soon as possible regarding General Smith. I knew it would be about Smith's request to extend his tour of duty in Saudi. What occurred at my meeting with the Secretary was the most unusual experience that I ever imagined would happen to me in my air Force experience to date. The usually lighthearted Robbie was serious this time and in a hurry to get down to business. The Defense Secretary was waiting for my input and to my shock, the Pentagon's response to Smith's request pretty much depended on what I was to say. Robinson told me that the Saudis, meaning Prince Sultan of course, had put in a strong endorsement supporting Smith. The State Department was willing to go along with that to please the Saudis. But Defense was inclined to get rid of Smith. Now it was up to what I said. I felt awful about being put into this position. While I felt that Smith was really a poor excuse for a General, and often treated me unfairly, he was my boss and he expected and deserved my loyalty. Now I was asked to choose between which entities I owed loyalty to the most. General Smith, my supervisor and the man I was expected to serve with total obedience or the Department of the Air Force, for whom I also worked and should give my best. Robinson pressed me for an answer, and I gave him my reply. I told him that I felt it would be best for the Air Force if Smith was not granted his request for an extension. It was time for a change of command at USMTM and I believed an AF General would be the best choice to replace him. Robinson agreed and immediately phoned the SECDEF with my recommendation. He hung up with a broad grin on his face. "Smith is gone." he said, "The SECDEF sends his thanks."

That did not make me feel any better. I have never felt great about what I had done that day as, despite Smith's craziness and wild idiosyncrasies, in the long run he had been good to me. Robinson and I had lunch before I departed to the airport for Spokane. Again, he told me I should come to work for him at the Pentagon. Again, I thanked him for the offer, but demurred. (I realized, of course, that his offer could have led to another promotion.) I now felt free and anxious to get home to Alice Marie, family and beautiful Lake Pend Oreille. I had about three weeks of leave before heading to my assignment in Alaska. I had reached the light at the end of the tunnel. My exile in the desert was over.

The following letter was written to General Smith by General Alzuhair upon my departure in June 1973. I did not see it until it was forwarded to me after I arrived at Alaskan Air Command. It arrived at the same time that I was awarded the Legion of Merit for my service in Saudi. I was touched that Alzuhair took time to write the letter, and I believe it was sincere. He and I had some interesting times together and I supported him in some touchy issues including the question of F-15s for RSAF and the battle over the KC-130s with Lockheed. He truly wished that I would extend my tour in Saudi, but separation from family was the overriding issue on that one. In many ways, I grew to enjoy my relationship with the Saudis. I realized what a difficult position they were in situated in a truly muddled part of the world, beset with fanatics, unbridled hatred.

Kingdom of Saudi Arabia
Ministry of Defense and Aviation
Royal Saudi Air Force

BG Olin E. Smith
Chief United States Military Training Mission
Dhahran, Saudi Arabia

18 June 1973

I am very pleased to express my deep appreciation for the close cooperation, productive advice and fruitful efforts given the RSAF by Colonel Eugene P. Cunneely, Chief USAFS, who helped us in many ways to overcome the administrative and technical problems which we are facing due to the rapid technological growth of the RSAF.

In these days when Colonel Cunneely's residence with us was very short, we see the result of his fruitful efforts and his close cooperation which brought both the RSAF and USMTM to very close viewpoints in analyzing administrative and technical problems and in selecting the best means and methods to solve them. These efforts are especially appreciated at this time as the RSAF seeks to renew and improve its present status in order to reach advanced Air Force levels.

Colonel Cunneely's command and leadership and high spirit left a very good impression on those with whom he made contact. This was reflected in our close cooperation in all areas between RSAF and USMTM.

I am stressing my deepest appreciation and thanks to Colonel Cunneely and wish him progress and success.

Please accept my best regards.
Asaad Alzuhair, MG, RSAF
Commander

Coming Home... An Uncertain Future

In June 1973, I was 44 years old. I had just completed 22 years of active duty in the Air Force. Alice Marie and I had closed in on 18 years of wedded life. We had five children: Gene Jr. age 17; Jimmy age 15; Mary age 14; Michael age 12; and Kathleen age 8. I began to reflect about the number of days, nights, weeks and months that I had to work separated from them. I admitted that I was not different from thousands of other military career people who suffered similar heartaches as a consequence of choosing the military life. But now, I had to admit, I wanted to do what I could to bring about some change. My decision to forego an assignment to the Pentagon, while unwise from a career standpoint, was the first step in my decision to close the gap in my life that I felt was hurtful to me and my family, especially my children. Their teen years were years when the presence of a father was critical, and I intended to do a better job of being there for them. My accepting an assignment to Alaska, I hoped, would be helpful in that direction.

These were the thoughts that entered my mind as I flew from Washington, D.C. to Spokane and a happy reunion with my wife and children. It had been a long year in a position that tested my patience and challenged my experience and judgment. It was a lonely year in an environment and culture with which I was totally unfamiliar. In addition, I was tormented by an Army General who seemed to delight in humiliating his underlings and overriding their personal feelings and professional decisions. For a year I was on the other side of the world from my loved ones mostly out of communication with them, and worried about their welfare and security. The days there dragged for me. But now it was over, and I had come out of it in fine shape. I was determined to look ahead and do a better job of enjoying each day God gave me. I was especially looking forward to spending some days at our cabin on Lake Pend Oreille before heading to my next assignment. Since I had left for Saudi Arabia, a new room had been added to the cabin that I had not yet seen. After a year in the desert, the thoughts of lounging on our deck and viewing the blue green waters of this pristine lake filled my head with delight. I thought about buying a new boat so the kids could again water ski and we could explore the lake for new coves and beaches. As the hours went by and my flights brought me closer to home, I began to settle down. Life now seemed to become better and better. I wanted to get home as soon as possible. I was determined to devote more of whatever time I had left to the things that were most important to me. And foremost of these were Alice Marie and my children. The more I thought about it, the better I felt about it and the decisions I had made.

PART VI:
MY FINAL ASSIGNMENT

CHAPTER THIRTY-FOUR: A NEW BEGINNING—A HUGE SETBACK

I am sore wounded, but not slain;
I will lay me down and bleed a while;
And then rise up to fight again.

-John Dryden

Family separations are the worst experiences you face in military life; family reunions are the best. As bad as being away from our loved ones for long periods is, coming home again makes it all worthwhile, once you are there. You are refreshed. You have done your duty. You served your country and the Air Force to the best of your ability, and now you are back. You made it. It seemed like the past year would never end, but it did. It is now behind you. Coming off that plane in Spokane and being greeted by Alice Marie and my children was such a great moment—the moment I had thought about for so many days and nights on the other side of the planet. It was exactly where I wanted to be. All I wanted now was to spend as much time with them as I could now and into the future, wherever that would be.

Alice Marie looked great to me, and my children looked terrific. I was amazed at how they had progressed while I was away, and I could only again wonder at what a splendid woman I had taken for a wife who could perform all the works Alice had during my year's absence. I found it incredible that she had accomplished from the day I had put her and the five children and a dog on a train in Pennsylvania a year ago. She had to do it all, every day, seven days a week, and she did. But it had been tough on the kids who had been pulled out of their comfortable situation at Dover and taken across the country to new schools, a new home and a totally new situation. It was rough on them, especially the older ones, and now they would soon be moved again to Alaska where they would have to do it all over again. The year I was gone was hard on Mary, who had difficulty making friends at Marycliff High School. She found the girls there to be cliquish and they left her out of their activities. Nevertheless. Mary did make a few friends, and this caused her to be very negative about moving to Alaska. Alice and I did not think highly of her friends. The boys were more positive about the move, and I was certain that they would enjoy what Alaska had to offer.

We Launch a New Boat

In the interim we had much work to do. First, we had to move out of the house we had rented. That entailed separating what we were going to take with us to Alaska from what we were going to place in storage. Our base housing at Elmendorf would be fully furnished, including washer/dryer, bedding and all other furniture. We could only take some kitchen things such as dishes, utensils, our clothing and personal articles. Lots of decisions had to be made, with much packing and agonizing. Another brutal military move was upon us in less than a year. Finally, it was off to the lake cabin for my last weeks before I had to report in. I had been dreaming about this for over a year and how I wanted to see Lake Pend Oreille after a year of desert sand and scorching days. The plan was for me to leave ahead of the family who could enjoy a summer at the lake until the end of August. But what a surprise I had awaiting me when I arrived at the lake, thanks to Alice Marie. We had a brand new 23-foot Fiberform boat.

Wow! It was a beauty and we put it to work right away. We named it the Alice Marie after the lady who went out and bought it on her own (with a little help from Sam Grashio and Father Gubbins). The kids had a great time water skiing and exploring the lake in the "Alice Marie," and I had to learn how to operate it without crashing it into docks and beaches. Dad came out and spent a few weeks and he, the boys and I built a new deck at the door of our new room that was added to our cabin during my absence. The room turned out nicely. We had it carpeted and thereafter it became the sitting room for the family with its picture window overlooking the lake and also Alice and my sleeping room. There was a lot of work to be done in and around the cabin and our days there were filled with both work and play. There was always something to paint or repair, and fortunately, the boys enjoyed helping out. Gene, Jim, and even young Mike liked to build things and I noticed that on each project their skills with tools improved. All of them were good swimmers and I never had to worry much about them around the dock or out in the boat. Our days and nights at the lake were blissful and brought me back to earth again. After a warm busy day of toil and fun, we had great BBQ suppers and then enjoyed ourselves around a campfire on the beach. The clear night skies of northern Idaho were cool and through them we saw the different constellations and occasionally the northern lights. I don't think I ever enjoyed myself and my family more than those days and nights I spent at our little humble cabin originally constructed by the indomitable Jack Chambers on the shores of Lake Pend Oreille.

Before leaving, I had plans finalized to have a permanent dock built and a boat hoist installed the following summer similar to the one that Sam Grashio had next door. His was built by that infamous river rat Freddie Fannin, who operated a tavern across the lake at Sunnyside with his wife Viola but was suspected of draining off at least half the profits of this enterprise by his own consumption. Freddie, in addition to his half ownership in the bar and grill, employed a group broken down scoundrels who performed services up and down the river putting in docks including his own invention, an electric boat hoist suspended on pilings which Freddie nailed into the lake bottom near the shore with his barge carried pile driver that was something to behold. (think Rube Goldberg). Freddie was a great character who provided Sam, my dad, and my family many hours of laughs watching him approach his work and hearing him cuss when things just did not go as planned; that is, if they were indeed planned at all.

My days at the lake went quickly. I began to count them down and once again I faced the inevitable. I would be leaving my family and reporting in for a new assignment and leaving them behind without me. However, at least I knew that they were safe, and that this separation would be relatively short. I would be flying to Alaska, securing our new home, and preparing it for my family's arrival in August. I was looking forward to my new assignment in Operations at AAC and seeing my next Commander Gen. Charles Carson, whom I knew from my days at McChord in 1956-57. I was also looking forward to flying again on a regular basis. I had always enjoyed flying in Alaska. It afforded absolutely fantastic geography over which to view the top of world from glaciers to rain forests to Arctic

Oceans and snowcapped peaks. The military life, for which I had volunteered, never left you bored for very long. I had no right to complain, but from my current standpoint, I was getting somewhat anxious.

Sorrowful News

During your military flying career, you become somewhat accustomed to tragic news. Flying is a dangerous business and accidents happen, even to the very best pilots. Doug Brenner was one of the very best. He was an ex-Thunderbird pilot and had completed a combat tour in Vietnam in the F-105 Thunderchief (Thud). I flew with him and against him in air combat training in the Delta Dart. He was my assistant ops officer in the 318th FIS and a special friend. When I took over the 95th FIS at Dover, shortly after, he took command of the F-106 squadron at Griffiss AFB, Rome, NY. We saw each other several times during that period and spoke often on the phone comparing notes. Several months after I departed for Saudi, Doug was assigned as Base Commander at Thule AFB, Greenland. Not an assignment he volunteered for, I'm sure. Doug had also been promoted to full Colonel. One night, he walked into the Officer's Club at Thule and fell dead of a heart attack. Heart problems had plagued his family and his brother, a former Navy aviator and an airline pilot, had also died a similar death. Alice and I were grief stricken at the news. We had lost a great friend and a true American hero.

Time to Say Goodbye Again

Our plan was for me to go ahead of the family and report into Elmendorf AFB and take up my duties and move into our new quarters on base. Alice and the kids would remain at the lake cabin for another month and enjoy some more warm days of summer in northern Idaho. There I felt they would be safe with Sam and Dee Grashio next door. Alice would then drive the family to Tacoma, turn in our car for shipment to Alaska, and then fly on an Air Force C-141 from McChord to Elmendorf where I would meet them and take them to their new home. My orders required me to report to Alaskan Air Command no later than 18 July 1973. This included eight travel days and 30 days leave. But this time went quickly, very quickly, and I was soon packing my belongings and saying farewell to the family, the *Alice Marie* and Lake Pend Oreille. This wasn't easy. But at least, it would be a short separation and, as a family together, we would be starting out on a new phase of our lives in the great state of Alaska. I prayed that it would be a happy and fruitful time of our lives. I was feeling optimistic about it.

My Orders Are Changed

Everything went well until I arrived at Elmendorf. My hopes to begin work in AAC Operations and resume flying again on a regular basis were dashed. I was met by Col. Anderson, who was the Director for Manpower for AAC. He greeted me at Base Ops and gave me the usual warm welcome to the Command. But I was confused as to why he was greeting me in place of someone from Operations. He then broke the news to me that my orders had been changed over a month ago while I was en route from Saudi to the USA. I had been reassigned from Operations to become the Deputy Chief of Staff for Plans and Programs for AAC. Colonel Anderson was surprised that I had not been informed of the change in orders. He was greeting me because the Directorate of Manpower came under the supervision of the Deputy for Plans. I made no attempt to hide my anger and frustration at this news. I did not care for this alteration in my assignment, and I recognized immediately that it probably meant that I would be dropped from the flying rolls. With the Air Force budget dropping annually, only Colonels and Generals who were in positions where their job description required them to fly in combat aircraft, were permitted to fly and receive flight pay. The Deputy for Plans would not fit that requirement. My initiation to AAC became a negative one and when I reported into Headquarters, I asked what the reasoning was for this change, who made this decision, and why I was not informed. No one would give me a clear answer. General Charles Carson, who was the AAC Commander was someone I had known from my days at McChord in 1956-1957. We were on good terms, but when I asked him these questions, he did not give me any good answers. His response was that the Deputy position I had was a good place for my talent and that he was not going to make any changes. He did not know why I had not been told beforehand. I ascertained that the Operations Deputy, Casper Bierman, who a junior to me, was the villain. He told me that they tried to contact me but were unable to locate me. This was baloney, as my home address was on my order. But he gave me no reason for the change being made. Of course, he was the beneficiary of the change in my assignment. I believed that Bierman finagled the change for himself after General Smith left the Command due to illness. Despite my complaint the way this was handled, I was helpless to effect a change. It deposited a bad taste in my mouth concerning the AAC command and some of the staff that I was going to have to work with. It was certainly not a great start for my career as a Deputy Chief of Staff at a Major Air Command. This episode also gave me reason to again be disappointed in how decisions were made at high levels in the military. Too often personal politics overrode fairness and good judgment. When I first entered the service as a young lieutenant, I had nothing but admiration and deep respect for senior officers, especially Colonels and Generals. However, as I advanced through the ranks, I began to observe too many occasions where men of high rank failed to uphold the high standards their grade and experience demanded. Much too often, careerism and politics came into play. When I saw this happen, it was always painful to behold. The results were that men were promoted to high grades that were not qualified and ahead of others who should have been. In the days to come, during my assignment at Alaskan Air Command, I would observe the results of these faults in many instances.

Our New Home

Our family was assigned a great old house on Colonel's row. These were two story duplex homes with full basements. Ours overlooked the base parade grounds and a terrific view of the Chugach Mountains, which were snow covered year-round. The house was large and comfortable, with plenty of room for our growing family, and I hoped that Alice and the kids would like their new digs. The house was within walking distance of the AAC Headquarters where I would be working, and as it turned out, I ended up walking to work most of the time during my tour of duty. Elmendorf was close to downtown Anchorage and outdoor recreation was prevalent both on base and off. The base had a primary school to which Katie would be enrolled and was very close to the house. Close to the base, a brand-new Middle/High School was opening to which Gene, Jim, Mary and Mike would attend. It looked to be a great facility, but as it turned out, Bartlett-Begich School was huge disappointment as a learning facility. Unfortunately, there were no Catholic schools in Anchorage. That was too bad because, up to this point, our kids had been lucky to have attended excellent schools. One thing I truly concentrated on was doing all I could to give my kids a good education. For us, that meant Catholic school when it was available. We were lucky at McChord where our boys attended St. Francis Cabrini and in Colorado Springs also. Even in North Bay, Ontario and Dover, Delaware, Gene, Jim, and Mary were able to obtain good schooling in the parochial schools. For the one year in Spokane, Gene, Jim and Mary attended Catholic High Schools, but, alas, Anchorage offered no Catholic schools for my children. That hurt them.

At this point I purchased a Volkswagen Bug for my transportation. It was a great little car and permitted me to get around the base and into town to eat and see the sights. It had been several years since I had last been to Alaska and I was amazed to see the growth that had taken place. Our personal baggage had arrived, and I moved those boxes into the house, and I selected much of our living room, dining room and bedroom furniture from the base warehouse and had that moved into the house. But I waited for Alice to arrive to select the other pieces to round off the living room and other furniture we needed for our living quarters. I envied them still enjoying the warm summer on Lake Pend Oreille. The weather in Anchorage was by no means summery by the standards in the lower 48, mostly in the lower 50's and cloudy. I began my work in Plans and Programs, an area in which I had very little experience and much to learn. I was given a nice office in the AAC Headquarters building and started doing a great deal of reading and listening to those who had more experience than I. I did most of my eating in the Officer's Club, sometimes venturing downtown on the weekends. It was back to living a bachelor's life again and writing letters to home, (we had no telephone at the lake cabin). I was still depressed about my change in assignment and began to wonder if I had erred in turning down the Pentagon assignment or taking my chances with Colonel Branch. Maybe I would have been lucky to have gotten a flying job after all. However, I could not change anything now, and I was determined to make the best of it and do whatever I could to enjoy this assignment in Alaska with my family. My kids were growing old fast on me and soon they would be leaving the nest before my very eyes. By God, I was going to finish the job of raising them and I had better get about doing it before it was too late.

In late August, Alice and the kids arrived via C-141 from McChord. I was there to meet them, and I was the happiest man in Alaska that day. They looked so great to me, and they all seemed thrilled with their flight on an Air Force aircraft over the Inland Passage and above some fantastic scenery into Anchorage. That is all except Mary. Mary definitely was not happy and made no effort to hide it. She wanted no part of Alaska. Alice and I could see problems ahead. But the boys and Katie were excited with the adventure of being in the new state and with the surrounding scenery; they couldn't believe it when they saw the sun go down over the snow-covered Chugach Mountains outside our front door. Alice loved our big new home and saw lots of potential in it and that made me feel good. Overall, it was a good beginning for our three-year tour of duty. I began to feel optimistic and uplifted. Despite the setback about my assignment, I was again united with my family, the people I loved the most. I told myself: I absolutely will not complain about anything. I will do my best, at home and on my job. I'll let God take care of the rest of it. Amen!

Alaska Command

I learned something about the command-and-control situation in Alaska that I did not know. I knew that Alaskan Air Command was a part of Alaskan NORAD Command, which was part of the entire NORAD system. But I did not know about the Alaska Command. This was a third joint service, composed of Air Force, Army and Navy units, whose mission it was to provide military forces to defend Alaska from attack in the event of a ground invasion by enemy forces. Thus, many of us assigned to AAC wore three hats. As members of AAC, we trained and provided USAF air defense units. As members of Alaska Command, we trained and provided USAF tactical fighter and cargo units in the event of an attack by enemy ground forces. As members of NORAD, we provided early warning and defense forces for Alaska in the event of an air attack on North America. The number of personnel at headquarters manning both AAC and Alaska Command exceeded 800. I was surprised to learn of this and shortly after taking up my duties, which included Manpower, budgetary restraints from higher headquarters began to rear its head. I knew right away that I would be in for a rough ride. I began to study up on Manpower, an important area of military budgeting that I not been involved in prior. I also had to delve into all of the Operational Plans, which my Deputate was responsible for preparing and updating. These plans had to coincide with the latest intelligence threat against Alaska and be in line with our capability to thwart that threat. I noted that most of our plans needed to be updated. However, I was pleased that the majority of personnel assigned to plans and manpower struck me as experienced. What they needed was some guidance.

Col. Dave Creekman

Col. Anderson, who had greeted me on my arrival, was rotating to the lower 48. I needed to replace him as the Director of Manpower. This was going to be a very important replacement as I foresaw a great deal of activity in this area with the large budget cuts coming, along with my inexperience in this field. Several personnel folders were sent to me, but none impressed me too much. However, one was working under an old friend, Col. Dave Winn. I had worked and flown with Winn at ADC Headquarters. Winn had been shot down in Vietnam and imprisoned in the Hanoi Hilton for five years before being repatriated in 1973. I called Dave about Col. Creekman, who was working for Winn. Creekman had a great record, but as a Personnel Officer, not in Manpower. Winn gave Creekman a high recommendation saying, “Hire him, you won't be sorry.” That was enough for me. Winn's assessment was right on as I knew it would be. Getting Creekman on board to handle the Manpower situation that we faced from 1973 through 1977 was the best decision I made. Together we were forced by budgetary considerations to take the Command in Alaska from over 800 slots to slightly over 350 positions. We did away with Alaska Command completely. Every position we deleted was a battle that ended up in the Commander's office in a verbal dogfight between our office and the Commander or Deputy whose unit was going to lose positions. Our office had to have its position completely supported with facts and figures that were backed up by studies and logic. The bottom line was money. Naturally, our deputate was not popular throughout the Command. But we had our work to do, and we received our orders from the Pentagon with deadlines. It was not a fun era.

Dave Creekman was a pillar of strength, and he ran his directorate expertly. He lifted that off my back and permitted me to concentrate on bringing the Command's plans up to speed with the current intelligence situation. Dave and his wife Joyce became close friends to Alice and my family. This was a friendship that lasted many, many years after we both retired from the USAF to the state of Washington and beyond.

Gen. Jack Gamble

General Carson left AAC after a year and was replaced by Gen. Jack Gamble. Gamble was my former Commander when I commanded the 95th FIS at Dover AFB in 1970-1972. He had since commanded the 25th Air Division at McChord AFB. Now I was working for him again in Alaska. Gamble was a man I truly admired as a good human being as well as an effective military commander. He was operationally oriented, a hard worker and fair in his judgments. I was pleased to see him take over AAC. He knew me well; he knew my knowledge of air defense and he respected my opinions. He gave my deputate some added respect at the staff table. Due to the circumstances of my reassignment prior to my reporting in for duty, there was considerable tension between Col. Bierman and the operations people and myself. This soon spilled over to the people on my staff. When the manpower cuts arrived, and Operations received their fair share, naturally Bierman screamed that they were being discriminated against. This resulted in some heated battles that ended in the Commander's office for him to decide. I had no doubt Gamble would be fair minded and would follow the rules that were pretty well dictated by the management studies conducted by our manpower experts under Dave Creekman. When the decisions went against them, relations became worse. Plans and Programs were not the favorite organization in AAC, but in every case, we studied carefully what we were doing, whether what we were doing was critical to the mission, what happened if we stopped doing it or transferred that task to some other unit, and how many positions would that action save the Air Force. It was a brutally cold study in which our management experts actually followed the daily flow of paperwork, reports, meetings, and individual workload of every office in the Command. The results of these studies showed that the Command was manned excessively in relation to its workload and that a large portion of the staff work being done was unnecessary and redundant. The manpower cuts we recommended and that eventually were put into effect were startling. Predictions that the Command would be unable to meet its mission requirements with these reductions in place proved totally wrong. No sorties were ever aborted; no alert requirement was ever missed. The Commands had been excessively manned and, to put it bluntly, we had been wasting the taxpayer's money. General Gamble did not particularly enjoy this drawdown as no commander likes to see his command decrease in number. But the concept of having a leaner and meaner Air Force intrigued me. I saw a lot of waste in many areas, and I began to write down some ideas to work on that could save the service even more dollars that could be put to better use. Unfortunately, some of these ideas did not find favor with Gamble nor with several other members of the AAC staff

The Oil Embargo

Although I did not at all miss my past assignment in Saudi with all the problems associated with it, I did follow all of the news emanating from that part of the world. And there was much of it happening affecting us in the USA. One was the Yom Kippur War against Israel in 1973 in which the Egyptians and Syrians launched a surprise attack. However, the Israelis recovered quickly and pretty well trounced the Arab nations with much help coming from the United States. In retaliation, the OPEC nations quadrupled the crude oil prices to the western nations. I recall that gas prices in Spokane that winter were about 27 cents per gallon. But when the price war began, the prices went sky high. The Elmendorf Base Commander opened the motor pool gas pumps to military and retired government employees. These prices were far below what civilians were paying at downtown gas stations. I brought this to the attention of the AAC Staff and recommended that this practice be terminated in order to avoid a strong public relations backlash. The military was still undergoing a time of unpopularity as a residue of the Vietnam conflict and there were many groups who were overseeing what we were doing and seeking out anything that they could publicly criticize. But I was the one criticized as my proposal met with strong opposition by a majority of my colleagues on the staff. The Commander rejected the idea. However, within a few weeks, the daily newspaper printed an article exposing the base's policy showing photos of civilian RV's filling up on gas, plus some extra gas cans, at a price way below what the downtown Anchorage prices averaged. The next day a TWX from the Defense Department ordered us to shut down the motor pool operation. I made it a point not to bring the subject up at any ensuing staff meeting and no one bothered to mention it to me outside of my own people in Plans. But we had a good chuckle about it inside our own office.

Colonel Ron Linder

There were some cliques in the AAC staff, a sort of an old boy's network. The leader was Bierman, the Deputy for Ops. Bierman was a big chubby guy, a loudmouth, who often spoke without thinking. His background was in tactical fighters, but he had very little experience in air defense, and this showed whenever we were involved in NORAD exercises. The Headquarters building had within it a Combat Control Center from where the Commander and his staff directed the air battle. But the center was an old manual center with the vertical plotting board. Plotters on ladders behind the board, wearing earphones, and using colored markers, would depict on a back lighted vertical boards the position of the target and the interceptor aircraft. This system was antiquated and had long ago been put out of use in favor of the SAGE (semi-automated ground environment) system now used in the lower 48 by ADC. This system was computer driven and much faster and better able to keep pace with the swift air battle of the latest bomber and interceptor forces. AAC acquired a new officer who was assigned to Operations by the name of Ron Linder. Linder was a graduate of one of the military academies and had a splendid record which included combat sorties in Vietnam in SAC bombers and a few below the zone promotions, the last one to Colonel. Only a short time in his presence was enough to convince me, here was a brilliant guy. Any organization would have been delighted to acquire such a person, but not AAC Operations. The truth was he was too smart for Bierman. From the beginning, Bierman realized that Linder would not fit in because whenever Bierman said something dumb, Linder would correct him. And this was often. Too often, so Bierman decided to get Linder out of Operations.

Linder was a short man, maybe five feet, nine inches, but his weight was close to 200 pounds. He was round. The Air Force required its officers to meet height/weight standards and checked it annually. General Gamble became worried about Linder's health and suggested to Bierman that he get Linder into the weight control program. But Bierman wrote Linder a letter telling him to present himself to the Flight Surgeon for a weight evaluation. Linder responded that he had already consulted with the Flight Surgeon and was now on a special diet and that he had met with the hospital dietician and received excellent instruction with his wife on the purchase and preparation of the foods for the diet. This was in December 1974. Shortly after, General Gamble called me into his office. He told me that he was reassigning Ron Linder to my office. When I reminded the General that I had no position for him, he told me that he knew that. When I asked him why Linder was being reassigned from Operations, he told me it was because Bierman could not handle him, and I could. End of the matter. I saluted. Little did I realize that the Ron Linder story had a long way still to go before it would end. Shortly after Linder reported to me for duty in XP, I received a short note from Bierman informing me about Linder's weight control self-imposed program and his progress. Attached was the letter he had sent Linder and Linder's response. I put the note in a drawer and forgot about it as Linder appeared to me as progressing well in his program. I told him to keep me informed on his progress. I was pleased to have Linder on board in Plans as it was always good to be surrounded with smart people. They come in handy.

Major Adjustments

As time passed, I slowly came to accept the situation I found myself in. As a Deputy Chief of Staff of a major air command, although a small command, I held a responsible position where I could wield considerable influence and do good things career wise. But I was no longer in a flying position (also lost my flight pay), which decreased my chances of obtaining a command assignment. Further promotion was unlikely in an Air Force that was winding down in numbers and flying organizations. For the first time I was beginning to seriously think about my future both within and outside of the military. But these were just passing moments. As long as I had this job, I would give it my best shot. I perceived many areas in the command that I recognized should be reviewed and modified in the interest of cost savings and improved mission effectiveness. At the same time, I knew that to take on these changes would require a great deal of hard work and would face a huge amount of opposition, much of it right inside my own command. And there would be no certainty of success. Would it be worth the battles? That was the question I would have to answer within myself before embarking on any crusade for change. There were two major concepts that I identified after my first year that 1 believed should be looked at to study for change.

One was to replace the military personnel at our remote radar sites in Alaska with civilian contractors. This had been done many years prior at the DEW line sites, (distant early warning). Over two hundred military were sent to the Alaskan sites for remote one-year tours of duty. Before their deployment, they were given medical and dental exams, and during their year tour they received a 30-day leave. Transporting these personnel back and forth to these sites remotely located throughout Alaska was costly. Each summer these sites received its annual provisions of food and supplies by barge from Seattle towed up the Inland Passage and through the Yukon River. Throughout the year, C-130s from Elmendorf supplied the radar sites its mail and perishable food items and brought personnel in and out. The logistical cost of operating these radar sites was enormous. It was also hazardous. The landing strips at many of these radar installations were short, unpaved, on hillsides and often snow or ice covered. Many days in winter it was impossible to get into these places due to the weather conditions. All in all, keeping these old 1950's radar sites operating was draining on the Air Force budget, on personnel and was a real morale buster.

When I approached General Gamble with my proposal to overhaul this system and turn the maintenance and supply of the ACW sites over to civilian contractors, his reaction was furious. Despite my argument that this proposal would save the USAF millions of dollars per annum once the change was in effect, Gamble lectured me on what this would do the Alaskan Air Command as a military organization. I did not agree with his point because I did not see how permitting civilians to keep the radars operating in the cold arctic weather and having their employers pay them and transport them, etc., instead of the AAC doing those things, would substantively change our mission. Air Force military personnel, not civilians, would still scramble the fighters and direct them to intercept the enemy bombers. In short, AAC would retain the combat mission. But General Gamble would hear no more of it. My first brainstorm to make things better for hundreds of Air Force airmen did not get very far. I was shot down on takeoff.

I did have one other idea that lurked in the back of my mind. I came up with an idea again concerning the Alaskan radar sites which ate up so much manpower and logistics. These sites were also archaic and should be upgraded. AAC's command and control system was still utilizing the old vertical plotting board system that was slow and costly in manpower. It was time for this system to be overhauled and brought up to the standards now employed in the NORAD SAGE system used in the lower 48. I was determined to discover whether this was a feasible concept or not. It was not easy and not too many people I spoke with were congenial to my idea. But once I had proof that this concept was feasible and cost effective, I was certain I could sell this idea at the highest levels. I made this idea my personal goal. But I needed to have some research done first before I brought this one into daylight. Without telling anyone, I had my staff begin researching this problem both from a technical side and from a manpower side. I wanted to know were there radars out there that would meet the specs I needed. If we could find them and procure them, it would change AAC in a big way.

The AAC Staff and Me

Outside of Col. Bierman and the operations people, I and my staff in XP got along pretty well with the remainder of the AAC staff. But Bierman continued to complain about how manpower had hurt his section by cutting too many positions. He also headed up a clique of people that included the Intelligence chief and some others on the staff such as the Judge Advocate General and the Command Surgeon. He also complained to the Wing and Base Commanders telling them the XP was hurting them and the Command. I did not pay much attention to him because his complaints were easily refuted when they surfaced. And while he made a lot of noise at the staff meetings, it became dear that several members of the staff paid him little heed. After I squelched him a few times at staff, he then rarely brought up his beefs at those meetings, preferring to get his digs in against me behind my back to his pals at the bar or in their poker sessions.

But a real squabble broke out when the Director of Intelligence chief circulated a paper updating the enemy threat to Alaska. This secret document purported to outline a real threat existed that the Soviet Union could attack Alaska with ground forces crossing the Bering Sea in winter when the sea is frozen. The Soviets, using their experienced cold weather army forces, plus tanks, could invade Alaska and take it over quite easily due to the absence of sufficient US ground forces capable of defending such a widespread force. The Soviet Union could take over the valuable oil resources developed on the north slopes and put out of commission the DEW line and other air defense units in Alaska belonging to NORAD thereby opening a clear path should the Soviet Union follow with an air attack on the North American continent. The paper went on with a scenario supporting that the Soviet Union had shown signs of placing such forces in its eastern provinces. This paper had already circulated through the AAC staff for coordination prior to it being sent to AAC Plans. Should I sign off on it, it would proceed to the Commander for final signature and transmittal to higher headquarters, meaning the Pentagon. Presumably, the purpose of this document was to support additional forces being allocated to Alaska.

When I read the paper, I almost broke out laughing. I could find no rationale to support why the Soviet Union would ever consider such an outlandish military action. I wondered if the intelligence chief had ever heard of the battle of Stalingrad or Leningrad. I circulated the paper through my staff and found no one supported its premise. I called the Intelligence chief and asked him what evidence he had from his intel sources that backed up his conclusions. He had none. In my position as Deputy of Plans, I was cleared for quarterly briefings that were of a higher classification than Top Secret because the source of the data covered was so sensitive. These briefings were given in the basement of the Headquarters building in a vaulted room and were limited to only a few high-level staff officers. None of these briefings ever touched on any possible threat of a land attack on Alaska. I refused to sign off on this paper. My refusal to do so caused quite a stir with my fellow deputies (not all of them), and a few warned me that General Hill would be quite upset with me about it. But I never heard from Hill about it. My opinion was that the General agreed that the paper was absurd and would be received at the Pentagon with total disbelief.

Col. Harry Goodall

With the realignment of the commands following the deep manpower cuts, Lt. Gen. James Hill replaced Gen. Gamble as commander of AAC and Colonel Harry Goodall came in as vice-commander. Hill came in from the Pentagon and was a completely different personality from Gamble.

Goodall, I knew from ADC. In fact, he had been an F-106 pilot in the 318th FIS at a period when I was in ADC Headquarters. Goodall was on a fast track. He was junior in rank to most of us on the AAC staff. His reputation was that of a careerist and he would trample over anybody who stood in his way to becoming a general officer. I was appointed to be his host on his arrival on base and I did my best to please him and his wife. I could tell that he was going to be a pain to all of us during his tenure in the command. However, I knew him well enough not to let his eccentricities bother me. In fact, I was beginning to come to grips with my situation and giving more thought about my personal future. Determined not to let things bother me as I did in the past, I concentrated on spending more time with my family and enjoying Alaska. At the same time, I decided to do the best job I could to make a difference in AAC before I moved on. I began to do a lot less worrying about my evaluations and use my time into making AAC a more effective and productive command. And I would do this with or without the approval of General Hill or Harry Goodall. Both of these men had other aspirations of their own. I soon found that Bierman and some other members of the staff were sniping at me and my staff behind my back. Much of this was due to envy as several of our plans and programs were well received and incorporated by the Pentagon. I also put Col. Linder's great brain power to good use assigning him some studies that he produced that were outstanding for their thoroughness and depth. The man was a genius and was certainly under used by the Command. But Ron Linder was about to get himself and me in a pot of trouble.

Attitude

As I settled into my work as a high staff officer, I began to put a lot of things into place that I had not done before. In my previous assignments mostly as a flying officer, I was engrossed in the day-to-day operations that took place on the tarmac where mechanics and armament guys and aircrews worked to launch planes into the air, on time, and return them to the ground after completing their missions successfully, and safely, in any kinds of weather, day and night. That to me was the Air Force. But now I wore an eagle on my shoulder and my duty station was a desk. I had a secretary, and my workday was meetings, and TWX's and letters and briefings and phone calls and lots of reading. And there was much politics and working with Colonels and even Generals many of whom had less experience in flying planes and knowing what it took to make life and death decisions on the ground and in the air. But they had power and they had ambition. And they often made awful judgments. I had no choice but to live with this and work with it; I had put myself here and now I had to make the best of it. It would depend on me and the attitude I brought to my work and to ensure that none of this would hurt my relationship with my family. After all, I took this assignment primarily so I would have more time with them and by God, that's exactly what I intended to do. I owed that to them and more, and my frustrations at work would not deter me from that goal. Charles Swindoll said it all:

> The longer I live, the more I realize the impact of attitude on life. Attitude to me, is more important than facts. It is more important than the past, than education, than success, than what other people think or say or do. It is more important than appearance, giftedness or skill. It will make or break a company… a church… a home. The remarkable thing is we have a choice every day regarding the attitude we will embrace for that day. We cannot change our past… we cannot change the inevitable. The only thing we can do is play the one string we have, and that is our attitude. I am convinced that life is 10 percent what happens to me and 90 percent how I react to it.

And so, I made a decision to bring about a life change in how I would look at things and react to them. I made an inventory of what were the more important things in my life and what were the less so. In 1974 I was 45 years old, with five children of whom four were in their teen years. My flying days were probably over, and I had to begin seriously thinking about post Air Force days... doing what? Living where? Gene Jr. and Jim were coming up to college age soon and Alice and I had to put some dollars away for that too and beyond. So, I started to turn my mind inward to affairs that were close to my own heart, those of my family and my own future. And this meant a change in attitude. I still loved the Air Force, and I always would. My attitude toward doing my best at my job would not change, but from here on, my priorities were clearly rearranged. Alice Marie and my children moved to the top of my list where they probably should have been all along. As a family we did spend more time together and as a family we began to love Alaska and what it had to offer in both the summer and winter months. The boys immediately got into hockey and skiing. Michael picked up some friends as did Mary and Katie.

There was lots of recreation on base and off base, including a pool, a ski slope, tennis courts, and a golf course. Arctic weather did not curtail our enjoyment. And I began to unwind. Alice and I had great close friends in Dave and Joyce Creekman. With a few minor exceptions, life on the home front, became better. Four of my children were in their adolescent years with the many difficulties that come with that period of life. I was pleased that I would be home with them much more so than in the past several years to be of whatever help a Dad could be. The boys were kept busy with their sports and with their schoolwork Unfortunately, their newly constructed school was a failure. Bartlett-Begich was a disorganized school that had a multitude of problems in all areas, and Alice and I were very disappointed. Up to this point, our children had attended excellent schools and had received the best education. But this was a disaster for them, and we had no choice but to wait it out until the system improved and things in the classroom got better.

My daughter Mary had the most difficult time adjusting to our move to Alaska. Mary was in her sophomore year in high school and again she found herself having to make new friends in a strange place. And the school did not help. When she finally did find some friends, Alice and I thought they were not the most wholesome and when we let Mary know our thoughts about that, she, of course, rebelled. It was a tough time in her life for the first few years there and we worked hard to keep close to our family. Ironically, Mary ended up loving Alaska and almost refusing to leave when our time was up. After a year of college at Gonzaga, Mary returned to Anchorage and lived and worked there for a few years. She did some growing up there and finally finished college at Eastern Washington University. The military life was hard on Mary during those years. But, looking back, I believe my children also gained much from the several moves we made. They did see much of the US including Alaska, which they truly enjoyed. They also made many friends from all over and they became pretty independent and flexible. In my new staff job, I was home most every night for dinner and weekends. I was able to watch the boys in their sports and we returned to skiing again as a family. We also took several trips around Alaska, visiting Denali National Park (Mt. McKinley), Seward (salmon fishing), Homer on the Kenai Peninsula, and Valdez, the drop off of the Alaskan pipeline. In the summer, the boys and I made great use of the base golf course and the tennis courts. I also took my annual leave each August and the family and I (including our dog Brandy) traveled to Lake Pend Oreille where we opened the cabin and spent three great weeks. It was a great time for us, and we seemed always to enjoy great warm sunny days in northern Idaho. We gave the *Alice Marie* a terrific workout, exploring this beautiful lake with lots of water skiing. My dad joined us and he and the boys and I did some projects on old Jack's Cabin. In the summer of 1974 Freddie Fanin and his crew built a dock and a boat hoist for the *Alice Marie.* Watching Freddie's crew at work provided all of us hours of laughs. But they got the job done to the surprise of everyone. It was always a heart break to depart from the lake and return to Alaska. But this was what I wanted and needed. I was now closer to my children than ever, and Alice and I saw them inching nearer to adulthood. We cherished the hours and days we were able to be with them and guide them. We realized that the road ahead for them would bring many hard decisions and all of them would not be the perfect ones, but we would be there to do whatever it took to get them through as best we were able to do. Despite my disappointments with my personal career outcome, I was feeling much better about more things than I had ever felt in the past. And about myself too.

Deep Personal Losses

During my early years in Alaska, both Doug Brenner and Harold (Yogi) Van Daele died. Both men were in their forties. Both had served their country in war and peace, and both were fine family men, strong fathers and faithful loving husbands. Both served under me and gave me tremendous personal support which I will never ever forget. I flew with them, and they were both fearless intrepid fighter pilots of the highest order. Highly skilled and always ready to fly, they enter my thoughts and prayers daily even to this day.

CHAPTER THIRTY-FIVE: DAD JOINS OUR FAMILY

Half the world is squirrels, the other half is nuts.

- Joe Cunneely, Sr.

When our family departed Dover for Alaska, this was not good news for my dad and Alice's family. During our two years on the east coast, we were able to see our families again and visited them in New York many times. Alice's mother came down to Dover and my brother and Dad did also. Dad had been long retired from the LIRR and still living in E. Patchogue, Long Island. But his two sisters, Anna and Lillian had died, and he was alone. My brother Joe and his wife Pat were living in a condo they had bought in Lynbrook and only came down to Patchogue on the weekends where they had a small weekend home close to Dad's. He worried about Dad living alone and having to do so much yard work, and he helped Dad out a lot by mowing the lawn for him and other chores about the house. My brother had a bad back and so, this was not too easy for him either. Then my brother Joe saw a great real estate opportunity there in E. Patchogue. He approached our father to go in with him and purchase a home that was up for sale on the water. Reluctantly, Dad consented, and they both sold their former Patchogue homes and bought the waterfront house for $36,000 with my dad putting up half the purchase price. However, Dad asked my brother to provide him a promissory note for $18,000 in the event he would have to leave the house for some reason. Now in 1974, things were not going well between my dad and his daughter-in-law. I began to get wind of this from both Dad and then my brother. Being in Alaska, and pretty far from the scene, I was frustrated in trying to get a handle on the problem. In November of 1974, I received a letter from Joe that spelled out the problem in dire terms.

It was clear that he was going to have to ask his father to leave the house. It was something that he clearly did not want to do and never ever expected that he would ever have to do. When I saw my brother and Pat when they visited us at the Lake cabin in the summer, I spoke to them about the problem. My brother was deeply depressed about the situation. Relations between his father and his wife had deteriorated to the point they were barely civil to each other. He was at his wits end on solving this problem. Any hopes that I had of reaching some mutual solution to this problem proved hopeless. When I talked to Dad, it was just as bad from his side. Dad was still spending his winters in Lakeland, Florida, and that helped. But when he returned in the spring, the problem was still there.

It came to an end in the spring of 1975. During this period, I was involved in several programs that required my visiting the Pentagon. I took advantage of these TDYs to visit Dad and Joe at Patchogue. I would take the Amtrak from Union Station in Washington, D.C. to Penn Station and from there the URR to Patchogue where either Dad or Joe would pick me up. These visits were not all pleasant as there was tension in the house. It was during one of these visits that I mentioned the idea of Dad coming to live with me. Joe was dubious about this idea, but he was, at the same time, desperate for a solution. I told him I would have to talk this over with Alice Marie. On my return, Alice and I discussed the matter. She, of course, realized what a change this would render on our household. But she loved my father, and she knew that our kids also loved him. She consented to the idea. Again, my wife demonstrated what a terrific person she was. I then mentioned the idea to Dad. He had been thinking of getting his own apartment in Patchogue or perhaps moving to Florida. But I opposed those ideas. The previous year, while driving to Florida, he rolled his car over somewhere in the Carolinas totally destroying the car but miraculously escaping serious injury. Both my brother and I did not feel that he should live alone, nor should he continue to drive a car as he was no longer a safe operator of a vehicle. I finally convinced Dad that he should come live with us. Once he decided, he was anxious to move out as the situation in the house with his daughter-in-law was extremely bad. In July, I had another business trip scheduled at the Pentagon. I added to it my three weeks leave. Alice and the boys flew down to the lake. After concluding my business in Washington, I took the train to New York and then to Patchogue. Though I arrived fairly late at night, Dad was packed and ready to go right then and there. I told him he would have to wait until morning as I needed to get some rest before starting out on a five-day drive to Idaho.

That night was a terrible night for my brother and I. Joe was visibly upset at the way things had turned out between his father and himself. This was certainly not the way he had planned it. No civil words were spoken between them. After Dad had gone to bed, I tried to reason with his wife, Pat, but I got nowhere. She was furious that my father had demanded that they repay him the $18,000 due on the promissory note. My brother did not object, but he was overridden by Pat. He told me that he would pay my father the money due.

The next morning, my father and I packed the car with his meager belongings, and we set out for Idaho in his Oldsmobile. The goodbye between my father and my brother was heartbreaking to behold. It was a family tragedy. It should have never happened. My brother was nearly in tears and my father was furious with anger, mostly against his daughter-in-law. Before we left, he gave me his pocket watch, the one he used over the years that he carried in his vest pocket with his Elk's tooth on the opposite end of the chain. He told me to give it to my brother Joe, which I did. This was a most emotionally moving moment for both of them and for me as well. It was not right that their relationship should end like this. Here was a father who had devoted his life to raising his sons when their mother walked out on the family. He never remarried. He sweated out and prayed for the safe return of his son during the great, terrible war. And now he felt that his son was throwing him out of his home, a home he had helped him purchase when he had asked him to. As my father often said: *"There's no justice."*

Our trip across the country was uneventful. Dad talked on and on about his problems with Pat and I tried to listen patiently without throwing any more fuel on the fire. I also attempted to defend my brother as a husband caught in an untenable position between his wife and his father. I also discovered that my father's driving was worse than I realized. At one point I permitted him to take the wheel to give me a spell. Within minutes I realized what a mistake I had made. Before I knew what had happened, the car had sped over 90 mph. From that moment on, I knew that Dad's driving days were over. When we reached Lake Pend Oreille, I was exhausted. I told Dad he was finished driving and we put his car up for sale. The kids were delighted to see him. They truly loved their grandfather, and he truly enjoyed their company. Being around them revived him and helped him forget his personal problems from back east. He enjoyed the lake, Alice's great home cooking, the company of Sam and Dee Grashio, and a whole new way of life. Alice, who took on the greatest burden in welcoming another person into the household, once again demonstrated what a tremendous person she was in all ways. She now had another mouth to feed, more laundry to do, another bed to make, and since Dad no longer drove, she also had to cater to his travel needs. She took it all in stride. But, the Cunneely family had made a major change that was very significant and would become more burdensome as the years rolled on. That was for sure.

Dad and His Grandchildren

A huge benefit of my father moving in with us was the close relationship he developed with our children. They had always loved him, but they had only seen him on his visits at Christmases and in the summers. We also had spent some terrific times with Dad at his Patchogue home on Anthony Street when we visited him there when we were stationed at North Bay and at Dover AFB. The kids loved his marvelous sense of humor and his unique, wild sayings that he repeated to their delight. Dad loved golfing and he frequently played golf with the boys. By no means was he an accomplished golfer, but he enjoyed the game and playing with his grandsons. Elmendorf had a great course, the first course in Alaska and I had a family season pass for a ridiculously low price. Dad and the boys played that course every summer sometimes three or four times a week. In the fall, Dad went off to Lakeland, Florida, where he spent the winter months, returning in the spring. There he met up with his retired cronies, played some more golf and enjoyed the warm weather. He also joined us in the summer when I took my leave and traveled to Lake Pend Oreille. There we opened the cabin and spent three great weeks at our own modest little log house that Jack Chambers built. Alice and I were amazed at how Dad accommodated himself to these travels and different conditions. He loved the lake and enjoyed our trips across the lake on the Alice Marie to the Litehouse Restaurant in Hope. There we enjoyed sumptuous steak dinners and delectable desserts, then an adventurous, moonlight boat ride back to our cabin.

Following my retirement and our move to Spokane, Dad came with us and lived with us there, continuing his annual routine of spending his winters in Florida until he had an unexplained run in with his landlord, an old pal down there. This caused him to return early, and he no longer returned to Florida. At that time, I was working for Gonzaga University and the boys were in college. It was in the early 1980's and we began to notice signs of dementia in my father that became more worrisome as time went on. It was becoming more and more difficult to take care of him. His memory began to fail, and he began to wander from the house and get lost. When son Gene got married in Hawaii, Dad traveled with us to Honolulu to attend the wedding. During the week we were there, he wandered off several times, once setting off a city-wide search. Thanks to the Honolulu police, he was found and returned to us unharmed. His confusion became worse and worse. Not long after, we made the decision to place Dad in a retirement home near Gonzaga. It was a comfortable apartment where he was reasonably secure and well taken care of from my job, I was able to visit him almost daily and ensure that he was okay.

But Dad was not happy there and his confusion got worse. Often, he was not sure who I was, and I found that heart breaking. Then he began to wander away from the retirement home and get lost. I had placed a wrist band on Dad providing his name and my phone number in case he became lost. This wrist band came in handy several times. Often, I would receive a call from someone who had found Dad. They were enjoying his company, finding him to be delightful and funny. I would hop in the car and retrieve him. Many nights I would stay the night with him in his apartment to reassure him. My concerns for him were growing greater daily as he was getting worse mentally and physically.

Disaster then Death

In 1984, Dad again left the home and fell in the parking lot breaking his hip. Fixing his hip was impossible and he was confined to a wheelchair from then on. I had him moved to a nursing home in the Spokane Valley. I knew then that he would never leave there alive. Seeing my father in this situation was the most heart-rending situation I could ever imagine, or I would ever experience. To see him as he was, was almost more than I could bear. When I visited him, he most often did not know who I was. He would ask me, repeatedly, if his brothers and sisters were alive. Though I would answer him truthfully, he would ask me the same questions again and again. He did not know where he was. I visited him during meal hour as he could not feed himself and I did that for him. I usually left in tears. Thankfully, Alice and other members of the family also visited Dad often.

In 1985, my brother and his wife visited Spokane. I had been keeping Joe abreast of Dad's condition. I thought it would be right for him to visit Dad and see him before it was too late. I drove Joe over to the nursing home. It was a warm sunny day and we brought Dad outside in his wheelchair to sit in the sun. Joe was taken back when he saw Dad. He saw immediately that Dad was nearing the end. I believe that somehow, Dad did recognize Joe, but I will never be certain. It was an awkward meeting and Joe came away somewhat shaken.

On May 11, 1986, I was called by the nursing home at work. I was told that Dad was dying. I called Father O'Malley at St. Paschal Parish. I asked him to go to University Gardens Nursing Home and give my dad the last rites. This he did. I rushed over to the home. Dad was in a coma, resting comfortably and in no pain. I prayed for him. The next morning his heart stopped. My father, who stayed with his sons when our mother left us, had now passed on.

In accordance with Dad's wishes, I made certain that he was given a funeral mass in the church where he was an altar boy, St. Raphael's, in Woodside, New York. I called my brother Joe, and when I notified him of Dad's passing, I told him that he wanted to be buried in Calvary Cemetery where his father worked. Joe made the arrangements at the New York end. After a vigil service here in Spokane, Dad's body was flown to New York. There he was given a Mass at St. Raphael's at which I rendered a short eulogy. Dad was then buried next to his father and mother. I realized what a great guy he was. When his wife left, he brought his two sons back to his mother's house and ended up taking care of her and his two sisters and his two brothers as well. He served his country and loved his faith. He also loved his grandchildren and his daughter-in-law. He also loved my brother despite a breakup that hurt him deeply. His was a hard life. Still, he never lost this love for a strong drink and a good laugh. At his death, Dad was 93 years old. May he rest in peace.

At the time of our father's death, my brother Joe had been diagnosed with cancer. His situation was growing worse, and he was undergoing radiation and chemotherapy. The day of Dad's funeral he was in serious pain. I begged him not to attend, but he insisted. He said:

"Nothing will keep me from attending my father's funeral, nothing!" With that he crawled into the back of the limo and off we went. It was a very bad day.

CHAPTER THIRTY-SIX: I AM COURT MARTIALED... ALMOST

Do you swear to tell the truth, the whole truth, and nothing but the truth, so help you God?

Blindsided

Air Force Manual 120-3 outlines procedures by which a commander determines whether a criminal offense has been committed by someone under his jurisdiction. The manual sets forth procedures by which an officer or a board of officers conducts an inquiry to determine if sufficient evidence exists to convene a court martial or to have the commander exact summary punishment on the individual(s) found guilty by the investigating officer or board. In April 1976, this is the position I found myself in much to my great surprise and consternation.

This matter came as a complete shock to me. It resulted from the feud going on between Col. Linder, who had been shunted over to my Deputate by Col. Bierman a year earlier. I mentioned this in Chapter 34. Although Linder was a brilliant officer and a great addition to our section, he brought with him several personal problems that dragged me into a serious situation I really did not need. And this was not the last one of them.

Briefly, Bierman, at the instruction of General Gamble, had ordered Linder to put himself on the formal weight reduction program in accordance with AFR 50-49. This was done just prior to his transfer from Operations to Plans. However, Linder did not do that. Cleverly, he had previously seen the Flight Surgeon and together they had set up a weight reduction program for Linder without formally entering him into the AFR 50-49 program. Linder wisely appraised Bierman of that in a letter to him, but Bierman apparently did not understand completely what Linder had done. In the final OER that Bierman wrote on Linder prior to his transfer to Plans, he stated that Linder had been entered in the AFR 50-49 weight reduction program. This, of course, would be viewed as a derogatory comment on Linder. Linder appealed the OER claiming that he had never been formally entered into the AFR 50-49 program, which was a true statement. When Bierman got the OER back from USAF, he was furious. He felt that Linder had lied. He went to Col. Bergeron, the AAC DCS for Personnel. They began to search around for the paperwork which would support Bierman's claim that Linder had been entered the AFR 50-49 program. Bierman then remembered the Memo he gave to me. Following a morning staff meeting, Bergeron asked me if I had Bierman's letter concerning Linder's weight control program. I told him I was not certain I still had it and I may have disposed of it. I asked him why he would want it. Bergeron did not answer my question which left me puzzled. At that time, I had no inkling of what was taking place and I soon put it out of my mind. But a month later, on returning from a lengthy TDY, I was called into the office of Col. Roy Brunhart, the AAC Director of Communications. I was shocked to learn that I was accused of wrongfully destroying official Air Force records and failing to comply with AFR 50-49 by not forwarding records pertaining to Col. Linder's weight reduction program to the Deputy, Base Medical Services, when Col. Linder had completed the program. Brunhart asked me to provide sworn testimony on these charges. I refused to do because I had not been informed that I had been accused of any violations of AF regulations and I had no time to prepare to reply. I was astounded that what appeared to me as an offhand request by Bergeron for a copy of a year-old Memo from Bierman, without explanation, had resulted in my being charged by General James Hill with possibly serious crimes.

The Defense

After a brief interview with one of our AAC lawyers, I did my own research into AFR 50-49. I also found my copy of Bierman's Memo in the back of one my desk drawers. My memory had served me well. While I was not permitted to speak to Linder, who was also facing two charges more serious than mine, it soon became clear that neither he nor I were guilty of any charges that had been referred. This was a clear case of dislike and prejudice against Linder, and I happened to have gotten caught up in it because I was totally unaware that Linder had disputed the OER that Bierman had written on him and that the weight control program thing was at the center of the problem. Bergeron had blindsided me, but why they concluded that I could be found guilty of willfully destroying Air Force records pertaining to the entrance of Linder into the weight reduction program when he had clearly had never entered it was beyond my understanding.

On April 14th, 1976, I gave my sworn testimony to Col. Brunhart. The session took about one hour. I declined my rights to remain silent and to have a lawyer present. In answering Brunhart's questions I reaffirmed that in receiving Bierman's letter and Linder's response, it was clear to me that Col. Linder had not been formally entered into the AFR 50-49 weight reduction program. I also informed Brunhart that I had spoken to Linder about his self-imposed weight control program and informally counseled him about it. At the conclusion of his interrogation, I added my own statement. In my statement I reviewed the documents Bierman had given me which he claimed was an official Air Force record (a case file) of Col. Linder's AFR 50-49 weight control program. I showed him the letter and pointed out to him that the letter and Linder's memo was no such official file and why it was not. And that's why I did not treat it as such. I told Col. Bergeron I may have thrown it out because I wasn't sure I had retained it, although, as it turned out, I still had it in my desk. And I gave it to Brunhart. I then went further in demonstrating to Brunhart that the Bierman letter and the Linder memo came nowhere near forming a letter of notification that AFR 50-49 requires. In addition, Linder never did receive a medical evaluation from the Flight Surgeon required by AFR 50-49 for entrance into the weight control program. In short, despite what Bierman thought, Linder had never been formally entered into the AFR 50-49 program and the Bierman letter never constituted a case file.

Col. Linder was charged (1). with failing to obey instructions contained in Bierman's letter i.e., to report to the Base Medical Services Unit for a medical evaluation and placement for a medically supervised weight reduction program, and (2). Falsely stating in his letter, subject: "Correction of an Officer Effectiveness Report," dated 5 Feb 76, to AAC/DP, that he, Col. Linder, had never been placed on a weight control program. It was Linder's letter to Bergeron regarding his OER that touched off the explosion in AAC headquarters. I had never seen the OER, and I was unaware that Linder had written his letter appealing the OER. When Bergeron asked me for Bierman's letter, I had no idea why he was asking for a letter over a year old on an issue I had thought was a dead subject. When I told him I probably no longer had it, I did not realize the implications of my answer. Clearly, they thought I was trying to help Linder by destroying evidence. But the "evidence" they were seeking turned out to be proof that Linder was absolutely correct when he wrote that he had never been placed in the program. But it did not help him.

Along with and following my testimony, witnesses included Col. Bealor Rogers, the AAC Command Surgeon; Major Douglas Shirley, Flight Surgeon; Col. Charles Koeninger, Director of Air Operations, Elmendorf AFB; Col. George Bergeron, DCS Personnel, AAC; Col. Casper Bierman, DCS Operations, AAC; Col. Ron Linder, Director Studies and Analysis, AAC.

Col. Rogers testified that Col. Linder had indeed seen the Flight Surgeon regarding his weight problem as Linder told Bierman in his letter. He also confirmed that Linder had not been placed in the medically supervised weight control program in accordance with AFR 50-49 as both Linder and I indicated.

Major Shirley confirmed Linder's visit to him and his setting up an interview for Linder with the dietician. He also confirmed that he did not set up Linder for formal entrance into the AFR 50-49 program.

Col. Bergeron, the AAC Deputy for Personnel, and the one who had asked me for the Bierman letter, issued a statement that was somewhat critical of Linder's non-standard indorsement of Bierman's letter. And he also had to fault Bierman for failing to send the letter to the Director of Base Medical Services as required by the regulation. This failure prevented necessary action to be taken entering Linder into the AFR 50-49 program.

Bierman testified that he thought that he had complied with the regulation. He also spoke about transferring the "case file" over to XP (Col. Cunneely). His response to Brunhart's questions concerning Linder's indorsement that he had already seen the Flight Surgeon and was on his own weight control program, was vague and showed that he was clearly confused about what the regulation required.

Most interesting was Col. Linder's testimony. Linder outlined a history of his problem with Bierman going back to when he was assigned to Operations. His testimony became very personal describing Bierman as an ill-tempered individual, a heavy drinker, a man who had many personality conflicts. He also alluded to the fact that on command matters Bierman and Col. Cunneely did not get along and had many run-ins. When Linder became acting Chief of Staff, Bierman was incensed and demanded Linder address him as "Colonel" Bierman. They also had problems over OER's. Linder's description of the weight control incident clearly showed that he had outfoxed Bierman, Bierman not being as well educated on the regulation as Linder.

Brunhart's Facts and Discussion

The AFM 120-3 Report of Inquiry requires that the investigating officer provides the Commander with the facts he uncovers followed by an analysis of those facts. Brunhart's facts and discussion (as it is called) was fairly tepid. However, he did state that it was clear that Col. Linder had not been officially entered into the AFR 50-49 program due to omissions on the part of Col. Bierman and the hospital. He also stated that no case file had been established. (As I had been charged with destroying the case file, this was important.) Brunhart did accuse Linder of making a premeditated attempt to later confuse whether or not he had entered the program. Later in his analysis he came down hard on Linder in Linder's appeal of his OER where Linder stated that Bierman had included "false and incorrect statements." Brunhart called Linder's use of the unit commander's letter to support his claim to be "unsavory."

Regarding the charges against me, Brunhart admitted that they were less serious. He seemed to doubt that my response to Bergeron, that I did not know if I still had Bierman's letter, was truthful and he questioned my "lack of concern." He appeared to have overlooked my testimony that no one had informed me in the first place what the fuss was all about. The more I reviewed Brunhart's analysis, the more I was convinced that he was biased through his close association with many on the AAC staff who were Bierman' s drinking and poker buddies. Some of his conclusions did not appear to be supported by the facts presented by the testimony. While there was no doubt in my mind that Linder outfoxed Bierman, who was as dumb as they come, he certainly did nothing illegal.

The Verdicts Are Rendered

After a few weeks I was called to General Hill's office. I knew what this meeting was about. Everything inside me told me that there was no way Hill could find me guilty of either of the charges, but I wasn't positive. I had lost faith in Hill, and I had come to dislike Harry Goodall, the Vice commander. Both of these men, while extremely intelligent, lacked the level of leadership that I had grown to trust. Why neither of these men had seen through Bierman and his shallowness was beyond me and most of the staff. Bierman was a loudmouth who operated his department through bullying and threats. He lacked knowledge in air defense which was AAC's major mission, and he was not above lying to get out of embarrassing situations he had gotten himself into. He hated Linder, mostly because Linder was so much smarter, and he disliked me because Plans and Manpower did not put up with his nonsense and called him out when he brought out ideas that were totally unfeasible. He did most of his mouthing off against us at the Officer's Club bar.

After seating me, Hill told me that he had reviewed the investigation report and had come to his decision. He told me that the charges against me and Col. Linder were serious, but he was a fair man. He had concluded that I was exonerated on the charges against me. He asked me if I had anything to say. I had prepared myself. I told General Hill that I appreciated that he had found me innocent and that I felt that I was innocent from the beginning. I did feel that I was blindsided in that I was called to testify without being notified what the charges were against me and what the charges were against Col. Linder who worked for me. This seemed to surprise him, and he said he was unaware of that. I then told him that I did not agree that Linder was being fairly treated and that it looked to me that there was bias against him in the headquarters. Certainly, there was a bias against him in Operations. Hill did not agree with me on this one and cut me off abruptly. He told me that what Linder did was not proper and would be dealt with fairly. He then dismissed me. I left with a feeling that Linder was not going to get off lightly.

I knew then that Linder was going to receive a punishment for the charges against him. A few hours later, Ron told me he had received a letter of reprimand. The letter would be placed in his personnel file for a period of one year. A letter of reprimand is a death sentence for an officer as far as any further promotion or favorable assignment is concerned. Ron was very upset. He left with a look on his face that told me that this might not be the end of this matter as far as he was concerned. Ron Linder was not someone you wanted to trifle with.

Aftershock

This whole episode had a terrible effect on me. I saw it as a total failure in leadership at a high level. Generals and Colonels, with years of experience, should have handled these kinds of personnel matters in a much better fashion. I perceived that in this case they used just awful judgment. I also saw great gobs of time and dollars wasted. I saw the career of a most brilliant officer tragically thrown aside for a petty incident caused by a clash of personalities and ineptness on the part of an officer (Bierman) who was in way over his head and made worse by superior officers who either did not recognize it or did not care. I began to feel some disillusionment set in. Up to this point I had enjoyed such a wonderful and enjoyable Air Force career. I had been fortunate to have served alongside and under some truly superb officers and airmen. I loved the challenges, the camaraderie and the successes. Everywhere I had been stationed, I had enjoyed the people and the community. But Alaska had become a difficult and discouraging situation. I had been asked to come to Alaska by an old friend, General Donavon Smith, who promised me a job in Operations, where I would put my air defense knowledge to work, and I could continue flying. But that went by the wayside when General Smith went down with a brain tumor and a medical retirement. That opened the door for Bierman to have my assignment changed to Plans before I even reported in and without my having been informed. Thus, my flying days were ended. I found myself in a job I did not want, and I also found myself working in a situation that put me and my staff in conflict with the rest of the command. Huge budget cuts caused by the Vietnam War years inflicted large manpower decreases that XP had to manage. Alaska Command (ALCOM), a joint command consisting of Air Force and Army personnel was to be deactivated. Within a year, XP was tasked to identify nearly 500 USAF positions for elimination. Col. Dave Creekman and his Management Engineering Team went to work, and we did the job, but in doing so we ended up engaging in fierce verbal and paper battles with commanders and staff officers up and down the line who fought tooth and nail to save their organizations from any cuts. It was exhausting work that brought us many enemies. In 1973, we had over 800 USAF positions in Alaska. By the end of 1975, we were down below 350. Yet, we never missed a scramble or a support mission. In place of ALCOM, we had the Joint Task Force Alaska which utilized existing AAC and Alaskan Army personnel to defend Alaska in the event of an attack. The cost savings were enormous.

Reflection

I determined that, as I was stuck in the position I was in, I would continue to perform my duty the best I could until the end of my service. In 1975, I was completing 24 years of active duty, and I began to think about my next assignment and maybe even retirement. Despite my unhappiness at the workplace, I really made up for it at home where the hours and days I was able to spend with my wife and children were providing me a satisfaction and a contentment that I had not had since our first years of marriage. I was golfing, skiing, watching my sons play hockey, and just doing all kinds of family things that for many years I was not able to do. I became a Lay Eucharistic Minister at church and even taught Sunday school. When Cardinal Cooke visited Elmendorf, I was his escort. What a great experience that was for me. All was not bad, thanks to Alice Marie, my beautiful wife and my five great kids. And don't forget Dad who lived with us and was always good for a laugh. Thank God, the kids loved him too.

Linder Fights On

Ron Linder received a Letter of Reprimand. He was found guilty of disobeying Bierman's instructions to enter himself into the AFR 50-49 program. The letter stated that Linder purposely maneuvered to avoid getting out of entering the program and then falsely appealed the OER Bierman had written on him. The Letter was to be entered into Linder's personnel file for a period of one year. This reprimand was a kiss of death on Linder's Air Force career. I felt it was a most unfair verdict. It resulted directly due to Bierman's ignorance of the AF Regulation that prescribed what needed to be done for proper entrance of an individual into the weight control program. Then he wrote it into the officer's OER. He clearly did not know the definition of a "case file" as he told the court he had given one to me on

Linder's weight control program. I was then accused of destroying an official Air Force document (the case file) which never did exist.

But this did not end the matter, not by a long shot. General Hill and Colonel Goodall did not know who they were dealing with when they took on Ron Linder. Again, without my knowledge, Linder appealed the verdict. And he did so with a brilliant rebuttal of the charges against him and the verdict outlined in the Reprimand. Using legal logic that was unchallengeable, Linder tore apart the case against him and the verdict. He asked that the verdict be set aside, the Letter of Reprimand be removed from his file, and the OER, written by Bierman and endorsed by Goodall and Hill, be expunged. It was a devastating shot back at the Command leadership.

Soon after the Linder's letter had started its path up through the Adjutant General channels, I was paid a surprise visit to my office by Harry Goodall. I could never remember Goodall coming to my office before. It was also the first time I had seen Goodall looking so somber. Usually Harry was haughty, all business, and patronizing. Not this day. He took a seat on my leather couch. I asked him "to what great honor was it that I enjoyed the privilege of his visit?" Goodall asked me if I would talk to Ron Linder about his rebuttal. He wanted me to talk Ron out of pressing his appeal all the way to the Pentagon. I asked Harry why I would do that. I believed Ron was innocent and should never been charged in the first place. I added the problem was Bierman, who did not know what the hell he was doing in the first place. I thought Goodall would argue these points with me, but he didn't. I saw then that he was truly worried about Linder's letter and what it might do to his own career. Goodall told me to tell Ron he was willing to pull the Letter of Reprimand out of Linder's file as soon as Ron retracted his letter of appeal. I ended the conversation. I advised Goodall to go speak to Linder himself.

Goodall left my office a worried and unhappy man. He was not in the least concerned that an injustice had been perpetrated on someone else. He was worried that Linder's rebuttal would somehow reverberate on him. He had endorsed Bierman's OER on Linder and concurred in the verdict and might even had written the reprimand himself. Goodall was the consummate careerist. His main worry in life was his own professional future and his dedication to get those stars on his shoulders. That came before everything else even his family life. His wife was alcoholic and I'm sure there was a reason for that. Incidentally, General Hill's wife also had a serious drinking problem which became a widespread point of gossip on the base, especially after a series of embarrassing episodes in the Officer's Club. Fortunately for them, Linder came to see me shortly after Goodall's visit. He informed me that he had been diagnosed with stomach cancer and it was a serious situation. The Base Hospital had assigned a civilian doctor from Anchorage to take over his case. Linder would have to begin chemotherapy shortly and it might take him off from duty quite often. He also told me that he would be retracting his appeal of his Letter of Reprimand as he would not have the time or the energy to follow through with it to its conclusion. I told him how sorry I was about that decision. Frankly, I was looking forward to his victory over Hill, Goodall and Bierman in the courts. In fact, I was assured of it. So was he.

In 1976, there was still quite a lot to be learned about cancer treatment. Chemotherapy and radiation were being used but were still in their early years. Cancer patients had a short span of years left in most cases. In a case like Linder's, five years was often quoted. But Ron, as I had expected, was determined to battle it hard. He had a wife and three young teenage kids, and he wanted to live as long as he could to raise them and support them. He was a good husband and father, and I told him that we would be behind him all the way in every way. Whatever he wanted he would get from me. And that was a promise that I kept. But in keeping that promise, I ended up in another major squabble with Harry Goodall, General Hill and the Command Surgeon—a difference of opinion that almost cost me my job and maybe another day in court.

Life in the Air Force was not getting any easier even with those eagles on my shoulders. What made it worse was the problems were coming from colleagues on the AAC staff and from my own superiors. But fortunately, I was now engaging in one of the most interesting and challenging projects of my career. In the beginning, I received no command support or encouragement. But in early 1975, I had come up with an idea that I felt needed to be worked on and developed with or without that support.

And the more I thought about it and the more I looked into it and the more I talked to people about it... people who knew more than I did about the technology... the more excited I became. It was a concept that, if it came to fruition, it would change Alaskan Air Command from what it was to something totally different. It might even mean the end of Alaskan Air Command as we knew it. But it would be better!

And so, I commenced work on a project that I had not as yet shared with General Hill or the rest of the AAC staff. I put my own people to work to do the research and analysis. Using Dave Creekman's management experts in Manpower and the air defense officers in my Plans section, we discreetly began to formulate a concept that would completely overhaul the current AAC ground control system. As yet we had given it no name, but slowly but surely, I began to see where I wanted to go with it. And the more I studied it and looked at what my staff people gave me, the more enthused I became about its possibilities. Soon I would have to brief the Commander and the staff about my proposal. But I had to be certain that I had my facts and figures absolutely right.

CHAPTER THIRTY-SEVEN: SEEK IGLOO

A journey of a thousand miles
must begin with the first step.

-Lao-Tzu

There is no great genius without
some touch of madness.

-Seneca the Younger

Just prior to my arrival at Elmendorf, General Ryan, the AF Chief of Staff visited the Command. He was quoted as saying that "there should be a re-look up there—either we are going to fix up those installations or we ought to be closing them." Therefore, one of the first jobs I was given was to oversee the cleanup of Galena. General Ryan had put his finger on it, we ought to do something about the sites or close them down.

From the very beginning of my tour of duty in Alaskan Air Command I was shocked at the cost and the waste associated with how the air defense mission was performed. It was back in the late 1950s that the ground control system in Air Defense Command (the lower 48 states) was upgraded to a digitized system. The SAGE (semi-automatic ground environment) system came on board and replaced the old manual system where aircraft controllers stationed at radar stations located across the country, some in remote areas and on mountain tops, directed fighters to intercept unidentified aircraft and defend the sovereignty of American airspace to a possible enemy bomber attack. With SAGE, the radars transmitted their radar digitized signals back to the SAGE center where the signals were displayed on screens to the controllers who then directed the fighters to intercept any aircraft that were not identified as friendly. This system was much quicker, more efficient and in the end, more productive. But AAC was still in the old manual system with radar sites (13 of them) located in very remote areas, snowbound for a good part of the year, difficult and costly to resupply and requiring to be manned by personnel on one-year remote tours of duty. And there was no plan or program in the works to modify or upgrade this system. Alaska, which in the event of an enemy bomber or missile attack on the USA, would feel the first line of the enemy's thrust, should have the most modern air defense and warning system available. But Alaska had the oldest and least efficient one, and our enemies knew it. The Russian Air Force frequently probed the Alaskan ADIZ, testing our warning and reaction systems going back to the 1960s when I was in the 318th FIS and the College Shoes Program and we sent our F-106s to AAC to meet that challenge. These things gave me lots to think about and I became determined that while I was in the job I was in, Deputy for Plans for Alaska Air Command, just maybe I could bring about a change that was long overdue.

In early 1975, I lead the program for AAC's participation in the Joint Surveillance System resulting in approval by the Air Force Council. I then decided to bring my proposal to modernize the AAC's ground control system to General Gamble and the AAC staff. My staff and I worked long and hard preparing this briefing and it came off fairly well. While the briefing was well received, the criticisms centered around the impact the results would have on the structure of the command and the feasibility of the technology proposed. However, General Gamble, with some reluctance, gave me the go ahead to proceed ahead with the project. From this point, I took my proposal to Colorado Springs, where I briefed both the Air Defense Command staffs and the NORAD Commander, General "Chappie" James and his staff. These briefings were vital, as without the approval of both of these commands, we would have no chance to sell the program at the Pentagon. Having many old friends at both ADC and NORAD helped me a great deal and I received some excellent feedback from these people who were the air defense experts. NORAD, of course, was Alaskan NORAD Command's air defense wartime control agency and had much to say about our proposal with these two commands on board, I was now prepared to approach Headquarters Air Force with the AAC program to modernize. Thanks to Col. Ron Linder's imaginative mind, we named it the MAR program, an acronym for "minimally attended radars." This was because, under our proposal, we intended to draw down the average number of personnel at each of the radar sites from over 100 to 150, as they were currently manned, to six.

For the next 12 months I was on a marathon, back and forth to the Pentagon. Working with air staff officers, we began to put in shape the AAC proposal for presentation to the Air Force Council and the Air Staff Success would have it inserted into a future USAF budget for implementation. The project was given the name Seek Igloo, identifying it as an Alaskan Air Command program dealing with its radar systems. Prior to briefing the Air Force Council, the project had to be reviewed and analyzed by the Rand Corporation, a think tank. This took a few months while their experts, mostly ex-USAF personnel, determined whether the project was technologically feasible and cost effective. I was fairly certain that their analysis would confirm what my own XP staff had determined and I was correct. In fact, their conclusions were more favorable with regard to the cost savings potential, coming up with a figure of an annual $40 million USAF budget reduction. With a number like that which I could now insert into my briefings to the high-level boards coming up, I felt much better about our chances. But nothing was certain in the Pentagon.

The Seek Igloo project became a grind. My trips to the Pentagon became more frequent and stressful. The flights back and forth were bad enough. Washington, D.C. was a five-hour time zone difference from Anchorage. Leaving Anchorage on the 8:00 AM flight to Chicago, it was already 1:00 PM in Washington. I would arrive in Washington at 9:00 PM that night. But that would only be four in the afternoon Anchorage time. I had a heck-of-a-time getting to sleep before two or three in the morning, even though I normally was due at the Pentagon at 8:00 AM for meetings or briefings. My biorhythms were all off kilter and I had to struggle to keep alert and stay awake during the days. My workdays at the Pentagon were spent constantly reviewing the material and checking my facts and figures for the presentations to the staff. My first briefings were given to a board of relatively junior officers, Majors, Lieut. Colonels, and chaired by a Colonel. These were the most difficult briefings. These officers were extremely bright and aggressive. Their questions were pointed and usually designed to find weaknesses in the proposal I was presenting. Their questions were meant to trip me up and many of them did just that. Sometimes, I had to answer honestly that I did not know the answer and I would have to do some research. I never tried to bluff my way through. In the long run, these briefings to the junior boards prepared me for my session with the Air Force Board composed of full Colonels and General Officers.

My briefing to the Air Force Board went well. I had been assigned a sergeant who assisted me with my slides, and we practiced together. He knew my speech by heart and my slides came on the screen in a timely and professional way. They had been designed so they were easily understood and emphasized the advantages of the program.

I Brief the Air Force Staff on Seek Igloo

A few weeks later, I was scheduled to brief the Air Force Staff. This was the highest-level USAF board composed of three- and four-star generals and chaired by the vice-Chief of Staff. I was a little bit nervous about speaking to such an imposing group of high-ranking USAF leaders. This briefing would be held in the "tank" located in the bowels of the Pentagon, a place where not too many have ever been, including me. But this briefing turned out to be my best and the easiest. The air staff response to my presentation was positive and the questions they did ask were not difficult. Early on I sensed that they were on board to a proposal that would save the AF some money and not the other way around. As I got into my briefing, my nervousness faded, and my confidence began to grow. And this was because what I was proposing would change the structure of the Alaskan Air Command in a most significant way and save the Air Force and the taxpayers a huge amount of dollars. And I was most proud of the fact that while at the same time I was proposing a restructuring of AAC that would save dollars, I was also proposing a program that would modernize AAC's air defense system and make it more effective in accomplishing its vital mission of protecting the country.

At the conclusion of my briefing, there were very few questions. Most of the comments were positive and supportive. And I was pleased that several of the Generals complimented me on my presentation, stating that it was thorough and comprehensive and left little to be concerned about. After returning to the office where I was working with the air defense staffers, two of the Generals called to congratulate me for my work that morning. I was feeling pretty high and relieved that the long ordeal that had started over two years ago, had ended so successfully. I called General Hill to give him the news. He was pleased although not that congratulatory.

Seek Igloo still had a long way to go. Approved by the Air Force Staff in 1976, the program then went to Electronic Systems Division, at Hanscom AFB, MA for implementation. I visited Hanscom and met with the project engineers. I had the honor of meeting with General Wilbur Creech, who was the Commander of ESD. General Creech asked me to his office, and we visited for an hour. He was a great wartime leader and a fighter pilot, and I found him to be a warm and personable human. He was certainly knowledgeable of the Seek Igloo program and complimented me for originating it. Meeting him was a special privilege for me and it remains a fond memory.

Seek Igloo Implemented

In August 1984, the Air Force awarded a contract worth $78.7 million to Sperry Corporation of Great Neck, NY to start design of new short-range radar as part of a plan to replace the 25-year-old early warning network against bomber or cruise missile attack across Alaska and northern Canada. There would be 39 of the short-range radar sets as well as 13 long range radars which will be minimally attended, which means that they will be visited by technicians only for maintenance. The long-range radar will be acquired under a separate contract with General Electric Co., Syracuse, NY. The Air Force estimated that the new system will be about $40 million cheaper a year than the present manned system. The new radars were due to start operating in about two years and the entire improved system would be in full use by 1992.

And so, seven years after I departed AAC and had retired from active duty, the Seek Igloo contract was finally awarded and eight years after that, the system finally came into fruition, a total of 15 years. Fifteen additional years that AAC continued to man those decrepit radar sites with officers and airmen on remote tours on snowcapped mountain tops in Alaska was way too long in my judgment and way too expensive. Looking back, I feel somewhat pleased and a little proud that I had a hand in bringing an end to a system that was so wasteful and ineffective and that should have been modernized many years before my time. I am also angered at how difficult it was to convince people who should have known better, to make the change, and to sell that change against strong opposition from those who wanted the status quo to go on forever I guess.

Contracting

Much of the opposition against the Seek Igloo program and some of the other proposals fostered from both XP and the Pentagon originated from General Gamble's and General Hill's opposition to contracting, i.e., turning what were current military positions over to civilians. In almost every case, contracting ended up in cost savings for the Air Force. But commanders saw a decreased military population as a decrease in their own command responsibility, and this they did not particularly want. But in the 1970's, budget concerns were pushing down hard on each command and Alaskan Air Command, being one of the smaller commands, was especially vulnerable. General Ryan's visit in July 1974 presaged that AAC would be pressured in the area of manpower cuts, and that was seen shortly after with the deactivation of ALCOM. Shortly after that, we were required to combine maintenance on our C-130s. Air Force manpower kept bringing up the subject of contracting more functions all of which AAC resisted. I tried to warn General Gamble that I felt that sooner or later we would be directed to contract the radar sites. General Hill, General Gamble's replacement, who had come to AAC from the Pentagon and was aware of the thinking going on there, was more willing to identify some things that would eliminate some of those remote tours, both at the radar sites and at the forward operating bases. So, what we were feeling from above and I was transmitting to General Hill and the AAC staff was, that we had better do something ourselves or it would be foisted on us by the air staff at the Pentagon. The Seek Igloo program, the contracting of more positions at the FOB's, and the elimination of as many excess positions as we could identify, was what helped save AAC as a major command for as long as it did. But I knew that AAC would not remain a major air command for long. And that came to pass.

Alaska Command (ALCOM), the command that I took part in deactivating in 1975, was resurrected again in August 1990 as a joint command with the Army to defend against a ground attack. Alaskan Air

Command was deactivated and became 11th Air Force under Pacific Air Force which was headquartered at Hickam AFB, Hawaii. Alaska NORAD Command continued to exist as a part of the NORAD system for the defense of North America against a missile or bomber attack. The modernized ground system installed in Alaska (the MAR sites) was now an integral component of that system. ANC (11th AF) was now equipped with the computerized joint operations center (JOCC) that was part of the Seek Igloo program and supported by the AWACs and the F-15 interceptors stationed at Elmendorf AFB. And so, the seeds that were sown in the 1970' s by the Seek Igloo program, a program that met serious opposition from within the AAC staff at the time, finally came into fruition. While Seek Igloo ended up in putting an end to AAC as a major command, it did preserve the Air Force's capability to continue its air defense mission in Alaska. And so, I look back on it as a major success. In pushing it through with little to no support from my own superiors and colleagues, I was determined that it was the right thing to do. I believe that time proved that I and those who supported me in XP and elsewhere were right. Our efforts against the odds resulted in my receiving the Legion of Merit awarded to me upon my retirement from active duty.

CITATION TO ACCOMPANY
THE AWARD OF THE LEGION OF MERIT
(FIRST OAK LEAF CLUSTER)
TO
EUGENE P. CUNNEELY

Colonel Eugene P. Cunneely distinguished himself by exceptionally meritorious conduct in the performance of outstanding service to Alaskan Air Command, Elmendorf Air Force Base, Alaska from 31 July 1973 to 31 August 1977.

In this important and demanding assignment, the leadership, exemplary foresight and ceaseless efforts consistently demonstrated by Colonel Cunneely resulted in significant contributions to the effectiveness and success of the programs which form the future military posture of Alaska.

The singularly distinctive accomplishments of Colonel Cunneely culminate a distinguished career in the service of his country and reflect great credit upon himself and the United States Air Force.

I have always considered that Seek Igloo was the best achievement of my Air Force career. And yet it came to me from a simple idea that was obvious and necessary from the time I came to the Command. It met with opposition from those who disliked and were suspicious of change. Those in charge then were enamored with the status quo despite growing evidence that the status quo could not continue. No matter how much you provided them facts and figures showing that unless we changed the configuration of our air defense system, making it more efficient and less costly, it would surely be downgraded, or worse still, deactivated. Once I was permitted to bring Seek Igloo out of the Command and present it to the Air Staff, I was confident that it would be successful.

In January 1986, General Joseph W. Ralston, commander of Air Combat Command, in his testimony before the Senate on his nomination to become vice chairman of the Joint Chiefs of Staff, had this to say:

> When I was in Alaska, we had three remote bases. We had 300 uniformed personnel at Galena. Their mission was to keep the runway clean, keep the barrier up in operation, keep the command post alive, keep the dining hall up and keep the billeting operations up. Today, instead of having 300 uniformed personnel at Galena, we have forty-seven contract employees. That is a tremendous savings for the taxpayer.
>
> Now you might ask the question: How could the Air Force need 300 people to do what forty-seven contractors do? Well, as we send the magnificent young Americans to Galena, one of the things that you expect us to do is provide them three meals a day when they are there. You expect us to provide them with a place to sleep and live. You would expect me to have some kind of morale, welfare and recreation activity for them there. And every time I add MWR people, I've got to add more cooks; I've got to add more billeting personnel. Pretty soon I'm at 300.
>
> So, I'm a personal believer that there are functions you can do more economically by contracting out.

I could not have said it any better.

In 1984, Ron Linder, who was then retired and living in Anchorage, sent me a copy of the *Anchorage Daily News* highlighting the implementation of the Seek Igloo contracts for the construction of the MAR sites. Ron noted that the enclosed article talks about the "tree you planted years ago to modernize the Alaska early warning radar net. Those who toil in the vineyard of AAC, probably never wonder how their system came to be- but we know- it was a gleam in your eye one day, that moment of conception, which you cannot remember exactly when it occurred. Oh well, it seems in life, that no one plants the tree under which he enjoys the cool shade."

He also added the old adage: *"Success has a thousand fathers, failure is an orphan!"*

CHAPTER THIRTY-EIGHT:
LIFE IN ALASKA—THE GOOD SIDE

There are strange things done in the midnight-sun,
by the men who moil for gold;

The Arctic trails have their secret tales,
that would make your blood run cold;

The Northern Lights have seen queer sights,
but the queerest they ever did see,

Was the night on the marge of Lake Lebarge
I cremated Sam McGee.

-Robert Service

In early 1975, I requested a one-year extension to my three tours of duty in Alaska. I had a reason for doing this, in fact, I had several reasons. The first was I was more and more toying with the idea of retiring from active duty at the end of my Alaskan tour. If my request for the extra year was approved, my tour would end in the summer of 1977 and that would give me 26 years of active-duty service. Twenty-six years was a significant milestone from the standpoint of retired pay, and I needed to consider that aspect with a large family to support. Secondly, I was engaged in the Seek Igloo project that was dependent on me to carry forward to the Pentagon for a successful completion. I was dedicated to staying on to get that program across the finish line. Third, I had lost my enthusiasm for the Air Force, especially since I had been dropped from flying. In Alaska I had also served under and with a group of commanders and colleagues for whom I had lost confidence and respect. Not all of them, of course, were sub-standard, but I found that the Command had more than its share of officers whom I believed did not measure up to what I was used to. Lastly, Alice and the children seemed to be enjoying their life in Alaska. We had great quarters, close friends and an active social and recreational life. The kids were especially happy, except for Mary who was going through adolescent difficulties not unusual for her age. Alice and I were most unhappy with the school which although it was brand new, was a real disaster from an educational standpoint. But Gene and Jim had a lot of great friends and were active on the school hockey team where they were star performers. Mike came up behind them and took advantage of the skiing, tennis and golf facilities that were available on the base. He also became an excellent skater and skier, who made many good friends. Katie attended school on base and loved horseback riding and accruing all kinds of pets.

Friday nights in the Officer Club was steak and Alaska King Crab leg dinner specials that our family enjoyed, and Sunday mornings, after Mass, it was our traditional family breakfast. Gene Jr. was now in his late teens, a handsome well-built young man who was a good student and an excellent athlete. Unfortunately, Bartlett High School was a failed educational institution, but he and Jimmy did well there as best they could. On graduation, Gene decided to attend the University of Alaska-Anchorage, where he also continued to play hockey. Jimmy was also a great hockey player at Bartlett, where he made some great long-lasting friendships. Mary did well scholastically at Bartlett although she really did not enjoy school and was more interested in going to work and making money. Mary had various jobs while attending school and Alice and I finally agreed to her leaving school early with her high school diploma on condition that she continue her education at Community College, which she did. Mike continued in school through our tour in Anchorage. Mike was more interested in sports than studies and his grades reflected it. He enjoyed Alaska and all that it had to offer. During the four years we had there, Mike returned to the lower 48 greatly improved in golf, tennis, skiing, backpacking and motorcycling. Katie also had a great time in Alaska with many friends and pets. Katie loved animals and she was a good student in all of her subjects. But Katie developed problems with her kidney that required medical attention of a serious nature. Fortunately, we had a great urinary specialist at the Elmendorf Hospital, Dr. Bias, who was able to diagnose Katie's problem. Katie was operated on and the operation seemed to have been a success. But in later years, her problem resurfaced, and her kidney troubles returned. This required a second operation which again failed to solve her problem. Over the years, Katie had three major operations on her kidney, one of them removing over a third of it. It became a lifelong problem that she had to endure causing her much pain and anguish. None of the many urologists who took on Katie's case was able to resolve her kidney difficulties.

During our years in Alaska, our family took many trips including a great week to Fairbanks and Denali National Park. We also visited Seward where we fished for salmon and toured the Port of Valdez, the southern terminal of the Alaska pipeline. Alice and I also vacationed in Juneau, Alaska's capital city and then had a wonderful trip into Glacier Bay National Park. Living in Alaska was a terrific period for our family, and we left it with such fond memories. Alaska is such a fabulous place that appeals to one's natural instincts. You cannot live in Alaska and take in its beauty and variety of scenic landscapes truly breathtaking in their vastness, color, enormity, and frightening depth without being moved as you had never been before. I know that whatever I may have missed career wise by accepting the assignment to Alaska was more than made up by the days and hours I was able to spend with my wife and children in that great part of our country. We also made many lifelong friends there, most notably Dave and Joyce Creekman with whom we shared many great evenings during our days in Alaska and later in retirement. The kids also fostered lifelong friends in Alaska with whom they remained in close contact in the years after we retired to the lower 48. All in all, I know that Alice and I and our kids will never forget those years.

Tennis

Surprisingly, one of the side benefits of my year in Saudi was my introduction to the game of tennis. Tennis was a sport which I enjoyed watching and I followed it to some extent having living some of my early years in the shadow of Forest Hills, the home of the U.S. Tennis Association and the site of its stadium. But I never really played it. However, while in Saudi that was the game. The Shubert's had a court and on the weekends, I played there with my friends mostly in doubles and mixed doubles competition that was not too competitive. There were several courts in the various corporate compounds around Riyadh and lots of players, and it was a great way to while away the afternoon hours and get in some good exercise. I bought myself a good racket and when I returned home, I got the boys interested in the game. It did not take long before Gene and Jim were cleaning my clock on the court. Col. Stockman, the vice-commander, was also a tennis player and together we were able to connive and get the base to put in place two indoor tennis courts in one of the empty hangars. These courts, composed of plastic tiles fitted together, were laid in one of the unused hangars on the west side of the base. Reservations were made by calling the MWR Center in advance, but the place was available 24 hours a day and it wasn't unusual for the Cunneely boys to be down there until 3 o'clock in the morning playing tennis while an Alaskan blizzard was blowing wildly outside. I can truly state that the Cunneely's had a great time in Alaska developing their skills in golf, tennis, skiing and all other outdoor recreational activities. In addition, in August we flew down to Spokane and spent three weeks at our cabin on Lake Pend Oreille. There we did some cabin projects, drove the *Alice Marie* all over that vast pristine lake, played tennis at the Sandpoint Beach park courts and very much enjoyed life together as a family. Despite the problems I faced in my last years on active duty and my concerns as to what the future held for me, I made it a point to enjoy every minute of it while I could. These were minutes that only come once and do not return. I realized that and I made certain that I was not going to let them slip by. To this day, I am happy that I did just that. I put my future into God's hands.

The Family Begins to Separate

When Gene Jr. graduated from high school in 1974, he decided to stay in Alaska and attend the University of Alaska-Anchorage. That was great because he continued to live at home, and we had him with us. He commuted to school which was in downtown Anchorage. He became a member of the UAA hockey team and was one of their high scorers. When Jim graduated a year later, he opted to attend Gonzaga University in Spokane. Alice and I were pleased at his choice. GU was a Jesuit school with a good academic reputation. Having graduated myself from a Jesuit university, I was pleased that my son Jim would be receiving the benefits of an education providing the benefits of the Ignatian traditions. Further, Gonzaga was reasonably close to us, and Jim already was familiar with Spokane having spent much time there already while I was in Saudi and on our vacations to our lake cabin in nearby Idaho. Some of Jim's classmates from Bartlett High School were also going to Gonzaga so he would have some Alaska friends attending with him. Sam Grashio, who was now retired from the Air Force, was now working for Gonzaga University and he would be another close friend to rely on while in Spokane. Gene Jr. would join Jim at Gonzaga a year later, a decision that Alice and I were both pleased with.

Mary continued to work in Anchorage and attend Community College. Mary, who was so opposed to coming to Alaska in the first place, was now a full-fledged "sourdough." Of all our children, Mary had the most difficult time adjusting to our move to Alaska. She was bitter about having to move from Spokane and change schools (even though she did not like Marycliff) and make new friends. The friends she did make in Anchorage were not the most wholesome, especially one young man who had a bad record. He was the son of a sergeant and I worked with him to ensure that Mary and his son did not get into any serious trouble together. Mary was quite upset with me and my interference with her social life and my efforts to warn her about the danger she was in. I finally enlisted the advice of a nun who was an expert in such matters involving teenagers. The Sister gave me excellent suggestions and I followed them. Slowly Mary responded and we gradually developed an arrangement under which tensions eased. Two tragic things occurred that helped her attitude. Mary's boyfriend was arrested and jailed. Then her close friend, Katie Simonson, the daughter of our command Adjutant General, after lying to her parents about where she would be going, was nearly killed riding on the back of her boyfriend's motorcycle speeding down an Alaskan highway when it collided with another vehicle at close to 100 mph. He was killed and she was badly crippled for life. Such realities struck home.

However, one rather nice highlight for Mary and me, was Mary joining several other young ladies, who were the daughters of base officers in a debutantes' ball at the Officer's Club. I had the honor of presenting her and I had to admit she was the prettiest of the bunch.

Alice and I worked hard raising our five children. We found that it was not an easy proposition, as each one was different from the others, and that each one needed to be loved and nourished in a special way. We also found out that each one was a marvelous gift from God and that He offered us a responsibility as well as a blessing with each one. I concluded that regardless of whatever else I did with my life, what other accomplishments or awards or commendations I might achieve, none would be as important nor as self-fulfilling nor satisfying as what I might bring forth as a parent. I became aware that what I did right as a father would be what God will judge me on when we come face to face.

Retirement

In the fall of 1976, after much discussion with Alice Marie, I finally made a decision to retire from active duty. A great many factors entered into my decision. One was that I realized that I had received my last promotion when I turned down Secretary Robinson's offer to work for him in the Pentagon on my return from Saudi Arabia. I pretty much knew that I might likely not receive another assignment of significant importance that would qualify me for consideration for a General officer billet. But General Smith's offer to me to come to AAC operations opened the possibility of a returning to a flying position and perhaps a Wing Commander' job in the future. But that did not come to pass. Also against me was my failure to serve in Vietnam although I had volunteered for it twice. The war ended in 1973 while I was in Saudi Arabia. I recall how great the news of that terrible war's ending was received especially seeing the magnificent return of our POWs some of whom I knew personally. It had become a wasteful and fruitless war one which our leaders had decided not to win and thus had it become a tragic waste of lives and a divisive issue that would not heal for decades. Mostly, I was disillusioned over being grounded. My decision to become a Regular officer in 1955 and to make the Air Force a career was due to my love of flying. Now that it was fairly certain that my flying days were over, I no longer had that motivation to remain on active duty as a desk officer. And lastly, more and more, I began to desire to do something different with what talents I had for the remainder of my working days whatever that might be. I was approaching 48 years of age and I deemed it wise to be under 50 years of age when I begin looking for a new line of work in the civilian world.

This was all a bit frightening, and I had to admit I had no idea what kind of work in civilian life I could do successfully. But I did know that I would need more income on top of my retired pay to raise my family and pay for their education. Ron Linder, my brilliant Colonel friend helped me by working up some figure that showed me how much I would need in terms of additional income and investment. And so, with much trepidation, I put in my papers requesting retirement from active service effective September 1, 1977. I further requested that I be discharged from Fairchild AFB Spokane, Washington. I requested that location as Alice and I decided that we would retire in Spokane, a city and locale we had grown to truly like and that was close to our lake cabin in Sandpoint, Idaho. It was also where Jimmy and Gene were going to college. All the kids knew Spokane and liked that area, so we did not receive any negative feedback on that selection, except from Mary who wanted to stay and live in Anchorage.

My retirement request was granted and so I began my last year of active service with lots of things on my mind. I was still in the midst of the Seek Igloo project that had been approved and funded by USAF and DOD. But I was now in the business of turning the project over to Electronics Systems Division who would be engineering the program and eventually awarding the contracts. There were many other important items that I wanted to have finished before my departure from the command. Many of our wartime plans needed to be revised and updated due to the deactivation of ALCOM and the origination of the Joint Task Force. The command (AAC) was also going to receive a new commander as General Hill was due to leave for a position in SAC. As you have read prior, my relationship with Hill had its ups and downs. He almost had me court-martialed when he mysteriously supported Col. Bierman in an absurd case where Bierman failed to follow AF Regulations in placing Linder into the "fat boys" program and then tried to place the blame on Linder when Linder pointed out his errors. Hill and Goodall had to eat crow and rescind the reprimand given to Linder when Linder appealed to the AF Adjutant General showing clearly that he was totally correct all along. I was not unhappy to see Hill depart. But another episode prior to his departure brought me into serious conflict with General Hill. Once again, this disagreement involving Col. Linder, left me breathless with disillusion about how some officers in our wonderful Air Force could possibly be given three stars.

CHAPTER THIRTY-NINE: I LEAVE THE AIR FORCE

Old Soldiers Never Die, They Just Fade Away.

-General Douglas MacArthur

In the fall of 1976, I began my last year of active duty. It was my decision, made after much thought and long meditation. It was a hard choice, and I knew that it would bring a huge change into my life and the life of my family. I had dedicated the past 25 years of my life to the Air Force and looking back, I had to admit I had found it mostly enjoyable and rewarding. It was certainly not what I had planned for or expected. It came to me suddenly, beginning in the summer of 1950 when the North Korean army unexpectedly invaded the south and President Truman thrust our nation into what became known as the "Korean police action." Being an Air Force ROTC senior, I was duty bound to active duty upon graduation in June 1951. It did not occur to me that I would ever remain on active service for a quarter century from that date. But I did. On an impulse I challenged myself to becoming a military pilot as we entered the jet age, and the love of flying fighter aircraft took over from there. I survived through God's good graces and the prayers of many, especially those of my wife Alice Marie. And now in my 47th year, I had come to the end of my flying days and was preparing to move into civilian life in which I lacked experience. It gave me a lot to think about. I had two sons in college, a daughter in community college and working, a son still in high school and the youngest daughter in grade school. We would have to move, find a home, find a job and begin a new life in Spokane. Alice and I had lots of work cut out for us to meet an uncertain future.

Illegitimus non carborundum
(Don't let the bastards grind you down.)

-General 'Vinegar' Joe

I'm in Trouble Again

Despite my large successes with the deactivation of ALCOM, the Seek Igloo program, and several other manpower and planning projects that brought many kudos to my deputate, General Hill and Colonel Goodall never gave me or my section the credit it deserved. At least, that was my opinion. Much of that was because of the feud which continued between the operations side of the house led by Col. Bierman which opposed almost every program that Plans and Manpower proposed. Unfortunately, our proposals were in synch with the current DOD trends that featured curtailing spending, cutting wasteful and inefficient programs, and where possible, decreasing manpower levels. Why Hill supported Bierman so much was hard for me to understand because he came to AAC from the Pentagon and surely understood what was going on there. His backing Bierman in the Linder case was absurd as Bierman clearly screwed up and Linder, who was a very smart cookie, caught him on it. When he had to rescind the letter of reprimand he had given to Linder, it had to have been an embarrassment.

In the fall of 1976, Col. Linder was advised by his civilian oncologist that he should receive a special x-ray of his liver using a new technology called a CAT scanner. The nearest military scanner was at Letterman Army General Hospital, California. Linder talked to Colonel Fred Conrad, who was the Hospital Commander at Elmendorf and also the AAC Surgeon General Conrad, who was himself a cancer specialist, argued with Linder about the value of the CAT scan and also about funding the costs of the TDY to California and back. He even hinted that Linder should pay his own way down to California and back. Linder then spoke to me and told me of his conversation with Conrad. I was appalled at Conrad's attitude. I told Linder that I would find some TDY travel funds to pay his way down to Letterman and back for his CAT scan. It seemed incomprehensible to me that the Air Force would not provide funding for one of its own to receive medical treatment that had been recommended by a physician.

Conrad Pays a Call

Shortly after, I received a visit from Dr. Conrad. Conrad was fairly new to the command and came with a great record. But his visit was not cordial. In fact, he was in a huge dither. He started in with a fierce diatribe against Linder, calling him a “manipulator.” This did not set in too well with me as I don't appreciate anyone criticizing my subordinates unless they have good reason, and Conrad did not. He clearly was biased and had already come under the “Bierman” influence. In a short time, Conrad and I were in a full swinging debate about Linder's case. At the end, Conrad stated that he was totally against Linder receiving TDY funds to go to California for the CAT scan and that he, in his capacity as Command Surgeon, was totally against Linder obtaining a CAT scan in the first place. Then he left. Shortly after, I received a call from Col. Balilies, the command Finance Officer telling me that I would not be authorized to use my TDY funds for Linder's trip to Letterman. Conrad had done his dirty work in a hurry. More was still to come.

With all this going on I had to leave on a business trip myself in conjunction with Seek Igloo. While away I received a call from Dave Creekman, who was in charge in my absence. Linder had decided to pay his own way to Letterman and requested that he receive “permissive TDY” from Elmendorf to Letterman for purposes of receiving medical diagnostic testing. I immediately granted the TDY. That decision was to cause me a great deal of anxiety and perplexity. The TDY orders were issued and Linder, paying for his own travel expenses, flew down to California and received the CAT scan at Letterman Army Hospital. When Conrad heard that I agreed to let Linder go to Letterman for the CAT scan he went to see General Hill. He told Hill that Cunneely had overrode the medical decision of the Command Surgeon (himself) in permitting Linder to go to Letterman and receive a treatment that Conrad had advised he did not need and that the Air Force should not support. When I arrived back at Elmendorf, Hill called me into his office and he dressed me down real-good. He asked me who I thought I was to override the medical advice of the Command

Surgeon and a cancer specialist to boot. My efforts to explain to Hill all the circumstances surrounding this complex case were fruitless. He would hear none of it. What saved me from another court-martial was Hill was departing the command the next day for his new assignment. General Marion Boswell, Hill's replacement, would be arriving from the Pentagon in another two weeks and Hill would brief him on this matter and Boswell would deal with me. Hill then dismissed me from his office and his sight forever. Inside myself, I said, good riddance. I left his office not feeling humiliated or angry but feeling totally ashamed of a general officer who was so completely inhumane. In this situation we were dealing with a fellow officer who was diagnosed with a terminal case of cancer. He was literally fighting for his life. The doctor assigned to him by the Air Force had recommended he receive this diagnostic treatment (the CAT scan), but the command refused not only to reimburse him for this treatment, it refused to let him receive the treatment at all. In this instance, Col. Linder did what he felt he had to do keep himself alive for his family and future. And I, as his supervising officer, supported him one hundred percent.

Col. Linder's Memo

Unbeknown to me, Ron Linder wrote a Memo to Col. Conrad recapping his interactions with Conrad, inducting Conrad's refusal to permit him to get the CAT scan, nor to allocate the TDY fund, nor to permit me to allocate XP's TDY funds to pay for his trip to Letterman Hospital and much more concerning Conrad's negative attitude toward him (calling him a "manipulator") and placing him as a cancer patient in a most difficult position that was dearly not supportive. The Memo was devastating, almost incriminating. In addition, Linder sent copies of his Memo to the Surgeon General of the USAF at the Pentagon and the Commander, AAC. This Memo would be seen by the incoming Commander, General Boswell, who would see it before I did. The effect of this Memo would be devastating because it accused Conrad of intervening between Linder and his appointed physician in handling his illness, a most serious and debilitating case of cancer that might bring on his death, and eventually did.

I Get a Summons from General Boswell

General Boswell arrived at Elmendorf and took command of AAC near the end of October 1976. He was a tall and serious looking officer with a wide and experienced background. Word was that he had suffered a major heart attack a few years ago, but he certainly appeared fit and healthy. He had flown B-17s in WWII and later piloted F-4Ds in Vietnam. His previous assignment was Asst. Vice Chief of Staff of the Air Force at the Pentagon. Although I knew I was going to catch hell from him about the Linder case, I was still impressed with the manner in which he took command and control at the first staff meetings he chaired. It was clear that he had done a lot of homework about the Command before his arrival and was up to speed on the projects that were ongoing including Seek Igloo. About 10 days after his arrival, I received a call from his secretary to report to his office. My secretary, Betty Martinez, gave me a look of gloom as she and I both knew what this summons was about.

After reporting to Boswell in a military manner, the General told me to have a seat. The look on his face was not sociable. His first words were: "Jim Hill briefed me on this Linder case. He told me that you believe that you know more about medicine than the Command Surgeon and you overrode his professional decision… Is that correct?" I answered in a voice that was probably a little shaky: "No sir, that is not exactly correct. I am not a doctor, and I made no decisions that had anything to do with medical treatment. All I did was try to do what I felt was right for an officer who worked for me and who is very sick." Boswell replied: "That's not what Bud Conrad says." To which, with a little more bravado I said: "General, may I tell you what happened here that caused this problem. I'll be as brief as I can." Boswell said, "Go ahead." And I did. And the story I gave him clearly was much different from what General Hill and Col. Conrad gave him. While I was speaking, Boswell took out Linder's Memo and placed it in front of him. When I explained that all I did was give Linder permissive TDY to travel to California to get a CAT scan that the oncologist, who the AF had assigned to him, had suggested he obtain, Boswell became more silent. He began to question me on various points because my narration clearly did not jibe with what he had been told. He then began to read Linder's Memo again which, of course, supported my description of what had taken place. I ended my discourse by restating that my decision to support Linder's trip to Letterman was because this officer's life was on the line, and I believe the Air Force should do what it could to help him in this time when he needed backing.

Boswell, surprisingly, spoke about his heart attack a few years back when he was at the Pentagon and how frightful a situation that was. I then felt that I had made some headway. With that, Boswell turned and using the telephone bank behind his desk, he called the Pentagon. He spoke to the Air Force Surgeon General regarding Col. Linder's Memo. It was clear that Boswell and the Surgeon General had already spoken prior about the Memo. Boswell told the SG that he was "on top" of the situation and there was no need for the SG to take any action. Boswell told the SG that he (Boswell) would handle it from here. After that, Boswell called Conrad and told him to come to his office immediately. His tone of voice was quite stern. Then he turned his attention back to me. I had been sitting there in front of his desk quietly listening to the General's phone calls and wondering just how all of this was going down. Boswell finally said to me that he did not understand how such a situation as this could have been so badly screwed up. He said: "I know what it is like to be really sick, life-threatening sick, and you want and expect the Air Force to come through for you and your family. It seems like you were the only one who cared and tried."

We talked some more, and I told him how much I thought that Col. Linder's knowledge and talents had been underused. While I was still in Boswell's office, Conrad arrived. Boswell greeted him sternly and told him that he was upset about the handling of the Linder case. As things began to heat up, Boswell dismissed me, and I departed. I thanked the General, saluted and left. It was one of the best days of my last year in the USAF. Reports were that Conrad received an awful reaming out for his part in the Linder case. He and I never discussed it again. Any hopes he had for promotion ended that day when Boswell read Linder's Memo and spoke to the USAF Surgeon General. Boswell took to heart what I told him about Linder's talents, and he began to use Linder as his personal advisor on Joint Task Force exercise procedures and other political matters. This really upset Bierman and other staff members who despised Linder and were envious of his superior intelligence. I and other members of Plans and Manpower were delighted to see this fine officer finally get the recognition he was due.

Most of all, this experience encouraged me to regain some of the confidence I had lost in the commanding officers I had served under, especially General James Hill and Colonel Harry Goodall. Both of these officers went on from AAC to other commands. Hill gained his fourth star and commanded NORAD before his retirement. Goodall's ruthless pursuit of promotion to star rank was successful as he attained three stars prior to his retirement. While at the Pentagon, Goodall was called to testify before a Congressional committee on some matter. He was later accused of receiving an advanced copy of the questions he was to be asked from some staffer. It was a minor scandal, but it did not affect his career. But I never believed he was a happy man for all of his success. Hill divorced his alcoholic wife and married a former girlfriend from his school days. He retired in Colorado Springs where shortly after he died of a heart attack. Goodall retired in Waco, Texas. He was killed in an air crash in 2006 while taking off from Waco in his own airplane. Bud Conrad retired from the Air Force and became head of a hospital in Houston, Texas. He died as a result of being murdered in his office by some distraught woman. Col. Bierman retired in Anchorage and was employed by one of the oil companies doing security work on the Alaskan pipeline. By chance I ran into him on the beach of Waikiki, Hawaii in 1983. While standing near the water, Bierman walked past me. I recognized him and called out to him by his first name "Casper" several times. (He hated that name.) He looked around finally seeing me. We chatted for a long time on the sands of Waikiki about our days at AAC. It was a friendly reunion, but I felt that he was trying to justify some of the things he said and did while we served on the AAC staff. Less than a year later, Bierman dropped dead of a heart attack.

Ron Linder took a medical retirement from the Air Force and stayed in Anchorage. For some time, he was a member of the Alaska Utility Commission, but Linder again proved to be a very difficult person for the political class to work with. He was too smart and too honest. So, they eased him out. He suffered a terrible loss when his son died from drug overdose while at the University of Colorado. But Ron continued to bounce back. His brilliance was seen in his success in the stock market by which he easily supported his family. But unfortunately, the cancer came back in force and eventually took his life. He was a man to remember for sure.

I Become a Short Timer

Time was beginning to run out on my Air Force career, but I was as busy as ever. I was still winding up the loose ends of Seek Igloo and turning over all the engineering details to ESC. Thanks to General Boswell, my personnel problems were resolved and now it was time for me to concentrate on my future and the future of the Cunneely family that, incidentally, now included my dad. What I needed to do was find a job and find a new home. Alice Marie and I had decided that Spokane would be our new place of residence and so I now began to concentrate on that city. I needed to find us a home to live in and a job that would help me pay the mortgage and the many other bills we had. I now had two sons in college, and I took out some Alaska State loans to pay their tuitions. Mary was coming up next. In 1976, Alice Marie went back to work. She took a job at B and G, one of many companies supplying parts to the oil pipeline companies. I was flabbergasted when she decided to do this, but I knew that when my wife made up her mind about something, then that was that. Alice is a very organized woman. Before bed she had everything set up for the kid's breakfasts, their lunches and their clothes. The next night's dinner was planned, and the household ran pretty much like clockwork. As I expected, despite that she was the oldest employee in her section, she quickly established herself as the most reliable and the most accurate. The pay was good, and it helped to pay some of the school costs for the boys. Alice really enjoyed being a breadwinner again and I was very proud of her as always. Our kids were amazed that their Mother was now a working woman and still performing most of her household responsibilities as she had been doing. I could see that they now saw their Mother in a new light.

House and Job Hunting

In 1975, Father Bernard Coughlin, S.J. the new President of Gonzaga University visited Anchorage. His trip was arranged by Sam Grashio, and he asked me to host his visit. I arranged a reception for Father at the Officer's Club where Gonzaga Alumni and distinguished guests including Mayor Sullivan would meet Father. It was a successful event with an excellent turnout and my first opportunity to meet Coughlin. I recall telling Father that I was planning to retire in Spokane in a few years and would be looking for a job. As a graduate of a Jesuit college and appreciative of the education I had been given by the Jesuits, I told Father Coughlin that if there was anything I could do of value for Gonzaga I would certainly give it serious consideration. Coughlin told me to call on him when I came to Spokane. In the spring of 1977, I did just that. At that time two of my sons were students there. Coughlin and I talked, and he sent me to speak with Stan Fairhurst, who was the Vice-President for Finance. Fairhurst was at least six feet four inches in height and exceedingly large in every physical aspect. He was very cordial and invited me into his office where we chatted for well over an hour. Actually, he chatted over an hour, and I said very little. Fairhurst provided me the story of his life and career, in detail. I listened carefully, trying to find what the point he was getting to. Was he going to offer me a position or at least ask me what kind of talent or skills I could bring to the University? But that never happened. Finally, the one-sided monologue ended and we parted. I left totally bewildered. I wrote him a follow up letter thanking him for his time and ensuring him that I would be interested in knowing what he had in mind for me. But I never heard from him again.

More Job Searches

In job hunting in Spokane. I began with my old friend Sam Grashio who had retired in that city and knew lots of people there and gave me some key assists. First, I was able to stay with Sam and Dee and use his home as a headquarters for calls and his car for transport. Through one of Sam's old friends, I was able to interview with a few law firms for positions as a Firm Administrative Manager. These interviews started out promising as my resume had lots of administrative management experience in it and I felt that I presented myself well. But I ran into something that I had not anticipated. In both instances and in several to come, I noted an antipathy to hiring ex-military. With the lawyers, my high grade of Colonel clearly was an obstacle to the attorneys. I also found them disorganized and confused in just what they were seeking in the position they were trying to fill. And in one of the firms, not all of the partners were convinced they needed an administrative manager. I felt that as badly as I needed a job, working for attorneys may not be a happy situation. Sam introduced me to Father Schoenberg, S.J. Father was the Director of the Museum of Native American Cultures (MONAC). The museum was located on the north bank of

Spokane River adjacent to the Gonzaga campus. Father was a noted historian and had written the definitive history of Gonzaga University. He was also an expert on the history and culture of the Native Americans of the Pacific Northwest. The museum was looking to fill a position coming open of Development Director. The current Director, Joe McKinnon was leaving the position after several successful years to enter another career field. I was set up for an interview with Father Schoenberg, Joe McKinnon and Dr. Robert Carricker, a professor of history at Gonzaga University who was a MONAC Board Member. I thought the interview went well although I had my doubts that I would be able to meet the fund-raising goals that the organization clearly was hoping for. I was then asked to come back the next day to meet with the Chairman of the Board and some other Board members. I agreed. But the next afternoon when I arrived at the appointed time, the Board members did not appear. We waited over an hour before the Chairman and the other member arrived. It was evident that these men had been drinking heavily. The interview did not go well. Their questions quickly showed that they were concerned about my military background and lack of fundraising experience. Some of their questions bordered on rudeness. Father Schoenberg and Dr. Carricker who were present, tried their best to ease the tension in the room. That evening as I was at Spokane Airport leaving to return to Alaska, I called Carricker. I told him that I was withdrawing my candidacy for the position. He asked me why. I told him that first, I would not work for people like those I met today. Carricker agreed. Second, I told him that I would be very concerned about the future of MONAC. From what little I saw and heard I would be very worried about its financial stability. Dr. Carricker said it is a worry. He told me that he would pass on my message to the Board and wished me well.

I See a House I Like

Sam Grashio worked for a man named Curzon Fager who was the Vice-President for University Relations at Gonzaga University. Curzon owned a home in the near Spokane Valley, and he was thinking of selling it. Curzon was a guy who bought homes, fixed them up and then sold them and thus made a little profit. I drove out and took a look at Curzon's house. To get there you had to drive east on Mission to the WWP Building and then take Upriver Drive. It was a beautiful drive along the river, past the dam and the beach until you arrived at Pasadena Park. There you drove up a short hillside through some nicely kept residences and on the corner of Ella and Ely was Curzon's house.. I really liked the area and Curzon gave me a walk through and the price was okay. The one disadvantage—a small kitchen. But I would make no decisions until Alice Marie saw the place and I had looked at several others too. It struck me that here I am thinking of buying a house and I have no job yet, so I was truly getting ahead of myself for sure. I still was working for Uncle Sam, had several months to go and needed to find a job for sure before decisions on houses could be made. So, it was back to Alaska for the time being with some more planning to be done.

Winding Down

As my Air Force career started to wind down to a conclusion, I began to become very emotional. I had served over a quarter of a century in the uniform of my country, first putting on the blue suit with the gold bar during the Korean conflict in 1951. While my participation in that war was not huge, still I was proud of what little I did contribute and that I did do my share of damage on an enemy that was out to enslave its brother nation. Looking at the manner in which the country of South Korea has brought prosperity and freedom to its people in the years after 1953 made me pleased to think that I played a very small role in making that happen. (*In June 2000, the President of the Republic of South Korea presented me and thousands of others a Korean Service Medal in appreciation for contributing to safeguarding the Republic of Korea.)*

In the years following, during what is termed the Cold War, I devoted my talent and skills to the air defense of the United States. I recall hundreds of days and nights in various alert hangars situated at the end of runways across the United States, Canada and Greenland. I remember ear splitting and heart shattering scramble horns that propelled me and my trusty jet fighter aircraft into dark night skies to search out incoming unidentified aircraft that needed to be intercepted and made known at all costs and often at frightening risks. Later, as an operations officer and a squadron commander, I felt the burden of responsibility when the lives of others depended on my making right decisions every time, every day. Later, in higher staff positions that I was fortunate to attain, I put my experience to work to improve our combat systems making them more efficient in every way. And now, as time was running out on my military career, I could only look back with both good and bad memories. Over the years, in many different locations that the Air Force sent me, I met and worked with and for some great people and some not so great. But wherever I was, and for whomever I worked, I always gave my best effort because I was serving my country, the greatest nation in the world. I had taken an oath to do just that, and I took that oath seriously. During my years in the service, I was proud to wear the uniform and I would always be proud to have worn it. During my years in the Air Force, I met and worked with some of the greatest people I have ever known or will ever know. They will always be my lifelong friends because they shared the vicissitudes of military life and dedication to the preservation of our nation's freedoms. I now realized how difficult it was going to be taking off the uniform I had worn for so many years for the last time and entering a new lifestyle. It would be a challenge I had not faced before. And I already began to see that it would not be easy. But I no longer could waste my time looking backwards. The best I could do was say to myself, I did my best. Could I have done better? The answer, yes, for sure. And now it was time to move forward.

Retirement Nears

In June 1977, ACC issued orders transferring me to Fairchild AFB, Spokane, WA for duty as Deputy Commander of the 92 Combat Support Group. In the same month, we purchased the house at 4618 N. Ely Road. Curzon Fager sent photos of the house which Alice looked at. But for the most part she relied on my judgment which worried me greatly. So, we went ahead and bought the house which I believed was in a great location and had much promise.

We then began to clear out of our quarters at Elmendorf and I began to phase out of the Air Force. The personnel at XP gave us a very touching going away party held at the Anchorage airport. Alice and all of the family became very sad and nostalgic about having to leave Alaska which had become a favorite "home" to us. All of our kids grew to love the state and had acquired many close friends there. Alice and I had become very close to Dave and Joyce Creekman who we were going to miss dearly. It had been a very interesting four years that we had lived in this beautiful state, and we had enjoyed many of its awesome natural wonders. We were all going to miss living there. With a sad heart, we flew south to Seattle where we picked up our car, spent a few days, saw the Seattle Mariners in the new Kingdome and then drove across the state to Spokane. There Alice Marie got to see the home we purchased. She approved. As our household goods were in storage in Spokane, moving in would be fairly easy. Our intentions were to spend our summer at the lake cabin and move into the house in late August. I was required to report for duty at Fairchild until my retirement date of September 1. Jim and Mary stayed in Alaska to work during the summer until school began. Gene Jr., Mike and Kate were at the lake with Alice, and I joined them there on the weekends.

While performing my duties at Fairchild, I continued job hunting in Spokane and moving into the house on Ely Road. As a new member of Strategic Air Command, I got a taste of how this behemoth command operated. Almost all of the years that I served in the Air Force, SAC dominated the command structure. AF Chiefs of Staff mostly came from SAC and SAC budget priorities became AF budget priorities. However, Vietnam dissipated some of their strength. In the few months I served at Fairchild I saw enough to dislike the decision-making processes in place and their reliance on fear to inflict discipline on the ground and in the air. However, I did meet and enjoy most of the officers I worked with during my short time at Fairchild. My duties as deputy commander were not too strenuous, consisting mostly of inspections and paperwork. One lousy job I was given was conducting an investigation of a drowning at the Clear Lake Resort occurring in August. The Clear Lake Resort was a recreation area under the control of Fairchild AFB. The investigation was a tedious administrative headache involving many interviews of witnesses, plus taking testimony from medical, police and forensic personnel. The victim was a black civilian and some of the testimony hinted of racial bias. Concluding the investigation and putting the report together was time consuming and I was barely able to finish prior to my retirement date. Unfortunately, some of my conclusions did not please the Group Commander. Although the victim's death was clearly accidental, (he was a non-swimmer and was in the water after the conclusion of swimming hours), lifeguard personnel were slow to react to his plight and might have saved him had they continued to survey the swimming area while people were still in the water.

Prior to my retirement I was required to review a draft of an interview I gave to Grant Hales the Alaskan Air Command, Command Historian. He had interviewed me in June prior to my leaving Elmendorf. The interview was classified SECRET. I reviewed the draft and made such corrections as I deemed necessary and returned them to Mr. Hales. Being only a Deputy Commander at Fairchild and a temporary one at that, I had a hard time getting one of the secretaries to do the typing for me. Years later I was able to get a copy of that interview draft returned to me as it was by then unclassified. I was pleased that Grant Hales had noted in his letter the following compliment: "My only regret is that there are not more senior managers within the Air Force of your caliber who make my position the interesting pleasure it can be rather than just a job." Reviewing that interview after over 30 years brought back many poignant memories of my last days at AAC and the long-term changes that resulted due to the vision that I and my assistants had and through hard work placed into effect.

Job Hunting Continues

Each day I checked the classified ads for any leads, and Sam Grashio spoke to lots of his friends. I had written to Pat West who was a successful stockbroker about the possibility of entering that profession and he had answered me. He told me to come see him and that the field was open to those who were willing to put the time into learning it. I was about to call him as I was running out of patience, and I would soon be out of work. Then Sam called and told me that there was a possible opening coming up in University Relations where he worked as grant writer. He said that I should call Curzon, the man from whom we bought the house. I called Curzon and made an appointment to see him. Curzon was getting approval from Father Coughlin to hire someone to write requests for grants to foundations, trusts and corporations. He asked me about my writing experience. I told him that I was good writer but that most of my writing was for the military. He asked me for some samples, and I had a hard time finding some articles I had at home which I had written for some military magazines and newspapers. I could see that Curzon was not too impressed. He told me that he would think it over and besides, the position had not yet been approved and funded. I left feeling that it did not seem like a good fit for me. We never did talk about salary. When I saw Sam afterward, I found out that he did not like Curzon at all and was working to get him fired. I did not know what to make of this at all. On September 1, I would be without a job. I began to feel a tightness in my chest. Our income in August 1977 was over $3,000 a month. But that would drop in September to approximately $1,600 and change. Help! I now had three kids in college (Gonzaga) and a mortgage. I had taken out some Alaska State loans for Gene Jr. and Jim to pay their tuition at Gonzaga. So, I really needed additional income or begin to dip into what few savings Alice and I had been able to accrue, and that wasn't too much. My brilliant friend Ron Linder worked out a formula for me that indicated I would need a minimum of $15,000 yearly to get by. That did not sound like too much to me but time was running out.

Curzon Calls

Curzon called me and offered me the job on the condition that Father Coughlin approved my being hired. He also informed me that I would be on probation for six months due to my inexperience in the field of development should the President okay my being hired. Soon after, I met with Father Coughlin. I reminded him of our having met in Alaska, but I don't believe he remembered that occasion. I also told him that I was a Fordham alumnus and two of my sons and one daughter were students at Gonzaga. However, when I told him that I was meeting with him to get his approval to secure a position as a grant writer in the Development office, Father became completely confused. He mentioned that he thought that position was already filled by a Colonel Duffy, who had been a past ROTC Commandant. This, of course, confused me, and I left his office not knowing where I stood in the matter. When I passed this on to Curzon, he told me he would straighten the matter out and that Father had confused this position with another that had to do with writing proposals for government grants. I told Curzon if the position was still being offered, I would think it over and give him an answer in a few days. September 1, my official retirement date was near at hand, and I had to wind up my Air Force duties and administrative matters prior to that.

Retirement Ceremony

I was ceremoniously retired from active duty on the afternoon of August 31st, 1977, at the Officer's Club, Fairchild Air Force Base. On that date I had completed 26 years, two months, and 19 days of active service. I was 48 years and just under three months of age. Special Order No. AC-15900 was signed by General David C. Jones, USAF, Chief of Staff. In addition, I was awarded the Legion of Merit, with Oak Leaf Cluster. (Previously cited in chapter 37). In addition to my *Certificate of Retirement*, Alice Marie was awarded a *Certificate of Appreciation* for her support to me over those years without which I know that I could never have achieved any of what I did accomplish. Attending my retirement was Sam and Dee Grashio and my dad, and my daughter Kathleen. Afterwards we enjoyed a few drinks and a lot of laughs and fond recollections. It was a memorable occasion for me both sad and happy. Sad, because I knew that my life was changing and that I would no longer be wearing the uniform of my country again as I did for so many days and nights. Happy, because I would now be able to devote more of my days and nights to my family, especially my children who had grown so quickly on me and were now approaching adulthood. I realized how much time I had lost in being with them and would never be able to recoup. But now I would do all I could to recover those hours to the best of my ability. And so, Alice Marie and I faced an uncertain future, and we hoped it would become a good and fortunate one for the both of us and for our children. We had to take care of them, see them through their educational years and also take care of my dad as he was approaching his very senior years. Fortunately, he retained his good health and continued to spend his winters in Florida with his golfing cronies. He was a superb grandfather to our children who loved him dearly. He was generous to them, played games with them, (lots of golf with the boys), and shared all kinds of silly sayings and jokes with them which they still remember today.

And so, on September 1, 1977, the Cunneely Family became a civilian family with an unknown future ahead. But God had always blessed us, and we trusted Him to continue to watch over us. I know He had brought me through many dangers and home to my loved ones safely. For that I will always be thankful.

Following are excerpts from my last Air Force OER written by Col. Harry Goodall, Vice Commander of AAC and endorsed by Lieut. Gen. Marion Boswell, Commander:

> Col. Cunneely has been a strong contributor to the current status of Alaskan Air Command affairs. He has been the focal point for progress toward several important initiatives, i.e., the Joint Surveillance System/Region Operational Control Center (JSS/ROCC), Minimally Attended Radars (MAR), and contracting base operational support-all clearly beneficial to the Air Force. Other commands are currently examining these programs for applicability to their operations. The MAR concept, an entirely Alaskan initiative, holds great promise for eliminating remote tours for our people around the world. His deputate carried its share of the load ....
>
> *Regarding my personal performance factors:*
>
> Makes choices which are clearly beneficial to AAC and the Air Force. Not at all parochial, always considers new ideas. Outcome is usually better.
>
> Strong leader; can move people with or against issues. Doesn't seem to be flappable. As a matter of fact, seems to enjoy tension and uses it to purify and verify.
>
> A clear articulate speaker who is persuasive and candid. Doesn't mince words. Very convincing and logical positions are provided for my use in meetings and conferences.
>
> Polished image and professional competence are immediately apparent. He is a strong contributor to group efforts.
>
> Decision on force development and equipment configuration changes will save AAC manpower and O&M dollars. He is among AAC's strongest proponents for efficiencies in operation.

I could not have written it any better myself. Finale!

PART VII: WE RETIRE FROM THE AIR FORCE; GONZAGA UNIVERSITY, SPOKANE, WASHINGTON

CHAPTER FORTY: WE BEGIN A NEW LIFE

Lord, what wilt thou have me do?

-St. Paul

Waking up September 1, I realized I was no longer in the active Air Force. I no longer could wear the Class A blue or put on my flying suit and head for operations then take to the air. My days and nights around fighter aircraft and fighter pilots were over and would never return. I could only reminisce about the past and my experiences with the best people in the world and flying the fastest and most advanced combat aircraft in the world. My turn to do those things had unbelievably come and gone. I felt old. How could this have happened so quickly? I would have given anything to have turned back the clock and started over again. Oh, what fun it was and what a ride it was! And now what was in store for me in this new mysterious world that I was about to enter? I could not hide from it. The decision had been made and I had no choice but to face it head on and do my best to succeed.

Becoming a civilian again was a shocking experience. Not only did I need a job, I needed a wardrobe. For 26 years my daily costume was either a flight suit or a uniform. My civilian wardrobe, worn only for special occasions, consisted of one suit and a few ties and shirts. Alice would have to help me in that area. But more urgently, I had to nail down a job that would lead to that new career, and at this point I was at a loss what that would be. I began to feel old. I recall visiting with the Principal of Gonzaga Prep on one of my visits to Spokane. I asked him about my prospects for a teaching position at the school. He told me that teaching high school was a "young man's" job. As I was still in my early 40's, was in great shape and still flying fighter aircraft, I came away slightly put off. It was also becoming clear that achievements in my military career would not carry much weight nor duly impress potential employers who had very little idea of the kinds of things I had been doing in my years of service. In many cases my military service frightened away potential employers. In fact, it was a handicap.

On the days following my retirement, I checked the want ads and I again thought about calling Pat West regarding the brokerage business. Then I received a call from Curzon at Gonzaga. He had got things straightened out with Coughlin about the position in University Relations and he offered me the position at a salary of $14,000. This was below what I had hoped for but working for the University did have some perks that made it desirable in other aspects. Still, I was very unsure whether or not I should accept the offer. I told Curzon, a man I respected very much for his openness and integrity, that I would give him an answer in a few days. I wanted to talk to Alice Marie about it for sure, also I wanted to give it lots of thought and some prayer as this was a big decision.

Alice pretty much left the decision to me. She liked the idea of my working for the Jesuits in a Catholic college atmosphere. She was sure I would succeed at whatever it was I would be asked to do. She was far more certain than I was. One of the perks included tuition discounts for our children beginning in my second year at the University if I lasted that long. That would be most helpful as we were already in debt with school loans for Jimmy, Gene Jr. and Mary, with Mike coming up in a few years. Still, I thought that I would talk it over with Father Gubbins before making a final decision. We had met Father through Sam and Dee Grashio. We saw him each summer at the lake when he would be a guest at Sam's cabin and often read Mass for us on the beach there on Sunday morning. He was another Montana native, a gentle, wise priest who was just a great person to be with any time. I called Father and invited him to have lunch with me. I picked him up at Gonzaga and we drove to Frank Rotunda's Town and Country Restaurant on Trent Avenue. There, as we ate lunch, I asked Father's advice: "Should I take this position? I have absolutely no experience in fundraising or in grant writing. I do not want to fail the Jesuits." Father Gubbins' reply stunned me. He said: "Gene, don't worry about it. The Holy Spirit will take care of that."' I said, "Father, are you sure of that?" Father replied, "Gene, I'm very sure." And that was that. I went home and told Alice of my visit with Father Gubbins. She laughed and said, "Well, Father

Gubbins has a way of knowing what's best." (She was right.) The next day, I called Mr. Fager and told him I would be accepting the job. I still had grave doubts whether I was doing the right thing or not. But Sam Grashio was happy, and it was great that I would be working with him again, as well as in the same office with Father Gubbins. I knew that I had a big challenge in front of me and that I had much to learn in a short time. All I could do now was dedicate myself to the job I had accepted and to do my best as I had always done in my past. I was no longer an Air Force Colonel with eagles on my shoulders. Instead, I was a newly hired civilian employee at the bottom of the barrel. I was on probation to demonstrate I could learn a business that was new to me and could handle the responsibilities given me. I was in a brand-new set of circumstances totally foreign to the military world that I just came from, and I was the one that had to adjust. I first made it plain that I was not to be addressed as "Colonel." I was to be called "Gene," because military titles no longer were applicable or welcome. And so, I began my second career on an anxious and hopeful note. Little did I dream of what was to come my way, and quickly, that would test my flexibility and patience.

The Gonzaga Experience

I began my career at the university in September 1977 as a grant writer. When I arrived, I was the sixth development officer in University Relations. In addition to Curzon and me, there was Sam Grashio who was the Director of Annual Giving, Bill Dobner, the Director of Alumni Relations, Lee Norton, the Director of Public Relations and Communications, and Father Frank Gubbins, S.J., the Director for Planned Giving. We also had two secretaries and a lady who handled recording gifts. John D'Urbal, a former employee of the Spokesman-Review, was a part time worker who laid out the university newspaper, the SIGNUM. I was surprised that our workforce was so small. Our work area was in the basement of the University's huge administration building which had been cut up into small offices. During my indoctrination, it became clear that Father Coughlin and the Trustees were putting much pressure on University Relations to increase its fundraising. The financial condition of the University was quite shaky. Only a few years prior, Harry Magnuson, a Trustee and successful businessman, had come to Gonzaga's rescue. Couglin had been brought in

Gonzaga University Administrative Building (Spokane, WA)

as President, Stan Fairhurst as VP for Finance, and some stringent spending controls had been implemented. Several new Trustees were recruited who were expected to assist in bringing more financial security to the University's future. And University Relations was to play a major role in all of this. Having come from an institution where finances were taken care of by others, this was a new world to me. And early on I perceived that it was a world where pressure to succeed never existed day to day.

Troubles in the Office—We Lose Sam Grashio

There were several great advantages working at Gonzaga. Living along Upriver Driver afforded me an easy commute from home which Alice and I were really beginning to enjoy. I also enjoyed the people I was working with, and I found Curzon helpful in teaching me the elements of fundraising and the grants business. I also enjoyed being reunited with Sam Grashio, my old Air Force boss from ADC in Colorado Spring and North Bay. We were also neighbors on Lake Pend Oreille and our families were close. Alice and Dee Grashio were very close friends. Sam was helpful in introducing me to many of the Gonzaga people in my first days at the University. Sam, being an alumnus and a Spokane war hero, was a beloved figure in Spokane. And I was grateful to him for his sponsorship. However, I was surprised to discover that he and Curzon had a serious falling out. Curzon had suspected that Sam had been taking time off from work under the pretext of making donor calls. One day Curzon drove to Sam's house on Courtland Street and caught Sam there when he had checked out of the office stating he was going to make some calls. When Curzon confronted Sam about this incident, Sam was outraged claiming that Curzon had impugned his integrity. Having known Grashio for many years, I knew that in Sam's mind this was an unforgiveable breach by Curzon. From then on, Sam was out to get Curzon fired. This all took place shortly before I came on board. Each day when I arrived at the office, Sam would come into my room, close the door and start in on Curzon. Sam took his case about Fager to everyone in the University, including the VP's and the President, but he got nowhere. Most of us tried to calm him down but he would not do so. I felt confused because, as much as I liked Sam and tried to assuage him, I also liked and respected Curzon who had given me my job. Curzon had even apologized to Sam, but his apology had no effect. Things became worse and I began to worry about Sam. Clearly this affair was beginning to affect him mentally and physically. It was the fall of the year and Sam was busy getting the annual fund, the Great Teachers Program, underway. This was a big and important undertaking and Sam was the guy that had to do it. Although I was studying the grant writing business, Sam asked me to help him in some of the GTP matters which, of course, I did. Then, in November, things got worse, and Sam's temperament worsened. One Sunday afternoon, I received a call from Dee. Sam had a bad day and was complaining of chest pains. She wanted to take him to the hospital, but he refused to go. Then she told me that he had pains in his left arm. I then told her I was coming over and I was going to personally take Sam to the hospital, which I did. Soon after we arrived at Sacred Heart Hospital and Sam was admitted, he suffered a severe heart attack. We had arrived none too soon. A few weeks later he received a triple bypass operation. This was in 1977 and this type of operation was still fairly new and risky. The operation lasted well over three hours because the surgeons were unable to get Sam's heart restarted for something like 45 minutes. It was a most grueling affair for the Grashio Family and although Sam survived nicely many interfamily problems resulted.

Losing Sam in the office was a problem. Curzon then asked me to take over the Great Teachers Program. He knew, of course, that I had little to no experience in that program, but he had no choice. So, I jumped in, and I did the best I could. Fortunately, there were many GTP volunteers already out in the field making their calls and Sam had gotten the first mailing out to the alumni, so the program was somewhat underway. After Sam was in his recuperation phase I was able to consult with him and Curzon gave me his advice as well. As it turned out I ended up spending most of my time in my first year running the GTP than in writing grants. The GTP results were good, and we met the goal for the Annual Fund.

My First Year at Gonzaga

My first year as a civilian employee at an institution of higher education afforded me an education of itself. Coming from 26 years in the military, an institution geared to making decisions based on careful study, reasoned introspection and adherence to cost benefit analysis, the slow almost motionless manner in which the University identified and approached its problems I found exasperating. Against this, I also found the people I worked with to be exceptionally nice. If there was any group that I found difficult it was some members of the faculty. Many were liberal in their politics and definitely anti-military. Having spent the majority of their live within the walls of academia, they had resolved most of the world's problems right there without benefit of having to test any of their theories in the real world. While most of them possessed PhD's, their practical knowledge of what it took to make things happen and how to get things done outside of their classrooms was totally lacking. They were articulated, but ineffective. However, there were also some truly some great teachers and humanists that I got to meet and became friendly with. Men like Dean Jim McGivern who originating Gonzaga's Engineering School and Dean Bill Ilgen, who succeeded him. Dan Brajcich, Professor of Accounting, who was a legend in Gonzaga's Business School, who through his genius Gonzaga's accounting students set records for passing the state CPA tests. Others included Tom Rukavina, of the Philosophy Dept. Franz Schneider and John Sisk in the English Dept., John Firkins and Dan Hughes in the Math Dept., Kay Nakamaye and Don Thompson in Chemistry, Bud Hazel in Communications, and Bud Barnes, Eddy Birrer and Mark McDounell in the Business School and Bob Gilmore in Art. During my years at GU, I also worked with and associated with some truly inspiring Jesuits including Fathers Art Dussault, Pat Ford, Frank Gubbins, Clement Regimbal, Cliff Carron, Tony Lehman, and W. Schoenberg, all now RIP. Others still serving include Fathers Bernard Coughlin, Frank Costello, Ken Krall, Peter Ely, Kevin Waters, Steve Kuder, Bob Lyons, Michael Cook, Armand Nigro, John Mossi, Al Morrisette, Al Carroll, Tony Via, and non-Jesuit Father Charles Skok.

Harry Sladich

Of all the people I met early on my arrival at the University, one of them that stands out in my mind is Harry Sladich. Harry was a GU alum from Butte, Montana, one of the many Montana alumni that seemed to permeate the campus and found among the Jesuits, lay faculty, administrative staff, Trustees and Regents of Gonzaga. But Harry Sladich was unique. I soon realized that Harry was a source of information and assistance on which you always could rely. He was always accessible and friendly even though I could see he was a very busy man. And Harry was that way through all the years I was at GU and beyond. Knowing that Alice and I were new to the school and to Spokane, Harry and his darling wife Marguerite included us in invitations to their home where they hosted parties. These were great affairs and still afford super memories of warm affectionate evenings that afforded fine beverages, great food, and sparkling conversations mixed with humor and wondrous memories. Alice and I were very grateful to the Sladich's for they accepted us so early on and afforded us such friendship and hospitality. Harry had a marvelous career at GU. Twice he was acting President and was honored over and over again by the University for his accomplishments and his dedication. Tragically, he passed away in 2007 still in service to the University.

Curzon Resigns

In May 1978 Curzon handed in his resignation. This came as a shock to everyone. He said the pressure of the job was what caused him to resign. In addition, Lee Norton departed his position as Director of Communications to go back to teaching. It was determined that Sam would not be returning and Curzon after a long search had hired Maury Hickey to fill that position. Curzon was not pleased with that choice but was pushed into it as Hickey's wife was an employee and Maury was well known in the community. On the date Curzon left town for California, we had no replacement. Unbelievably, I was appointed interim VP. I had moved from a novice grant writer to Vice-President in less than nine months. I had a bad feeling that things were not going to get better. And they did. Father Gubbins came down with a serious case of lupus and had to give up his duties as Planned Giving Officer. Our staff was depleting by the minute. During the summer of 1978, I found myself helping Maury Hickey get the Great Teachers Program going, worked with John D'Urbal on writing articles for the SIGNUM, (which included the annual donor honor roll), and tried to do something with Father's schedule for Planned Giving lecture series that he planned for the farming community. I was truly meeting myself coming and going, making decisions on matters that I only had a smattering of knowledge about. My lake time that summer suffered badly. The administration was unable to settle on a new VP choice and in September Coughlin decided to ask Fager to come back for one more year. I thought that despite our shortage of an experienced leader, begging Curzon to come back was not a good idea. I made the mistake of telling that to Father Coughlin. He dismissed my suggestion abruptly. So, Curzon did come back (no doubt at a high price). So at least we now had an experienced leader, but our problems still did not disappear. Curzon and Hickey did not get along, so I had to help Hickey and, at the same time, run the Planned Giving Program in the absence of Father Gubbins. We did finally hire a Communications Director (Dale Goodwin), so that relieved me of that task. But Curzon then gave me another job. He was frustrated with our system of recording and acknowledging gifts. We had a lady in the back room with the responsibility of recording each gift that came in on the donor's record card and eventually into the university's computer system. Curzon thought that our lady was inefficient and probably should be replaced. After studying the situation for three days, I reported to Curzon that I thought the gal was a superwoman. The system was terrible. We needed to change immediately and utilize the computer completely or we would fall further behind. As I expected, Curzon put me in charge of the gift recording and acknowledgement system. I was moving further and further from grant writing and picking up more and more duties.

The Kids Adjust to Civilian Life

During the 1977-78 school year, Gene Jr., Jim, and Mary attended Gonzaga University. They lived in the dorms and Alice, and I saw them when they wanted us to see them, i.e., on certain weekends and holidays. They dropped by my office occasionally and I was pleased to see their progress. Gene and Jim loved GU and enjoyed their studies and their many friends. They also participated in ice hockey which was their favorite sport and through which they were reunited with several of their Alaskan teammates. But Mary was not happy at Gonzaga. Unhappily near the end of the school year she hurt her ankle. Nevertheless, she insisted on returning to Alaska. She and Jim took the ferry back to Alaska in the summer of 1978: Jimmy returned to work there for the summer, but Mary returned to Anchorage to stay and work. Alice and I were disappointed in her leaving college. Mike had attended Gonzaga Prep from which he graduated and applied and was accepted for Gonzaga University. Katie was attending St. Paschal Parish School. Katie was an excellent student and received first rate teaching there and she was doing well. So, with the exception of Mary, our kids were progressing well in their educational pursuits. Alice was busy with our new home, furnishing it and making it into a really comfortable place to live. We were finally in a house that we knew we were going to stay in as long as we wanted to, and we planned to make that a long time. No more moving every three or four years as we did in the military. So now, unlike in the past years of our married life, we could make long term plans. And that felt good for a change. We were now masters of our fate.

CHAPTER FORTY-ONE:
THE GONZAGA YEARS 1977-1992

Ad Majorem Dei Gloriam
(To God's Greater Glory)

-Motto of the Society of Jesus

When I began working for the University in September 1977, I was very unsure that I had made a right decision in job selection. As much as I wanted to work for the Jesuits, the great teaching society that provided me such a splendid education at Fordham, perhaps to repay in some small part what they had given me, I truly did not know whether I had the knowledge and background to do them any good. My military career was so totally unlike academia in the manner in which they approached their problems and formed their decisions. And the transformation would have to be made by me. Could I do it? I really wasn't sure. But I did make the decision to try. I respected Curzon Fager and I learned how to communicate with him and to help him deal with Father Coughlin and the Trustees. And in the 19778-79 school year, when we were still short of Directors, I did my best in filling in for Father Gubbins in Planned Giving and helping Maury Hickey in the Great Teachers Program. I also undertook a complete revamping of the gift acknowledgement and recording program. This led to full utilization of the University's IBM computer system, and a few years later to the acquisition of a new software system. I was the lead officer in the design and selection of this system and worked closely with the University's computer office in its installation. This was reminiscent of my days in AAC and the MAR system. Computerization of the gift recording system was a terrific advance for University Relations. Being able to accurately record gifts quickly and accurately, then acknowledge those gifts to the donor within a few days, greatly improved relations with donors.

The new system also greatly improved the way the University handled pledges. All in all, the computerization of the gift recording system was a huge step forward for the University.

Bill Wagner Comes to Gonzaga

Curzon had told Father Coughlin that he would only stay one year so the University needed to find a new VP. Several candidates were interviewed, and the staff was permitted to meet with them. One of those interviewed was Bill Wagner. He was an experienced development officer from California. And he was a real character. None of us knew what to make of him except that he was likeable and funny. We were surprised when we learned that he was hired. He was totally different from Fager. On my first meeting with him I was completely honest with him about my inexperience in the fundraising business and what I had been doing. Prior to Curzon's departure he had fired Hickey, so we were again without someone to run the very important Great Teachers Program. In my conversation with Wagner, I told him that I was running the Great Teachers Program again in the absence of a Director. He told me to go ahead. After our first meeting, I wasn't sure what Wagner's impressions were of me. I was not confident. But I liked Wagner. He was outgoing and fair-minded. He had the trust of the President and the Trustees, and they were supportive of his plans to bring on board experienced development people who would begin to meet the financial needs of the University. Within a short time, we hired a Planned Giving Officer, when it became clear that Father Gubbins would no longer be able to return to his duties. When Bill Dobner departed, he hired Marty Pujolar to take over as the Director of Alumni Relations, a post he held down successfully for over 20 years. Wagner also added additional staff in public relations, research, donor records, and administration. After the 1978-79 FY, Wagner gave me the Great Teachers Program to run and the title of Director of Annual Giving. Eventually, our office assumed the responsibility for all donations to the University's athletic programs. John Preston was hired as the Director for Athletic Development. John reported to me although his office was located in the Mamo Center and he worked closely with Dan Fitzgerald, who was now both the Men's Basketball Coach and Athletic Director. A few years later, University Relations was given the responsibility for the Law School development program. John

Power was brought on board to be the Director of Development for the Law School. John also reported to me and his office was located in the Law School Building adjacent to the Dean. I was also given an assistant position and I hired an excellent one who helped me with the GTP, Teri Mathis.

The years 1978 to 1992 were active and exciting as the University launched a Capital Campaign which required all of us in University Relations to work hard in planning and executing the programs necessary to raise the University's endowment. The campaign was a success and brought Gonzaga into a safety zone with some breathing room from where it had been in the 70's. In addition to the excellent work put in by Bill Wagner and our staff, much credit was due to Father Coughlin, who gradually honed his skills in asking for the big bucks. He also brought on board some outstanding Trustees and Regents who did yeoman work in both giving of their own funds and seeking out donor prospects. Gradually, I gained confidence in my ability to raise funds for Gonzaga. I put aside all bashfulness in going face to face with a prospective donor and asking him or her to make a contribution to the University. I attribute that to two factors. One was my capability to express my thoughts in a convincing manner that truly depicted my own sincere enthusiasm for Gonzaga. Second, I brought from my Air Force career many of my management and planning skills. These helped me in hiring some excellent people to assist me in the four areas of my responsibility. These were the gift recording and acknowledgement area, the Great Teachers Program, the Athletic Development Program (Bulldog Club), and the Law School Development. These folks included Phil Akers, John Preston, Rene Seibert, and John Power. Bill Wagner was a huge help in providing me guidance and encouragement and each year Annual Giving exceeded goals in the GTP and the Bulldog Club. The Law School was ever a tough sell. Nevertheless, we showed great improvements and even completed a capital campaign that eventually brought us a new law school building. Some of the new concepts our team originated included student tele funds, a President's Council Board, an annual Bulldog Club Golf Scramble, pre-game and halftime Bulldog Club socials, (featuring the ever-entertaining Dan Fitzgerald), resumption of the Red Mass for Law School alum and annual President Council Banquets. During my 15 years in University Relations, my directorate was responsible for raising over $27 million in annual gifts for the University. Twice during my time in University relations, I filled in as the acting vice-President while the University searched for a replacement. It was position I did not wish to fill permanently as I began to look forward to retirement years when I would be able to spend more time doing some things that were in my mind to do before time ran out. When Bill Wagner resigned, Gonzaga utilized a Headhunter firm to find a replacement. It took many months, and the Trustees went through several candidates before settling on a finalist named Ed Engel. Of all the candidates that our staff interviewed, he was the least liked. The consensus was that he was not a listener, did not understand the Gonzaga experience nor was interested in learning about it, and talked down to our staff as though he was coming in to save us from our past failures. We all felt that should he be hired disaster was in the works. Up to the very last minute I tried to talk Jack Butler, the Headhunter, into withdrawing Engel's name, but he refused. Just prior to his going to the Trustee meeting, I pleaded with Father Coughlin to delay hiring Engel and to look at some more candidates. Coughlin was departing for a trip to Burma in the morning and refused my request. I told him it was a huge mistake. Coughlin was furious with me, and Engel was hired. He lasted less than two years.

My Gonzaga Years

Father Gubbins' advice to me in the Town and Country Restaurant that the Holy Spirit would take care of things for me should I take the job offered to me by Curzon Fager in 1977, proved to be right on. The 15 years I labored for the University were golden. I certainly did not foresee that I would ever perform duties in the area of raising money for a Jesuit university after 26 years around fighter airplanes. Yet that is just what happened to me. That I was able to succeed to any extent could only have come about through divine intervention, of that, I am certain. In addition, several added good things accrued to me and my family.

Because of my employment at Gonzaga, after my first year of employment, Gene Jr., Jim and later Mike and Kate received tuition benefits, the boys at Gonzaga and Katie at Regis College. This helped Alice and me to pay off the loans we had taken in Alaska for the boy's college education. Mary, who had returned to Anchorage after a freshman year at Gonzaga, decided after a few years to return to Spokane and finish her education.

Her decision was to attend Eastern Washington University. After completing a year living at home and commuting, Mary moved to Cheney to finish her degree at EWU. After finishing Gonzaga Prep with excellent marks, Katie went off to Denver to get her degree at Regis. So, thanks in large part to my ties to Gonzaga, our three sons received their bachelors at Gonzaga, Katie received hers at Regis and Mary's at EWU. Alice and I were proud of all of them.

During my years at GU, I decided to obtain my master's degree. Utilizing my GI Bill privileges, I began attending classes in the Business School offered mostly in the evenings and on Saturdays. In 1983, I received my Master of Science Degree. I had great rapport with many of the professors especially those in the math department and the business school. Many of them had been with the University for many years. They really loved Gonzaga. They were not paid very much, but they loved their students, the Jesuits and the Catholic spirit of the school.

I Become a College Professor

In 1985, Bud Barnes, Dean of the School of Business Administration hired me to teach a course called *Business Communications*. It was a required course for business school majors. I had to arrange my schedule to teach it and receive permission to do so from Bill Wagner. I really enjoyed teaching although it was an added workload due to hours spent in class preparation and correcting papers, tests, etc. I taught this class for three years until the course was moved from the Business School to the English Department. (I believe this was a mistake.) Teaching gave me an opportunity to meet students in the classroom and to get a first-hand feel for their intellectual prowess. And so, for that three-year period, I was at the University as a teacher, a parent, an alumnus and a member of the staff.

Teaching college students in the 1980's was a shock to me. The students I had were juniors and seniors. But the majority of them were weak in spelling, grammar, and writing in general. This required me to spend the first three weeks of the class semester reviewing these fundamentals. In their oral presentations, I found that most were unable to speak to a group in a mature manner utilizing eye contact, strong opening and closing statements, and good use of vocal inflection and pitch control. In this area, the female students generally performed much better than the males. I included mock interviews in the curriculum and had the students prepare resumes which they could use for their post graduate job searching. I truly enjoyed my teaching years as it gave me a respite from the pressures of fund raising and an opportunity to meet and know the students. As I was an adjunct professor and my background exceptionally different from that of their other teachers, I believe that I was able to relate to them in a better and closer manner. Most of them were intrigued that a former fighter pilot was teaching them how to improve their writing and speaking skills preparing them for a successful business career. My student evaluations were for the most part excellent with the exception of a few dissident students who received grades lower than they hoped for. I would have liked to have stayed on as a teacher after my retirement.

Joining Gonzaga University after my Air Force career brought me in full circle. The Jesuits who taught me at Fordham were men of great stature and intellect. I remember with joy and fondness Jesuit Fathers Ignatius Cox, William Bier, Joseph Donceel and Thurston Davis, all brilliant minds dedicated to the education of Fordham's young men, and I compared them to the wondrous Art Dussault, Frank Gubbins, Cliff Carroll, Clement Regimbal, Frank Costello, Tony Lehmann and others, who likewise dedicated themselves to the young men and women of Gonzaga. How many lives have these men touched through their vocations? Now, through my second career, I was in a minor way able to give back a small amount of what they had given me, an amount I would not ever be able to fully repay. I truly found a home at Gonzaga. It was a place I enjoyed working for, mostly because of the people I worked with and the cause we worked for.

Another Family Loss

In 1986, my father died at the age of 93. He had lived his last years under our roof, and he was a great joy to our family. Our children came to know him and love him dearly and became endeared to his humor and his wisdom. In his last years dementia overtook his agile mind and we had to place him in a nearby retirement home. But he was uncomfortable there and wandered off often and walked great distances before being found by friendly strangers who called us and helped us retrieve him. But on one of his escapes, he fell and broke a hip and that required him to be confined to a wheelchair, and he quickly began to fade. It was a most frustrating period for me watching my wonderful father slowly begin his final days and I was powerless to do anything to prevent it.

My Brother Dies

Worse, my brother, at the same time my father was in his final days, was suffering with a terminal case of cancer. He was beginning to suffer much pain and neither chemotherapy nor radiation was effective. On the day of my father's funeral in New York, I pleaded with him not to attend, but he insisted. I believe his estrangement from his Dad for so many years made things worse for him. A little over a year after Dad's death, in August 1987 my brother Joseph, age 61, also died. We had been in constant touch during his last months, and I visited with him a few times, watching him slowly give in to the clutches of the onslaught of cancer. It was heartbreaking as the last of my immediate family passed on. My brother was my hero. He joined the Navy at 17, right out of high school and served his country honorably and bravely. In the midst of the Battle of Leyte Gulf, his ship was hit by Japanese suicide bombers (kamikaze), and it was so badly damaged that the Captain ordered all hands to abandon ship. They did, but when the carrier, the USS Kitkun Bay, refused to sink, the crew reboarded and sailed it through dangerous waters back to port for repairs. So, my big brother was away at war for many years during my teen years. When he returned, he married the girl who lived across the street from us in Jackson Heights, Patricia Halloran. He served 20 years as one of New York City's finest, (the NYPD) retiring as a Sergeant with a great record. He worked successfully on Wall Street and with his wife Pat, eventually settled in a beautiful home in East Patchogue, Long Island. Joe and Pat visited our family many times in Tacoma and Spokane, and we kept in touch constantly. Whenever I was able, I flew into Long Island, landing at what is now Gabreski Airport and visit with them. Unfortunately, they never had any children of their own, but my brother had great rapport with his nephews and nieces. Losing him at such any early age was very hard to accept. I always looked up to him as my "big brother" and I respected his judgment and how he handled life's adversities. He showed his bravery in the Big War and again when cancer hit him in the prime of his life and when he had attained financial security. He was the last remaining member of my family. His passing remains another one of life's terrible mysteries.

I Retire—Again

In 1992, I decided to leave Gonzaga. The two years under Engel were disturbing. We were in constant disagreement, and he proved to be a dishonest man. He attempted to fire John Power behind my back and when I confronted him about it, he had to back down. He promised to advance me to the position of Director of Development, a promotion which would place all fund raising under my supervision. However, he failed to follow through before he was cashiered. I was now 63 years old, and I had just had a hip replaced. I wanted to stay on as an adjunct teacher, but the job I was offered by Bud Hazel in Communications fell through at the last moment. I was disappointed, but not overwhelmed. Faculty promises were always tentative and unreliable. All my kids had finished college. Gene Jr. was in Hawaii, married with a son and teaching. Jim and Mike were living and working in Seattle. Mary was with the airlines as a Flight Attendant and doing well. Katie was married and had a son. Unfortunately, her marriage was shaky. Alice and I were planning to travel a little more and spend more time together doing more on our lake place and reading. We intended to stay close to Gonzaga, continue to be basketball season ticket holders, and attend social events there and with the Serra Club. We were going to try to be with the folks we enjoy the most. In other words, we were going to phase down the things we cared less about and phase up the things we cared more about. As long as we had our health and our energy, we would do what we could for our family, our friends and ourselves in the hope that it is what the Lord wants.

While my years at Gonzaga were most rewarding and beneficial, I have to admit that it was a far cry from the exciting and challenging years I spent in the Air Force. How often did I long to be back in my flying suit, head out to the tarmac, climb into the cockpit of a fighter and blast off into the sky on a mission that would take me far above the terrain, the traffic and the cares down on Mother Earth? I truly envied the young pilots I saw at Fairchild AFB when Alice and I went out to pick up groceries at the commissary. I saw myself in them and I recalled my days as a young Lieutenant, still wet behind the ears with lots still to learn about handling the latest jets I was flying. The years I spent in flying and working with dedicated men and women in the various locations we were sent, were no doubt the best years of my life, and nothing I would ever do again would match them. But time marches on, and the years take their toll, and I had to adjust and move on with them. I was fortunate to be employed by the University and I did enjoy my time there. I did the best I could to help Gonzaga financially and to provide it the best I could intellectually. The people I worked with were for the most part just super and many became lifelong friends. I utilized the campus facilities to continue the habits I brought with me from the service. I attended daily Mass in the campus Chapel and from there, skipping lunch, headed for the gym, participating in the lunch bunch basketball games. In many of those court conflicts, I brought my half century old wiles up against the likes of John Stockton, Eddie White, and several others of our varsity players. All in all, it was a great second career for me and one that I never foresaw. It was another example of the Holy Spirit at work, as Father Gubbins had told me.

Later Years

One of the advantages of retirement is the option to pick and choose what you want to do, then do it when you want to—for the most part. I began to pare down my participation in volunteer causes. I was a lifetime member of the Air Force Association, the Military Officers Association of America and the Order of the Daedalians. And I, of course continued my membership in the Serra Club, which at the time of my writing, has now surpassed 25 years. Although I no longer attend most of the monthly functions, I do attend the important activities which are intended to express appreciation to our priests, deacons, seminarians and women religious. I also am a member of the Foundation Board for the Immaculate Heart Retreat Center. For several years I was President of the Financial Commission of St. Paschal's Parish, (until I was fired by the Pastor who did not agree with the Commission's advisories). But I did assist St. Paschal's School in their annual appeal for over 20 years, as well as Mater Dei Institute, and Immaculate Heart Retreat Center, on whose Board of Directors I was a member for many years. I also gave some minor consulting advisories to my alma mater Chaminade High School. Best of all, I have been able to spend time with my grandchildren. I was able to watch Gene III play sports (baseball and football) at G-Prep, and Connor, Casey and Brendan McClary as well at Assumption, G-Prep and AAU. I was also able to see Kanani in her excellent volleyball career in Spokane and at G-Prep and now Grace and

Michael are beginning their sports as well. I have even been able to see Collin pitch baseball a few times and play basketball in Bellevue. Having time with our grandchildren and getting to know them and participate in their development to some degree is a valuable gift for Alice and me and we hope it is for them too. We see how difficult it is for the parents of today to raise their children in a Christian home trying to instill in them the right moral values that they must bring to adulthood in what today is a very secular society. It is by no means an easy task and they, the parents, must be diligent and watchful to ensure that their children are provided excellent example in the home and are guarded from the temptations that are so prevalent in our culture.

During our years at Gonzaga and then in retirement, we traveled to many foreign lands including Ireland, Italy, France, Spain, Switzerland, Germany, England, Netherlands, and Mexico. As well as Hawaii and several other great places in the beautiful USA. We also were able to spend long summers on Pend Oreille where the boys, Alice and I have made some great additions to the cabin. In the 80's we added a second story and new roof, repainted the interior and exterior and completely caulked the inside (downstairs) and chinked the outside of the cabin. The adventures and misfortunes we have gone through at the old cabin are legendary and provide us many hours of good sport and hearty laughs. A few years ago we tore down the old boat lift and purchased and installed a new one. In 2010, we installed some new TREX like decking in the front of the cabin. There is always work to be done at the lake cabin that will keep our family forever busy. Still, it's fun and our family never seems to tire of retreating there and enjoying the scenery that northern Idaho, the Selkirk Mountains and Lake Pend Oreille offer.

The proximity of the cabin was a major reason that Alice Marie and I decided to retire in Spokane. We believe that it was a good idea since our family felt and continues to feel that the cabin is their second home. It would be hard for us to perceive what our summers would be like without our days and nights at the lake, swimming from the dock, boating, tubing, barbequing, and working on the many maintenance tasks that are required to keep the place going. And now our grandchildren are enjoying it also. Thank God! As you make decisions in life, some are good, and some are not. You have to live with them and go on. You ask for God's help when you make the big ones. I have to say here and now, I believe God has answered my prayers more than I could ever hope for.

The Sunset Years

If someone had asked me when I was in my 20's and I was flying the fastest jet aircraft that men could build, what my chances were of living until I reached 80, I would have answered, not very good. There were so many opportunities that I put myself in to crash and burn, almost on a daily basis, that longevity was certainly not very feasible. And I saw so many truly fine fighter pilots, most much more skilled than I, "buy the farm." Strangely and unexpectedly, I survived pilot school, gunnery training, combat in Korea, and then numerous checkouts in advanced fighters (Mach II types), night intercepts, tanker hookups in clouds and turbulence, zero/zero weather scrambles and landings, all kinds of in-flight emergencies, lightning strikes, bird strikes, student pilot errors portending fatal consequences, not to mention completely ridiculous pilot errors of my own initiation that somehow, due to God's graciousness, I recovered from.

And so, I have reached my eighties, and with much thanks, my sweetheart of over 55 years of marriage, Alice Marie, is still with me. I spend many hours wondering why. Within the Lord's Prayer are the words, "Thy will be done, on earth, as it is in heaven." Surely it is for some reason God has willed that I have survived, and so it is up to me to determine, as best I can, what it is that He wants me to do with the days that He has given me. I believe that in these golden years each day is golden and that I am to use each one to serve His will in the best way I know how. Why else would God have spared me so long? And so, I do try to live my life in the way I can best serve others, especially my wife and my family, and then others who are in need of hope, sustenance, and love. Looking back on my life, I can see that I could have and should have done more of such things for others. But the awful truth is, I did not. So now, must be the time for me to make up for what I failed to do. Figuring out why each of us is brought into creation and then what our Creator wishes us to do with the wondrous gift of life, is the most important question we must put to ourselves. The earlier we answer this question and live by it, the sooner we can be at peace with ourselves with the realization that lasting happiness will not be found in this life.

In these times I have become devoted to the Prayer for Serenity, attributed to the Protestant theologian Reinhold Niebuhr:

God, grant me the serenity
to accept the things I cannot change,
courage to change the things I can,
and the wisdom to know the difference.

Living one day at a time,
enjoying one moment at a time,
accepting hardships as the pathway to peace,
taking as He did, this sinful world as it is,
not as I would have it.

Trusting that He will make all things right,
if I surrender to His will.
That I may be reasonably happy in this life
and supremely happy with Him forever.

I am now seeing many of my friends and former comrades in arms passing on. It is a reminder to me that we are made of dust and "unto dust thou shall return." I am so grateful for the gift of faith given to me from my Irish parents and developed through my Catholic education from the Nuns who taught me at Blessed Sacrament School in Jackson Heights, the Marianist Brothers and Priests at Chaminade High School and the marvelous, intellectual Jesuits at Fordham University, the Bronx, New York. Without the presence of the Catholic Church and its administration of the Sacraments to us struggling sinners, who are still waging the good fight against the temptations of the evil ones, I am not certain how we could fare.

Love and Marriage

In my own life, I also owe a debt of huge, unimaginable proportions to my wife Alice Marie. Her coming into my life at the time she did, completely changed everything for me for the better. Her love and dedication to me and then our children was a miraculous thing, and still is. I realize now, and have for many years, that my decision to propose marriage to her was an inspirational one and had to have come from above. I had no idea how significant an effect on my entire life, including my perspectives, habits and personality, she would have. In my bachelor years in the service, I had little time to devote to thinking about things outside of my professional career. I was being shifted from base to base, undergoing ten transfers within four years. I was mostly absorbed in training, learning to fly, then on to combat training, then transitioning into a new aircraft, etc. Living alone in BOQs and eating (and drinking) in Officer Clubs was getting old. I was now 26 years old, balding, with a cloudy future. I had no steady girlfriend. The closest female friend I had was Alice Marie who was still single but lived over 3,000 miles away and whom I saw only a few times a year when I was able to cross-country back to New York. The thoughts of marriage began to enter my mind more and more often. And Alice Marie was my only prospect. In August 1955, while I was on TDY at Tyndall AFB, Panama City, Florida, I proposed to her by phone. (I relate the story in Chapter 14.) I had decided that I was ready to move away from the single life and begin a new life with someone that would share the future with me. It was the best decision of my life. Alice, who was a working girl in New York City's publishing industry, thankfully accepted my proposal, gave up her job, and flew down to Florida where we were hitched. Of course, I did a faulty job in planning the wedding, but from the moment Alice came off the plane at the Panama City airport, I knew I had made the right call. We were married at a very simple Mass in the AFB Chapel with only the two witnesses, an altar boy and an unidentified lady in the rear of the church. No bridesmaids, flowers, limousines, rice, or photographers, et al. But my memory was that we were two very happy people. From that point, I could never conceive that I would ever consider being married to anyone other than Alice Marie or that we would ever separate. I believe history proved I was correct.

EPILOGUE

Epilogue can be defined as a closing section providing further comment. The concluding narrative of My Story (not counting the Appendix), ended with my retirement years through the 2000-2015 period. Little did I expect to remain in this world for such a long duration. Lord knows, God had thousands of opportunities, especially during my flying years, to call me into the great beyond. But for some reasons that only He could explain, I have continued to extend my days into the mid-eighties. As I write these words, I have now attained the ripe old age of 86.

Amazingly, I feel pretty good, suffering only the usual ailments of the geriatric set, hard of hearing, bladder urgencies, forgetfulness and slowdown in the energy department. I continue to exercise regularly through swimming. And I am blessed with a wife who keeps me busy with an endless list of projects to take care of, and who also makes certain that I am served nutritious meals that are also tasty. Alice and I celebrated our 60th wedding anniversary on September 17, 2015. Such benchmarks are always awesome and wondrous to capture and behold. They are important because they represent a constant that give an extra meaning to your life's work. In this case, our anniversary as this one was, reminds me to thank God for all the blessings He has bestowed on me, none of which I have earned. And the greatest blessing I received has been my wife Alice Marie, who over these many years has brought so much joy and love and peace into my life. I shudder just wondering what my life would have been had she not consented to share her life with me. Her decision to marry me was a most risky choice by her, and I can only wonder what prompted her to decide as she did. Thank God!

But even a good and happy life has its ups and downs. And we have had both for sure. When Jesus asked us to follow Him, we only had to see how His life on earth ended to realize that life is not going to be a bowl of cherries. For Alice and me, our five children were true blessings. They gave us such joy and we were so thankful that they arrived healthy and beautiful. Raising them was serious business, but also a wondrous pleasure and they gave us so much in the way of love as we watched them grow and mature. Our three boys are now great fathers and husbands and we are so proud of them. Our daughter Mary, after college, was a Flight Attendant for 17 years and then became a Registered Nurse. The other daughter, Katie, married and had three children, all sons, and also became a nurse. Unfortunately, her marriage failed and she had to raise her boys herself.

Altogether, our children gave us nine grandchildren, another great blessing. The vocation of marriage and family kept us so busy and involved and it brought us closer to each other as the years rolled by. From the day we were married in the little chapel at Tyndall AFB, Florida we always tried to keep Jesus near to us in our lives following the adage that "the family that prays together, stay together". It worked for us.

In 2015 my daughter Mary, the older of our girls, passed away unexpectedly. It was such a shocking event and a reminder that in the long run, God decides when your hour is up. And we must accept His will. It was not an easy acceptance. Trying to understand why I am left here after ninety-plus years of age, but my daughter at 56 and such a valuable member of our society was taken, is difficult to understand. During my Air Force years I dealt with death many times. Loss of comrades was never easy to take. But losing a child is worse by far. I don't believe I will overcome my grief soon at losing Mary, if ever. Mary was such a good person, strong in her Catholic faith, and always helping others in need. I am consoled that I know that she is with Jesus now. It is at times like these, when your inner strength is tested that your faith in the Savior comes in handy and sees you through. I can only pray that I will not be further tested in my lifetime.

It was a tough and heartbreaking year, 2015, as my second daughter Kathleen (Katie) was suffering from cancer. It was a serious situation and our family, especially her sons, were extremely worried lest they might lose their mother. And Alice and I were fearful lest we might lose our remaining daughter. So once again we place ourselves in God's merciful hands and ask for His compassion and spare Katie's life. But we must accept that God's will be done.

Katie eventually passed away in December of 2015. It was tragic again losing my second daughter. We have great memories of the girls. Plus, Katie left behind three handsome sons. God did not allow this void to continue. I did not know it at the time, but God would be sending me three new great granddaughters!

I am now into my 94th year of life. I have seriously decided to turn my memoirs into a book. But on November 30 of 2020 I came down with the now famous COVID-19 virus. I spent fifteen days at Valley General Hospital. After being discharged from the hospital, I had to recuperate at home with the help of my family. By the summer of 2021, I was out at the cabin on Lake Pend Oreille. It is a restful spot with lots of memories of family and friends for over 50 years. Again, the Good Lord interceded and allowed me more valuable time on earth with family and friends. I'm still pushing on with the book.

On October 1st, 2021, my granddaughter, Kanani and her husband Micah, brought a beautiful baby girl named Ellis Beckett Mahealani Duncan into the world—I am now a great-grandfather! My grandson Gene III and his wife Mallory followed shortly after with the birth of Quinn Rose Kawaimālie Cunneely on January 13th, 2022. On June 15th, 2023, my third great-granddaughter, Charlotte Avery Anuhea Duncan, was born. All girls so far, but a great-grandson is expected in October of 2023 to Gene III and Mallory! All are doing well and growing fast. My other seven grandkids are thriving as well. They are either gainfully employed or pursuing degrees from universities. I am so proud of them all!

APPENDIX

PART I:
FLYING STORIES

Insidious Onset

As if I did not have enough worries facing me, as whether I could handle the T-28's faster speed and the 360-degree overhead landing pattern, plus checking out in the T-33, the USAF's first jet, along came a trip to the pressure chamber. Bryan AFB had a building which contained a pressure chamber and before we student pilots began flying over 10,000 feet, we had to "fly" in the pressure chamber. And that's when we first heard the words, "insidious onset." There we were told that there was something else that could get us killed besides engine failures, compressor stalls, midair collisions, not to mention enemy fire, and that something was hypoxia.

Hypoxia, in short, is lack of oxygen which, as everyone knows, the body needs to survive and which there is a lack of at higher altitudes. And while pilots are provided oxygen for high altitude flying, like all other mechanical systems, oxygen systems sometimes fail. And when they do, it can cause death before a pilot realizes what has happened to him. When he begins to lose his oxygen intake, a pilot will slowly and usually without warning or understanding what is occurring to him, pass out and not recover. Thus, hypoxia gets its appellation. To help understand hypoxia, pilots and other aircrews who fly at high altitudes are required to receive altitude chamber refresher courses biennially. After a thorough briefing, the crews enter the chamber where they are slowly taken to high altitude by releasing air from the chamber. The crews have oxygen masks. After leveling off, each crew tests himself on how he would experience the effects of hypoxia. He does this by taking off his oxygen mask. As he begins to breathe without oxygen entering his system, the deleterious effects of hypoxia occur. Some pilots last longer than others, but all eventually will show deterioration mentally and physically. Their minds will become sleepy, and their coordination will rapidly decrease. They will begin to pass out. Only when their buddy hands them their mask and they begin to breathe oxygen do they soon recover. Watching this and then going through the exercise yourself is frightening because when you are flying in a single seat aircraft, should this occur, there is no one to help you. When

I left the Bryan chamber that day, I had another important thing to think about and to store away in mind for the future.

While I was at Bryan in the summer of 1952, we lost a student pilot due to what the accident board determined to be a case of hypoxia. He was on a weekend cross-country flight with his instructor and two other students. On one of the legs, I believe they were in the Midwest, he went radio silent and his T-33 began to stray from the formation. The instructor was unable to rouse him on the radio and he could see him slumped over in the cockpit. His plane finally went down into a field. The entire student body at Bryan Field was shocked at this accident which took the life of one our fellow students, and it was due to a cause that seemed so unnecessary and avoidable. But that was the nature of hypoxia, an insidious killer of aircrews.

Ed Gates and I Take a Cross Country

Years later, when stationed at McChord AFB in Tacoma I had my first close encounter with hypoxia. It occurred in 955 and I was with the 317th FIS. I was also the head T-33 instructor pilot for the 25th Air Division and had access to two T-33s and two F-80s which I often used on the weekends for cross country excursions if I was not on alert. (This was before I was married.) One of my fellow 317th buddies was a pilot named Ed Gates, and he agreed one weekend to go with me on a weekend T-33 trip to New York. We were able to leave on a Friday and after making a refueling stop in the Midwest we departed for Steward AFB. We were cruising nicely at about 35,000 feet over clouds in Pennsylvania when I noticed that Ed, who was in the rear seat, had not said anything to me in a long while. I then asked him how he was doing, but he did not answer me. A second time, and again no answer. Looking through the rearview mirror, I saw Ed slumped over with his helmet against the canopy. I began to shout at him through the interphone and he did not respond, and I knew instantly that he was hypoxic. With that, I went to emergency on my IFF responder and told ATC that I was declaring a MAYDAY and was immediately descending to 10,000 feet. I pushed the nose of that T-Bird over and went down through those clouds almost vertically on instruments. All the time I was yelling at Ed on the intercom to wake up. I had no idea how long he had been out. I had that T-33 up against the red line on speed as I leveled off at 10,000. I began to notice Ed's helmet moving a little and I asked ATC for a vector to the nearest field. Ed now began to moan and then talk a little as he started to come to. I was still in the clouds and ATC asked if I wanted to go to Syracuse. I told them I would like to head for Griffiss AFB which was close by and for a lower altitude. ATC gave me a heading for Griffiss and cleared me to a lower altitude which helped Ed come completely out of his daze. I then called off our MAYDAY and we landed at Griffiss.

After we landed I asked Ed if he wanted to see a Flight Surgeon, but he declined. He was perfectly normal as far as I could determine although minutes ago I was frightened that he might be dead or permanently injured. It was clear that I was more in a state of shock than he because he had been out of it for the most part. For my part, I was angry that I had permitted Ed to pass out on me without my realizing it until he was already unconscious for an unknown period of time. Thank God, it came out okay. It was a lesson for me, and Ed Gates too, on how insidious hypoxia is and how dangerous it is because it can creep up on you when you least expect it.

Hopeless

In 1971, when I was commanding the 95th FIS at Dover AFB, my operations officer, Lt. Col. Harold Van Daele, while on a training mission in an F-106 was diverted by ATC to intercept a US Navy fighter that was over the Atlantic Ocean and headed eastward. The aircraft was an A-6. When Van Daele's F-106 closed in he saw the pilot slumped over and obviously unconscious. All efforts to rouse him or to turn his aircraft back toward land failed. When Van Daele's fuel began to run low he had to leave the Navy aircraft and begin his return to Dover leaving the A-6 and its pilot to continue eastward to its eventual tragic ending. It was another victim of hypoxia's insidious dangers.

Helpless

During the Cuban missile crisis in the Fall of 1962, I was stationed at ADC Headquarters in Colorado Springs, Colorado. The entire USAF was placed on higher alert status and most of the squadron aircrews were busy on alert duty. All inspections were postponed and headquarters pilots, like me, were utilized to augment the units on alert duties. We were also used to ferry unit aircraft to the various depots for scheduled maintenance and systems modifications. I was scheduled to pick up an F-102 at a Midwestern airbase and ferry it to Keesler AFB, Mobile, Alabama. I arrived at Mobile after dark and encountered a thunderstorm in the area. This forced me to make an instrument approach that vectored me over Mobile Bay around the center of the cell that contained heavy rain squalls and lots of lightning. It was a bumpy approach and I was really happy to see those runway lights and get that baby on the ground. I was soaking wet by time I arrived at the BOQ, got a little something to eat and hit the sack. I was very tired as well.

The next morning, I received a note informing me that I should see a Captain at Base Ops who would transport me back to Peterson Field in an EB-57. EB-57s were utilized by ADC as target forces in their exercises. Stationed at Hill AFB, Ogden, Utah, these aircraft were excellent "target faker" aircraft because they had long range capacity as well as electronic countermeasures (ECM) capabilities, including chaff dispensers. The rear cockpit was manned by an Electronic-Countermeasures Warfare Officer (EWO), who worked the ECM gadgets during its target missions. However, the rear cockpit did not have any throttle or control stick. The occupant in that cockpit was for all purposes just a passenger.

On arriving at Base Ops I encountered the young Captain who would be my pilot, only to find that the aircraft had a mechanical problem. He was a junior officer, not yet a Senior Pilot, but I did not concern myself too much about it as I was very anxious to get home asap. After several hours of delay, we went out to the plane and loaded up to depart. But after starting up, the Captain shut the engines down due to a recurrence of the malfunction. This happened again later in the day and it was not until dusk that the problem was finally solved and we were able to depart Mobile for Colorado Springs, a flight of about two and a half hours. It was a clear, starry night with unlimited visibility, and I just settled back and relaxed for the long flight home. After being airborne for about an hour and a half with no apparent difficulties, the Captain advised me on the intercom that he was not feeling well. When I asked him what his problem was, he told me that he was "woozy" and that he thought that he might pass out. We were cruising at above 35,000 feet. I advised him to check his oxygen system and to immediately descend to 10,000 feet. I told him to continue talking to me and fight to stay conscious by all means that he could muster. I reminded him that without a throttle and controls in the rear seat, I was unable to assist him or land the airplane. I reminded him that I was helpless to save him and the plane if he went unconscious. He continued to talk to me, but intermittently told me that he felt weak and that he might pass out. I took over the radio, called ATC, and declared an emergency and a vector to the nearest AFB and a descent to 10,000 feet. ATC gave us a vector to Amarillo and approved our descent. But the Captain was still a problem. Even after we leveled off at 10,000 feet he complained of wooziness and now of poor vision. On contacting the control tower at Amarillo, I advised them of the pilot's condition. I requested that we be met with an ambulance and medics and that the runway lights be turned up to the highest intensity. I did have an altimeter in the back seat and an airspeed indicator which I was able to read out to a very woozy pilot. Even as I turned into a long straight in approach to the runway, he continued to tell me he was unable to read the gauges and I had to read the airspeed reading to him. Our EB-57 came down the final approach in a S curve. Each time the pilot told me he was passing out I shouted to him I would have to bail out on him. I never felt so helpless and frightened in an airplane in my entire flying career. As we came over the runway threshold, I told him to cut the throttles. The plane sank quickly and bounced heavily onto the runway, back into the air and down again, up again and down again, this time for good. I yelled into the intercom, "Brakes, brakes." Fire trucks raced after us and we finally came to a stop 500 feet short of the end of a 10,000-foot runway. My heart which had to have been racing at a record speed, finally began to slow down. I could not believe what had just occurred on what should have been a routine, boring cross-country flight.

They unstrapped the pilot and took him to the Base Hospital in the ambulance. A mechanic on a ladder helped me shut down the aircraft engines, a procedure I had no idea how to do. The EB-57 was then towed into the flight line. I gave the doctors my description of what took place in the air with the Captain. I also wrote out a statement of the incident for the Flight Safety personnel. I then went to Base Ops and made some phone calls to ADC and to Alice Marie letting them know I would not be home tonight. I did not tell my wife any details of why I would not be home tonight. By then I was totally worn out from a long, frustrating day in which I was scared out of my wits by circumstances that were totally out of my control, that could have easily taken two lives, one of them being my own. I checked in at the BOQ and tried to get some sleep. But I laid awake a good part of the night going over in mind the events of that flight. The more I recalled the details, the more I realized how close we had come to disaster. For one thing, I now realized that I was not even certain I knew how to execute a safe ejection from the rear seat of that EB-57, especially as we descended to very low altitudes on final approach. My mind was on other things, especially on attempting to keep the Captain awake and still in control of our plane. Nor was I certain that I would have made that fateful decision to leave the aircraft and the young pilot to certain death, even though I had threatened to do so many times over.

The next morning, I visited the Base Hospital. I saw the young Captain and he appeared to be back in good shape. I spoke with the Flight Surgeons. While they surmised that he was a victim of hypoxia, they also believed that it was brought on to some degree by the long hours he put in the day of the flight. He was also suffering from a cold and probably should not have been flying. The plane's oxygen system was working normally and that was not a factor. The Captain and the Doctors thanked me for the job I did in getting us down safely. ADC was sending a T-Bird in to fetch me, so I said goodbye to the young pilot and wished him well. I left him there in the hospital as the Flight Surgeons wanted to keep him there at least another day for observation. I never saw him again.

As you might guess, that was my closest call to having hypoxia claim me as a victim, and that would have been as an indirect victim. Remembering this episode, I recall most the horror of what my emotions would have been had I bailed out of that EB-57 and left the young Captain to go in with it and die. Could I have lived with that? And how would others react to my decision? Thank God it turned out as successfully as it did. It was so close to being a major tragedy.

I was extremely happy to return home the following day and see my family, and in cases like these, you appreciate them even more. I did not speak about this incident to Alice Marie, but the word got out because it became a topic of wide discussion at ADC Headquarters. All kinds of people congratulated me on the incident including several General Officers. General Bill Elder put me in for an Air Medal, but that was denied. The Flight Safety Office made it a subject of an article in ADC's safety periodical *Interceptor*. But to this day, I look back on this episode as another one of those many examples where God was looking out for me, and in His goodness and wisdom decided that He had other things in mind for me.

Death in the Afternoon

During my flying career, I was often called upon to serve in the capacity as Airdrome Officer or as it was sometimes called, Air Operations Officer. The AO served in place of the assigned unit's Operations Officer. His job was to oversee all flying operations, check all flight plans for safety and accuracy and to be available to assist in the case of airborne emergencies. The AO was located in the Operations Building and remained there throughout the period that flying operations were being conducted in and out of the airfield.

I was fulfilling this duty with the 80th Fighter Bomber Squadron at Suwon Air Base (K-13), Korea, in the fall of 1953. The war had ended the previous summer and those of us left were filling out our tours flying patrols along the DMZ and up along the coasts of the Korean peninsula letting the North Koreans know we were still here and ready for them lest they want to resume hostilities. On this particular day, our squadron had completed its missions early and no further flights were scheduled other than a test flight to be flown by my good friend Bob McDermott in one of our F-86s that had just come out of some engine maintenance. While sitting in our Ops control center reading *Stars and Stripes* with our faithful squadron mascot Hobo (dog) lying by, Bob called in with a problem. He was back in the landing pattern, but his right landing gear apparently would not come down. He tried several times to extend his wheels, but the result was the same. On a slow flyby, the tower controller verified that the right wheel did not extend. I then advised Bob to extend the gear and bounce the aircraft on the runway to see if that would somehow open the gear door and drop the wheel. McDermott was an excellent pilot, and he accomplished this tricky maneuver as well as it could be done, but it did not work.

By this time, word has spread throughout the base that we had an aircraft in trouble that might have to land with only one main landing gear and the nose wheel extended. I was now receiving calls from the Base Commander and other agencies on base regarding our aircraft. Bob was now getting low on fuel and would have to land soon. About this time, Bill Knuttel, one of our recently arrived pilots came running into Ops. He was excited about McDermott's having to make a crash landing and he had with him a movie camera with which he was going to record the accident. Bill was a guy that I had not quite taken to yet. I found him to be a loudmouth and a bit of a "know it all." I liked it better when new guys took their time in letting us know how good they were. Still, I was much too busy at this time to pay much attention to him. I advised Bob to bring his aircraft in and touch down at the slowest speed he could and hold his right wing up as long as he could and cut the throttle when the wing tip hit to avoid fire. (This I was reading from the Dash One). The fire trucks were in position and some foam had been placed on the runway. It was now up to Bob and the gods. Knuttel ran out the door with his camera to get into position to record the event. As McDermott came down final approach, I also left Ops and went out to watch his touchdown.

When I got outside, I climbed up on one of our revetments to better see the F-86 which was still about a half mile from touchdown. I then saw that Knuttel had gone out onto the infield about 6,000 feet down the runway to get his pictures. I saw that he was also fairly close to the runway, maybe only about 50 to 60 feet from the edge. It crossed my mind, Knuttel should not be there; he could get hit. Bob's F-86 made its touchdown about 1,000 feet down the runway on its left gear and he held the nose high, and the plane rolled for a good distance and then the right wing began to lower and then it gently struck the runway, and the aircraft began to turn into it and as it turned all of us who were watching saw that the F-86 was heading directly toward Bill Knuttel. What happened was like a slow-motion replay. Knuttel, who was aiming his camera at the F-86, realized that the aircraft was bearing down on him. He began to fall away and escape, but too late. McDermott, who was trying to hold on to an aircraft that was no longer responsive to any control, looked out to see a figure in front of his plane disappear beneath his right wing with a thud. As the wing rolled over Knuttel, his body flipped into the air and came down lifeless. The aircraft came to a halt and McDermott got out and ran back to see what had occurred. The ambulance arrived as did the fire trucks. It was an awful scene depicting a fatal tragedy that never should have happened. We had lost a pilot, a human being, who placed himself in a spot where he had no business being. The aircraft was not badly damaged and flew again in a short time. Bob McDermott was badly shaken. He had performed masterfully in bringing back his malfunctioning F-86 and doing it as well as it could have been done, but it ended up in killing one of his fellow squadron members. He was grounded for several days until he was able to adjust to what had taken place and accept the fact that it was not his fault. For Bill Knuttel, it was a useless way to die. I now felt terrible. Maybe I should have somehow prevented Bill from going out to the infield with that camera. I also felt badly that I had not gotten to know Knuttel better; perhaps he was a good person, better than I realized. It was my worst day as an Airdrome Officer.

My Best Day as AO

In 1959, I was assigned to AO at Peterson Field, Colorado Springs, Colorado. At that time, I was assigned to the ADC TACEVAL Team, Air Defense Command Headquarters. When I arrived for duty I realized that something unusual was up as some Secret Service men showed up. They wanted to search the place and go over some details regarding aircraft parking. It was not for a long time was I able to learn that Vice President Nixon was to arrive later in the morning. Nixon was flying down from Denver in an Air Force passenger plane and would be met by some local Republican bigwigs.

When Nixon's plane arrived, there were several limousines there to meet him with some dignitaries to welcome him. I was standing at the aircraft to ensure everything was in order. When Nixon came down he shook hands with all the bigwigs. It was a sunny day and Nixon was sprite and quite cordial. The pilot of the Air Force plane had his family there and Nixon saw them and had them stand with him and have a photo taken. He was handsome and friendly. Before getting into the limo, he shook everyone's hand including mine, saying, "I'm Dick Nixon." As if I didn't know. He would later become my Commander-in-Chief. Little did I realize what a future this man had to face, both good and bad.

Weather Tales

Only after returning from Korea and joining an all-weather squadron, did I realize the importance of becoming a master of instrument flying. I set about doing that at McChord AFB, Washington, an area of the country noted for fog, rain and low visibility situations. Although I had come from a squadron where I was the unit's Instrument Instructor Pilot, I quickly realized I had much to learn about that end of the flying business. And the only way to do it, was to do it. And at McChord, I had plenty of opportunity to practice. I made it routine to end all of my training missions with practice instrument approaches, both GCAs (Ground Control Approaches)and

ILSs (Instrument Landing System). Whenever I could, I practiced my instrument flying under the hood in the backseat of the T-33. I became proficient to take-off in the T-Bird once the aircraft was lined up on the runway and fly the entire flight under hood all the way down to round out. This concentration on flying the gages gave me enhanced confidence in my instrument flying and provided me lifesaving support on several occasions when, due to circumstances beyond my control, such as quickly changing weather conditions or air defense scrambles, I needed that confidence to get back on the ground safely. Two of these I recall vividly.

Cocked Pistol

Shortly after Alice Marie and I were married, the 317th FIS held a party in the Officer's Club at McChord. As was customary, the party began with a cocktail hour. My usual routine was a beer, followed by a couple of scotch and sodas to begin the evening. Alice enjoyed a glass of wine. However, as we were about to enter the dining room, a bell sounded and Major Williams, our Squadron Commander, called everyone to silence for an announcement. The 25th Air Division just called a "Cocked Pistol". This was a code word for a division wide air defense exercise. All aircrews were to report to the flight line and man their planes for scramble against an incoming "enemy" force that would be attacking the Division. Williams had been in argument with the Division, explaining that his squadron was in the midst of a unit party at the O club, much alcohol had already been consumed, and sending his aircrews into the air might be unsafe. However, the Air Division was adamant, the "Cocked Pistol" stayed in effect. Major Williams then had the pilots lined up and asked each one how many drinks he had consumed. He also gave them a breath test. When he got to me, I told the Commander I had a beer. He told me to head for the flight line. I told Alice that she had to get home by herself, which she did.

I hitched a ride to the flight line, suited up and went out to my assigned F-86D and set it up for a scramble by pre-flighting it, placing my parachute in the seat, and my helmet on the canopy. I returned to ops and waited. Most of the pilots, including myself, expected that the Division would back down and cancel the 317th's participation. I certainly hoped they would as Major Porter, the Ops Officer and I were the first two in line to go. I began to think that maybe I should have been more truthful with

Major Williams. I was hoping to get a cup of coffee in me when the scramble horn went off. Porter and I ran out to our birds and started up. It was a typical Tacoma night, low clouds and rain. We taxied to runway 34, hoping all the way that the scramble would be scratched, but no dice, off we went into a very black, rainy sky. We climbed up through thick layers of clouds and did not top out until we reached 30 thousand feet. We were vectored north over British Columbia where we were put into orbit waiting for any incoming targets. It was a real dark night, no moon, and I immediately got the worst case of vertigo. I was flying in trail of Major Porter, about a quarter of mile behind, but my senses kept telling me we were climbing, then diving, then turning, although when I checked my instruments, none of the above was true. Vertigo is a true enemy to a pilot because it is telling the pilot to forget what his instruments are saying, and to believe what his senses are telling him instead. And we are all more willing to believe our own senses first.

This went on for quite a while and my nerves were screaming. But I just decided that the gages were correct, and I would stay with them. After almost a half hour, we were finally given a target and were vectored to intercept an "enemy" T-33. We closed in, made a kill and were finally set up for recovery back to McChord. We were a long way from home and the weather had deteriorated. So had my spirits! Major Porter was a teetotaler and was in much better shape than I, for sure. We had to get into a holding pattern as there were several other aircraft ahead of us and now we were beginning to get low on fuel. The ceiling and visibility was now close to minimums as we began our approach to the field. In 1956, the letdown to McChord Field was a teardrop on a heading of 120 degrees from over the field, which aimed you directly at Mt. Rainier. At 12,000 feet, you began a starboard turn using a 20 degrees bank back toward the field, rolling out on a heading of 340 degrees. Leveling off at 5,000 feet, Approach Control should pick you up on radar and vector you and let you down, handing you to Ground Control Approach at 2,000 feet. Thank goodness, everything went like clockwork. Thanks to my hours of practice and built-up confidence, and despite my tiredness and bouts with vertigo, I held to the glide path and center line all the way down until I broke out about 200 feet above the ground, well high enough for a safe rollout and landing. I was exhausted and very happy to be back on Mother Earth. It was a flight that should never have been flown. Foolishness on the part of superiors at

Division level put pilots in the air at serious risk for an "exercise." Good performance by those pilots saved them (the Division Commander) from being sacked.

Never Believe a Good Forecast

Weather forecasts are necessary to file a flight plan and most of the time they prove valuable, that is, they are reasonably accurate. But not always! That is why, a pilot must always have an alternate airport on his flight plan that the forecasters say will have good weather and which will be reachable after the pilot makes a missed approach at his original destination. There is a good reason for that rule. Weather forecasters make lots of mistakes.

One of them that sticks in my memory occurred on a cross-country flight I took from McChord to New York. It was a flight I had made several times as I enjoyed going home and visiting the family when I could; also, to see Alice Marie.

On this trip, using a T-33, I was alone. On the first leg I filed for Ellsworth AFB, South Dakota. With good tailwinds and trimming back carefully, I planned to overfly Ellsworth and, should I have sufficient fuel and weather permitting, I would refile in the air and land at Sioux Falls AFB (now Joe Foss Field). If I could make Sioux Falls on the first hop, I could then fly into Mitchel Field on the next flight without having to stop anywhere else for refueling. So, I was really pleased when about 100 miles out of Ellsworth, I had conserved enough fuel to make Sioux Falls if the weather there permitted a VFR approach. I called ATC for a weather update, and I was given a good forecast. Sioux Falls was "clear with 10 miles visibility". With that, I refiled for Sioux Falls. But as I approached my destination I noticed cloud cover over the city and when I checked in with Approach Control I received a totally different picture. The field was now closed in with a cloud ceiling of 300 feet and visibility of two miles. My fuel remaining would be okay for a VFR approach but would be extremely tight for an instrument approach and would certainly not accommodate a missed approach. How did this happen? What a few moments ago appeared to be a no sweat deal was now a big sweat deal because a weather forecast I was given proved totally wrong or had changed so rapidly that it left me with no options other than to make a perfect approach to an airport to which I had never made an instrument approach before.

Fortunately, I broke out and landed although with very little room to spare. When I taxied in and shut down, my fuel supply was close to fumes only. My first stop was the weather office in Base Ops. I wanted to know what happened to the weather from the time I was told the field was wide open to the time I arrived at the field. The weather guy had no explanation other than ATC must have been using an outdated weather bulletin. Actually, the field had been socked in for the past hour. I was flabbergasted. I had changed my flight plan based on an erroneous weather advisory. The results could have been disastrous. Who was to blame? The aircraft commander, of course! Me!

After calming down and scribing this experience into my memory bank for future recall, I filed my flight plan for good old New York where I planned to see my family and Alice Marie. It occurred to me that I had better be more careful in the future and leave myself a little more room for error if I wanted to live a long life. It is said that "there are old pilots and bold pilots, but not many old, bold pilots."

A Sonic Boom That was a Real BOOM

Sonic booms are a real pain in the you know where to the Air Force. They scare the heck out of people, and in some cases they can cause real damage. Air crews are constantly briefed not to go supersonic over populated areas and flying at Mach One is usually restricted to designated training areas which are not well populated, frequently deserts or over the ocean. Breaking the sound barrier is never allowed over urban areas as sonic booms there frighten large numbers of the population, awaken people who are sleeping, and cause serious damage especially to windows. (Remember when a fighter pilot cleaned out the windows of the Chapel of the Air Force Academy doing a flyby.)

In 1962, as a member of the ADC ORI Team I was flying an evaluation flight with the 319th FIS, stationed at Bunker Hill AFB, Indiana. At the beginning of the "hostilities" I was flying an evaluation flight with one of the 319th pilots in an F-106B, a two-seater. I was in the front cockpit and the 319th pilot was under the hood in the rear. It was a bright clear, sunny afternoon when we were scrambled and headed north out of Bunker Hill under the control of Chicago SAGE Sector. The pilot in the rear was flying the aircraft using data link instructions from the Sector while I just monitored his performance and ensured we were clear of any other air traffic. Not too long after leveling off at flight level 350, data link instructed us to a heading of 360, descend to 25,000 feet and accelerate to Mach one. Obviously, we had been assigned a target. But there was just one problem, and it was not a small one. Right smack over the nose of our F-106 was the city of Chicago. Should we follow the instructions given us by Sector, we would be descending at supersonic speed, nose down, right over downtown Chicago. I instructed the rear seat pilot to hold up and that I would take control of the aircraft. I then broke radio silence and called Chicago Center.

I asked our Controller to confirm instructions for us to go Mach One. He confirmed our instructions immediately. I told him our position and that we would be directly over Chicago at Mach One. Again, he confirmed. He then told us to immediately comply. I then asked him to reconfirm with the Senior Controller. I could tell that our Controller was a little exasperated. A short time later he stated that the Senior Controller confirmed our Mach One instructions. I gave him a "Roger" and turned the aircraft back to my rear seater. The nose of the aircraft was pushed over, and the afterburner was lit. We were soon Mach One and the City of Chicago went quickly under our nose. The data link then directed us eastward over Lake Michigan and our target, a B-57, came into view. We had a quick lock-on and a 'Splash.' All in all, it was a very successful intercept and probably a kill just in time before the enemy aircraft reached its ground target. Shortly after, we returned to base, where my back seater made an instrument approach under the hood. But all the way back to base I had a worrisome feeling about flying supersonic over Chicago. And I was right. Immediately as we pulled into the parking area and shut down, I saw Col. Verne Stinson, the ORI Team Chief waiting for me. I knew what this was about. His face was red. He took me aside and told me that General Elder, the ADC Inspector General, who was somewhere in the Air Division, had already heard from the Pentagon about our booming Chicago and wanted to know what the hell I was thinking of by doing such a thing. It seems that our F-106B cut a swath of devastation through the middle of Chicago, smashing windows and frightening people. Some folks who were out riding their horses through the park were thrown from their mounts and a little old lady, somehow, fell out of her rocking chair. The damage costs would be very high. Stinson went on for some time before I was able to tell him my side of the story; although I don't believe he felt I had a strong defense.

I went into Ops and called General Elder. Again, he laid into me for my lack of judgment and grounded me before I could get in any words. He threatened to send me back to Colorado Springs. Finally, I told him how I verified twice, the second time with the Senior Controller, approval to go Mach One. This at last sank in and he told me he would get back to me. He never did. He did get back to Col. Stinson, who relayed that I was off the hook because the Senior Controller took responsibility for the decision. Which is what I said in the first place, and which was tape recorded.

So, it came out okay for me, but not for the Senior Controller, I guess, and not for many people in Chicago who were scared out of their wits and who suffered damage. There was quite a write up in the papers about it and the Air Force did not come out looking too good. I did not feel good about it either. The Senior Controller was trying to win his "war" on which his job was on the line and maybe he did not realize what sonic booms can do. I also did not feel too good about how my bosses were so ready to throw me under the bus. I felt that they did not show much faith in me and maybe were more concerned about their own rear ends first. If they could put the blame on a stupid Major, so much the better. This episode served as a lesson to me for later events in my career.

Air Force Heroes

During my years in service, I was fortunate to have met, and in some cases, to have served with some real war heroes. Checking quickly, I can count at least 14 that I got to know during my career who were designated Fighter Aces both during World War II or the Korean War, and in some cases, both. Of these, I would have to say the ones I most admired were the following, not necessarily in order:

Bud Mahurin: Bud had a total of 24.5 kills in WWII and Korea. He was a POW in Korea. When I met him after the war, he was a civilian working in the air defense industry, among those Northrop. I once had dinner at his home in Los Angeles.

Joe Foss: Joe had a total of 26 kills in WWII, and when on later to become the Governor of South Dakota.

Chuck Yeager: Chuck was the first man to break the sound barrier. I met him in Korea in 1968. I guess you would say he was an icon to all AF fighter pilots.

George Ceuleers: George was Commander of Hamilton AFB when I was ADC TACEVAL Team. He was a great friend to me on several occasions.

Bill Shomo: Bill was the 318th FIS Executive Officer in 1966-1968. Few of the 318th pilots knew that he had eight air victories in the big war.

Frank Gabreski: I met him when he commanded Suffolk Co. AFB. Frank was the leading WWII ace with 31 victories. A splendid man.

Eugene Paul Roberts: I met him after I retired in Spokane. He and I shared our first two names. He was a stockbroker and a wonderful man with a terrific record with nine aerial victories. He died in 2009.

Donovan Smith: General Smith was the one who asked me to come to Alaskan Air Command and take over Ops. He was a great man and I admired him for his brilliant mind and can-do attitude. But he acquired a fatal brain tumor and retired before I reported in. He had 5.5 kills in WWII.

Major Gen. John Alison: When I met him, he was vice president of Northrop and I had just checked out in the F-5 and was heading to Saudi Arabia, a major customer of Northrop's. Alison was a WWII ace in the Pacific Theater and later had great impact on USAF's programs. He was a modest man but with huge accomplishments.

Harrison Thyng: When I reported for duty at Northern NORAD Command, RCAF Station, North Bay, the Vice-Commander was Brig. Gen. Harrison Thyng. Gen. Thyng was a double ace, having shot down five enemy planes in both WWII and Korea. We first met when our family rented a bungalow on Lake Nippissing, very close to General Thyng's family. We were still seeking a permanent residence in North Bay and so we were enjoying living on the beach and enjoying the summer weather. The kids were having a ball and they soon got to know the General's daughters who at times babysat for us.

General Thyng had a reputation for being a pretty hard-nosed officer. And so, he was. The RCAF officers were much more laid back in their approach to their duties than the Americans and even I noticed this and had to adjust to it. However, Thyng was not accustomed to adjusting to a relaxed approach to problems, and when he passed this on to the RCAF staff, they did not like it at all. Whenever General Thyng sat in at a staff meeting for the Commander, Air Vice Marshal Harvey, and I happened to sit in as well, I could see that the atmosphere in the room was different. Thyng did not appreciate it when a staff member did not promptly and completely answer his questions and he let them know of his dissatisfaction. I recall following a minor air exercise I initiated against the Ottawa Sector shortly after they became operational. I did not flunk them but gave them a pass because it was clear that they were caught by surprise and needed some more time to get their act together. I felt that it would be more helpful to debrief them on their mistakes in a more positive way. Thyng called me into his office and reamed me out unmercifully for being a pussycat. (He used more indiscrete language.)

But General Thyng liked to fly with me, and he had garnered a USAF T-33 for us at North Bay on a permanent basis. (He did not like to fly the RCAF T-33s). I truly enjoyed flying with the General and felt honored that he wanted me to fly with him. But there was one flight with him that I will always remember. I don't recall from where our flight originated but our destination was McGuire AFB, New Jersey. It was a clear day, and our arrival was uneventful. Thyng was in the front seat and as we taxied in toward the parking ramp, he began to let out over the interphone a stream of nasty invectives, the cause of which I did not comprehend. Finally, I asked the General what the problem was. "Look at those G.D.M#$@! no good Nazi bastards parked on our air base." At that, I looked over to my left and saw a couple of what appeared to be German Air Force aircraft (Luftwaffe) parked near Base Ops. Thyng was furious that he should see this. It was clear that for him, WWII had not completely ended although it was now 1964 and WWII had ended in 1945, the year the Germans had surrendered. I was speechless.

Thyng went on for many more minutes on that subject, but I decided I would refrain from giving an opinion. But it made me think just the appearance of those black iron crosses on those aircraft brought back once again the latent hatred he held for an enemy he considered vile and one that had to be defeated at all costs. It was probably what made him such a potent and skilled fighter pilot in combat. Nineteen years after, those feelings were still there and surfaced.

Thyng was a tough and disciplined officer. While the RCAF did not admire him, I did. I thought that, when a war started, this is the kind of General I would want to serve under and fly with. Later, after he retired, he ran for the governorship of New Hampshire, but lost. I suspect he was too conservative and too rigid in his views for those folks. I believe the citizens of New Hampshire made a mistake.

A Historic Flight with Col. Grashio

While I had met Col. Sam Grashio when he has the commander at Paine AFB and I was at McChord in the late 1950s, I really got to know him well when he came to head up the ADC TAC/AVAL Team in 1959. I had been there a year already and Grashio soon became a well-received and popular leader of the Team. When the Team traveled to an exercise location, the ranking leaders flew in the T-33s, the remainder had to travel by C-54 which were slow, noisy and uncomfortable. That was the way I usually traveled, but on one occasion I was fortunate, or so I thought, to be scheduled to fly with Col. Grashio via T-33 to the east coast.

We hit a few bases where Col. Sam visited with the Commanders, most of whom he knew personally and so enjoyed friendly reunions with. My job was to handle the flight planning from each base to the next stop. We were at Suffolk County AFB on Long Island, where I had an opportunity to visit with my dad and my brother while the Colonel was doing his thing with the CO's. Our next base was Stewart Field, Newburgh, New York. We were scheduled to depart at about three in the afternoon. The weather was good, a sunny day, cloudless and it was a short hop to Stewart, a distance of about 70 air miles. I instructed the transient crew not to fill the external wing tanks as we would not need that much jet fuel to get to our destination. I then filed a VFR flight plan to Stewart (visual) estimating an airborne time of 35 minutes.

Grashio arrived a little late, but I had the aircraft pre-flighted and ready to go. He took the front seat, and I gave him a copy of the flight plan and we were off. I handled most of the radio calls and provided him the outbound headings and we headed over the north shore of Long Island and the Sound.

But as we climbed out, Col. Grashio became troubled that the forward visibility was hampered due to the afternoon haze. This is a common occurrence in the east as the sun begins its downward course. There is a lot of haze in the east coast skies on sunny afternoons from particles, smoke and what they used to call smog. Forward visibility is severely hampered, although if you look straight down, you can see the ground clearly. However, Grashio became unduly upset with not being able to see ahead of the aircraft. He asked me why I had not filed an IFR flight plan. I told him I did not think there was a need to as the forecast was for clear skies and our destination was wide open. Grashio had now climbed the T-33 to above 20,000 feet on hopes that the visibility would be better, which, of course, it did not. He then decided that we should obtain an IFR clearance. I told him that I thought we would be at Stewart Field before we would get one. My remark only infuriated Grashio. He then started to call New York Center.

He then asked me for our position. I gave him an approximate position off of the Yonkers Beacon. We started to circle and Grashio kept calling New York with no contact. I gave him some alternate frequencies to try, but he had no luck on those either. After several minutes, he realized that we were on internal fuel only. I had to own up that I did not have the wing tanks refueled because it was such a short flight. The Colonel was now nearing a state of panic. We were circling at 20,000 feet, somewhere northeast of New York City, in a jet aircraft that was quickly running out of fuel, and he could not see where he was or how to get where he wanted to go, nor could he raise radar control who could tell him those things. He made it clear that this was all due to the idiot he was flying with in the rear seat. By this time, I was wondering what I would have to do myself if Grashio doesn't stop panicking and shouting and let me get us down.

Finally, Grashio said, "Cunneely, do you know where we are?"

Idiot in Back Seat: "Yes, Sir."

Grashio: "Can you get us to Stewart?"

Idiot in Back Seat: "Yes, Sir."

Grashio: "OK, you better get us there, you Irish S.O.B., or I'll have your ass.

With that, I took the stick and rolled the airplane over so I could see the ground. Below was the Hudson River and I could plainly see the Tappan Zee Bridge. Banking starboard I headed north up the river and quickly spotted on my port the gray walls of West Point. Quickly ahead was the bridge at Beacon and I tuned in the Stewart Field beacon on my radio. I had already descended to 5,000 feet and the visibility at that level was better. I called Stewart Tower and asked for landing instructions. I then homed in on the outer beacon, leveled at 1,500 feet. By now I could see Newburgh and the airfield. I turned the T-33 onto the initial approach and asked Grashio if he had the runway in sight. He told me he had it, "you sonofabitch." Grashio landed without incident.

When we taxied in, a group of the TACEVAL guys were there waiting to meet us. Of course, they wanted to know why we were so late past our ETA. (It took us an hour to get from Suffolk Co. to Stewart.) They quickly noted tension between Grashio and me, and they all left with the Colonel leaving me to fill out the logbook and get myself to the BOQ. Grashio told them it would be the last time Cunneely would fly with him and that I almost got him killed. I decided my best strategy was to refrain from any comment. The fact is, over the years, I had landed at Stewart Field many times and I had flown in and out of New York many, many times. The flight to Stewart was an easy safe flight and there was no cause for concern. At no point in that flight was I not completely confident that I knew how to get to Stewart Field, safely.

So Long Geno!

If you fly a lot, you are bound to make lots of mistakes. Most are minor errors that don't end up killing you. But you know about them, and you correct them, and you learn from them. If you are wise, you don't make the same mistakes again. If you do, sooner or later you will end up in serious trouble with very serious consequences. I've made my share and, thankfully, I have lived to tell the tale. Creating good habits in the cockpit is essential and, of course, keeping your wits about you at all times is a necessity for a long life. Always being in good health and of sound mind is also important to being a formidable fighter pilot.

In my office at home is a drawing etched by my good colleague Major Ed Cleary. The title of the picture is 'So Long Geno.' This piece of art was given to me on my departure from my tour with the ADC TACEVAL Team, Ent AFB, Colorado Springs, Colorado, May 1963. Ed was a Scorpion pilot (F-89), but he also an accomplished artist who did much of the artwork for the ADC *Interceptor* magazine. The scene depicts me in a WWI SPAD flying behind a Luftwaffe fighter, presumably the Red Baron himself. The caption says: "*Now that's not Rickenbacker out there! I wish I could be a Team Chief instead of a Chase Pilot!*" It's a great picture and it truly represents one of my most embarrassing episodes in the air during my years with the ADC Team.

One of my assignments with the team was to evaluate the operational performance of the pilots in the ADC Squadrons we visited during our TACEVAL/ORI's. We did this by giving aircrews written exams, then performing individual flight evaluations in the two-seaters (F-106Bs, TF-102s or T-33s). Then during the aerial war phase, we acted as chase pilots. On a TACEVAL we were conducting on a squadron stationed at Truax AFB, Madison, WI. I was set up to chase an element of two F-102s. It was nighttime when we were scrambled. After heading north for some time, we were directed on an incoming target at 30,000 feet. It was a really dark night, and I was using my radar to keep locked on to my element. But I broke my Judy on the element and picked up the Bogey on my radar and locked on to the target in order to ensure I would know where it was. I followed my fighters to the target and went on in and got a Kill on the target myself. Breaking off the target I then tried to find my element and rejoin them. I could hear their dialogue with Ground Control and heard their instructions to Return to Base. But I had some problems in finding them. Finally, I picked them up on my radar and locked on. I gradually closed in on them in trail as we returned to Madison. However, as we got close to base, I became confused because the Flight Leaders calls to Ground Control did not correspond to our position. As we entered the initial approach for landing, I joined up on number two's wing and we landed. But as I taxied in, I realized something was askew. Indeed, it was. Somewhere in the flight, probably after coming off the target, I had left my original element and had joined up with another and had returned to base with a different flight from the one I had taken off with. To make matters worse, I was still confused until I got into the debriefing room with my original flight. The flight leader then told me that after breaking off the target he did not break off to the rear of the target but continued straight ahead. I, of course, did break to the rear and lost sight of my element. I then picked up the second element and joined it mistakenly all the way back to base.

It did not take long for my faux pas to become well circulated. I really had goofed in not noticing my error before returning to base with the wrong flight. I had to laugh at my own mistake. You can only imagine how much ribbing I suffered for that incident to the point that when I left the team, many months later, Ed Cleary memorialized it with his picture. I still have it on my wall. Whenever I look at it, I remember that night well and wonder again how quickly a mistake can happen when things are moving so fast. I also remember Ed Cleary, who is now passed away, a great friend and a talented artist. His depiction of my error over Wisconsin years ago still provides me a smile and a wonderful memory.

"Now that's not Rickenbacker out there! ...
I wish I could be a Team Chief instead of a chase pilot!"

Last Flight in the Delta Dart

My PCS orders were in, and I had already handed command of the 95th FIS over to Lt. Col. McDougal. It was March 17, 1972, St. Patrick's Day and I had just returned from Williams AFB, Arizona after completing my checkout in the F-5. That was a great experience. I found the F-5 to be a terrific aircraft to fly in every aspect. My course consisted of 10 flights that included dive bombing, strafing, skip bombing, firing on the airborne target, and low-level navigation. These were skills I had not performed since my Korean days in the F-86 Sabre. That happened to be 19 years ago. My instructors were amazed at my scores and so was I. But the F-5 was an easy plane to fly and position into the angle of attack needed to put the piper on the target and keep it there. I had a ball returning to the kind of fighter flying that I really enjoyed the most. It was also invigorating flying in the clear sunny skies of southern Arizona where there were no worries about GCA approaches, etc.

After returning to Dover, I visited with the family before my next group of TDY's to Washington, D.C. I then requested Yogi Van Daele for an F-106 flight, probably my last, as I had a long series of TDY's coming up until my leaving for Saudi. Yogi scheduled me for a flight with a couple of his new lieutenants on a routine intercept training flight. It was an overcast day, and we took off with me as Flight Lead and headed to the training area over the Atlantic east of Cape May. After about an hour and a half of high-altitude intercepts, we headed back to base under the control of 20th Air Division. However, weather at Dover had deteriorated. A snowstorm had hit the base and ceiling and visibility were now at minimums for approach and landing. At this point, word came from Col. Van Daele in the 95th FIS Combat Alert Center (CAC), that my two wingmen would proceed to and recover at our dispersal base Atlantic City. I passed the word to my two guys, and they peeled off and headed to the DOB where the weather was much better. I then proceeded back to Dover and made my approach. It had been well over a month and a half since I had flown the Six, still I never gave it a second thought. Visibility at the base was now down to a half mile (minimums) with heavy snow. I made a GCA (radar-controlled approach) and broke out and spotted the runway at just about a quarter of a mile and landed without incident. I knew as I taxied in that this was my last flight in the Delta Dart, an aircraft I had truly grown to love to fly. It was a most emotional moment for me.

My good friend Yogi (Van Daele) later confided to me how his junior pilots, whom he dispatched to Atlantic City, complained why they were sent to the DOB, but I was permitted to return to home base. He let them know, in very clear terms, that when they had that wreath around their wings, and a thousand hours in the Six, he might let them take a shot at landing at home base in minimums. That shut them up.

List of Military Training

(Date indicates completion date)

1950 July 28	Certificate of Training, Continental Air Command, completion of training in Administration and Logistics, Air Force Reserve Officers Training Corps Summer Camp, Stewart AFB, Newburgh, NY
1951 June 12	Advanced Course, Senior Division, Air Force Reserve Officers Training Corps, Fordham University, NY
1952 September 25	Pilot training, Class 52-G, Pinecastle AFB, Orlando FL. Training included 130 hours in the T-6 at Bainbridge Air Base, Georgia; 87 hours in the T-28 at Bryan AFB, TX; and 152 hours in the F-80/ T-33 at Pinecastle AFB. FL Ground school included 85 hours covering Aircraft Engineering, Principles of Flight, Radio Communications, Navigation, Flight Instruments, Electronics and Radar, Weather, Learning Techniques, Flying Safety, and Code, Aural and Visual.
1953 March 7	Fighter Training (F-86) Nellis AFB, NV

1953 April 19	ATRC Course ATRC Course Aircrew Familiarization (F-86-9), K-13 Air Base Suwon, Korea
1953 April 27	Fifth Air Force Survival School, APO 970, Korea
1954 September 8	Advanced Flying School (Interceptor) F86-D, Perrin AFB, TX
1955 September	Interceptor Weapons Instruction, (F-86D) Tyndall AFB, FL
1963 August 29	ADC SAGE Orientation Course
1964 July 31	The Defense Atomic Support Agency Orientation Course, Sandia Base, NM
1966 August 18	ADC Life Support Training School, Tyndall AFB, FL
1967 October 19	USAF Survival Training Course, Fairchild AFB, WA
1967 August 10	Aerial Combat Tactics Qualification (F-106), McChord AFB, WA
1967 September 18	Aerial Refueling Qualification (F106), McChord AFB, WA

1969 June 12	ADC Life Support Training, Perrin AFB, TX
1971 February 9	Chief of Maintenance/Maintenance Officer Course, Tyndall AFB, FL
1972 March 24	The Foreign Affairs Executive Seminar, Foreign Service Institute, Washington, D.C.
1972 April 7	USAF Special Operations School Military Assistance Advisory Course, Washington, D.C.
1972 April 25	ATC Training for MAAG/Mission Personnel, Randolph AFB, TX
1974 April 26	The National Cryptologic School, National Security Agency, Elmendorf AFB, AK

PART II:
OTHER THOUGHTS AND MEMORIES

It's a Great Life If You Don't Weaken

-Joe Cunneely, Sr.

As you get older, you get wiser—at least you are supposed to anyway. Not everyone follows that rule I know, because I certainly have met many old fools in my time. But I like to think that I was not one of them. I believe I inherited a lot of what we used to call "common sense" from my father, and also from my grandmother. But my dad was a beacon of common sense and I learned from an early age to trust his judgment in matters that called for decisions on things that were important. During my life I came into contact with people whose IQ's were well above the 100 mark, guys and gals you had to describe as brilliant, folks whose educational achievements and intellectual acumen just overwhelmed me and made me feel inferior. And yet, I saw so many of these same folks make such a mess of their lives, do such stupid things and hurt so many of their loved ones in the process. While they possessed intellectual brilliance, they were lacking in the one ingredient necessary to make the right choices in life to achieve success and contentment. Not that having common sense will guarantee a happy life in every case. It did not help my dad when my mother walked out of our family and left him with two small sons to raise on his own. Still, he managed as best he could, thanks to his mother, my grandmother. And he did a splendid job as a father, devoted son, a grandfather, and so forth. Those who knew him loved him for his great sense of humor, his energy and his faith. In addition to raising his sons, he took care of his mother, his sisters, and his brothers to the end of their lives. He was a wonderful grandfather to my five children who remember him fondly. He served his country in World War I and worked for the railroad throughout the difficult depression years, always able to support his family. He helped others as well who were in need during those terrible times. He loved his country and was dedicated worker and supporter of the Democratic Party, a political party that at that time gave much support to the struggling workers and immigrants of the nation. He sweated out my brother Joe's years away at war in the Navy during World War II and again during my time in Korea. He prayed for our safe return. He helped Joe Jr. financially purchase his home in Patchogue on the inlet, a place he could never have bought without Dad's help. He was generous to a fault. He was a family man, through and through. He loved his grandkids and loved being with them on Christmas and at the lake. He never missed Sunday Mass and prayed his Rosary often. When he died, he was buried in the Cunneely Family plot at Calvary Cemetery next to his mother, father, and sisters. According to his wishes, his Mass was celebrated at St. Raphael's Church where he was an altar boy in his youth. His father James was a Section Foreman at Calvary Cemetery. My father left Bryant High School when he was 15 years old to work for the railroad. He never even had a high school diploma, yet he was a whiz in mathematics. I have no idea what his IQ score was, but I do know that when it came to common sense, he was a genius. He was also the greatest man I ever knew, and I loved him dearly.

Love, Marriage and Children

The Lord God then built up into a woman
the rib that he had taken from the man.

When He had brought her to the man, the man said:
"This one at least, is bone of my bones, and flesh of my flesh; This one shall be called 'woman,' for out of her man this one has been taken."

That is why a man leaves his father and mother and
clings to his wife, and they will be two in one flesh.

-Genesis

At the time I am writing this piece, I am at the age of 83 and my first child born has attained the age of 56. I cannot persuade you how astounded I am at these numbers and the mystery of it all. When his mother and I married I must admit, the thoughts of having children were not foremost in my mind. In fact, my bride had warned me that due to the loss of one of her ovaries due to an emergency operation she had undergone years before that the doctor had believed would be an appendectomy, she might not be able to bear children. But little things like that hardly matter when love takes over and you are in a rush to the altar. However, seven months later, Alice Marie announced to me, out of the blue, she was with child. My transition from bachelorhood to fatherhood was breathlessly rapid. Our honeymoon days were coming to an end I couldn't believe it. But when I saw my first son, Eugene Jr., on his 1956 Memorial Day arrival, I had to believe it. I was a father.

Between the dark and the daylight
When the night is beginning to lower
Comes a pause in the day's occupations,
That is known as the children's hour.

-Longfellow

Since that day, I have had a great amount of time to think about what it means to be a father. Two more sons and two daughters arrived from the single ovaried Alice, and each arrival was greeted with awe and wonder at God's marvelous caprice. Unlike in today's society, our children were not planned, nor spaced, nor denied. They were the outcome of our love for each other in coordination with God's creative handiwork. Each was welcomed and loved equally from the first heartbeat. Though Alice Marie did the hard part of carrying the babe in her womb for many months, suffered the pangs of childbirth and the early days of feeding and diapers, etc., I tried to join in as much as possible. I truly enjoyed the nighttime feedings, the baths, diaper changes and all the other duties I was able to contribute to in those early days. I wondered long hours at the great mystery of conception and birth, and I thanked God over and over that our children came into the world healthy and well put together. I surprised myself how much I enjoyed being a father. And more than anything else, I hoped to be as good a father to my children as my father was to me and my brother. It also became very obvious that I married a woman who was destined to be a great mother. That became clear when I began to notice that my children moved into first place ahead of me on the priority list. I was amazed how she fit into that role so beautifully. It was something I could never complain about, ever. As I moved onward through my professional career, the reality hit me that the most significant thing I would ever accomplish in my life would center on how I would meet my responsibilities as a husband and a father. Nothing else I did would be more important, and on those areas, I would be judged most critically. Alice felt the same.

Looking back on my parenting years, I realize that I could have done a better job. One major problem was how much my work kept me away from home. And children do need the presence of their father, especially in their teen years. I found this to be particularly true with my daughters. When this came to my attention, I tried hard to be more attentive to them and to make certain that I gave them as much affection and closeness as I could muster. For someone who had spent most of his time in a masculine world, I had to be very careful and sensitive to them and show them in every way that they were equal to my sons. And, of course, that was indeed the case. Each of our children was a unique gift from God, to be received and blessed and nurtured in his/her own distinctive way. Our job as parents was to raise them to become responsible Christian adults prepared to accept the burdens of life that God gives them and to do their very best to earn His graces.

Alice and I believe strongly that the family is the basic unit of our civilization. A strong family begins with a father and a mother, who together, set the example for their children and looks after their full needs. Their needs cover not just food, housing, education, etc., but include moral training i.e., religious faith. As both Alice and I came from broken families, we worked to keep ours close and firm as much as possible. There is no question that raising children gave us the most challenging situations in our life. Each child is a different person with individual viewpoints and problems. You must give each one your special attention. And at times there can be frustration. But when there is love, there is perseverance, patience, and forgiveness. Alice and I are so proud and happy with our five children and what each has accomplished in their lives. Each has given back to us more than we deserve in love and respect. And now we have nine grandchildren to love, and watch grow. We are so blessed.

Heartbreak 2015

To be Irish is to know, that in the end,
the world will break your heart.

-Daniel Patrick Moynihan

One must live many years to appreciate the wisdom found in the above axiom.

At the time I am writing this, I have reached the ripe old age of 86. To be truthful, I never dreamed that I would be alive and kicking in my 10th decade. Not that I'm complaining. No one looks forward to dying early in life. And most of us usually are pointing to the future, not concerned too much with death, especially if you are living a happy life. And for the most part, that would describe me. Even in my late 80's, I am in good health and live an active life and participate in many activities that make my life enjoyable for the most part. Over the years, the Lord has bestowed many blessings on me, none of which I have earned. By far the greatest of those blessings is my wife Alice Marie, a wondrous woman who brought into my life so much love and joy and companionship and support for over 60 years. It is beyond my comprehension to understand what my life would have been without her. I am so thankful and fortunate for her coming into my life and I am certain that it was the result of divine intervention.

So, to repeat what I have already stated, this man (me) has absolutely no cause to complain. Nevertheless, the words of Daniel Patrick Moynihan do ring true. No one escapes this planet without heartaches, and we must be prepared to expect them to come our way and to accept them. "The Lord gives and the Lord takes," is applicable to everyone; still it hurts when it comes our way.

And so, they did. In 2015 our family received a double blow. The previous deaths in the Cunneely clan came in August 1987 when my brother Joseph died after a long difficult struggle from cancer. This occurred a little over a year after our father died in May 1986. However, in December 2014, our daughter Kathleen was diagnosed with cancer in the bladder. Katie, since the age of eight, was plagued with kidney problems. These were uncovered when we were stationed in Alaska, and she was attended by an excellent urologist who eventually operated on Katie at the Elmendorf AB hospital. But, for the rest of her life, Katie suffered from recurring painful kidney infections. These infections were painful beyond the norm, requiring frequent visits to emergency rooms where she suffered long waits before relief and often faulty diagnosis. Doctors usually prescribed antibiotics but declined to give her needed pain medicines. Katie went through several operations on her kidney; one was to extract the kidney completely, but the surgeon only took out a third of the kidney, believing he had done what was needed to cure her problem. It did not. The leading urologist in Spokane advised taking out the kidney. After months of tests, the surgeon chosen to operate, declined to do the surgery stating he felt that Katie's kidney should not be taken out because it was unnecessary. I was infuriated. Her life continued to be hellish due to continuing kidney infections of the worst degree. Her dependence on painkilling medicines increased. Failure on the part of her surgeon to take out the second urethra she was born with and was causing reflux of her urine, eventually resulted in cancer of the bladder. The result of continuing only to treat her symptoms and not the fundamental cause of her kidney problem was disastrous for her.

In the following six months, her cancer moved so rapidly that they were unable to operate to extract the bladder despite chemotherapy. She was then given additional chemo plus radiation. We drove her to Seattle to visit with the experts at the Freddie Hutchinson Cancer Center. They were very helpful and we then tried to enter her into a clinical trial. Bill Curry, our longtime family friend was also involved in procuring Dr. Steve Anthony in Katie's case. Katie fought hard and received tremendous support from her friends, from the Gonzaga Prep days, and also from the folks at Dr. Chaundry's office who helped her get ready for the clinical trial. She was given terrific love and support from her sons, especially Casey who gave up his job and devoted himself day and night to take care of his mother's needs. Her sister Mary came over from Seattle to be with her as much as she could and was a great source of love and dedication. She tried to arrange a cruise with Katie and their cousin Colleen; however, they failed in that effort as both (Katie and Colleen) were too ill to make it. Another wonderful pair were Patty and Dr. Jim Doyle who gave unselfishly of themselves to help Katie and our family right to the end.

On the afternoon of July 8, Alice was alone in the house when she received a call from Seattle. On the phone was a person identified as King County police officer. He gently told Alice that our daughter Mary was found dead in her apartment. When I arrived home shortly after, I found my wife totally distraught. We had not been able to reach Mary by phone for a few days which was unusual and we had even called Jimmy to check in with her. He had no luck either. Mary had passed away at least two days previous. Cause was unknown. You can imagine what a shock this was to us and to everyone as Mary had no health problems we were aware of and had completed a visit with us and Katie only a few weeks previous to her passing. Having to tell Katie this news was the most difficult act I ever performed in my life. Alice and I had been concentrating and completely focused on doing everything in our power to saving our daughter Kate, and now this happens. To put it mildly, our world started to collapse. Our family started to crumble. We were now totally involved in bringing home our darling Mary and laying her to rest. This while her sister is fighting for her life.

On July 18, 2015, we laid our dearest Mary to rest in St. Joseph's Cemetery. I was still in state of denial. Mary was such a great blessing to our family. The number of people who sent in their praise of Mary as a human being who gave to others and who was so beloved by her patients and by her co-workers was staggering. Her colleagues at Swedish Providence Hospital held a memorial ceremony in her honor that brought tears to my eyes.

Katie lost her battle on December 29, 2015. Her last thoughts were about her three sons and how they would be after she died. And her sons were with her right to the end at the Hospice of Spokane where she passed away peacefully and finally without pain. She was laid to rest at St. Joseph's next to her sister Mary. Our wish is that they are again joined where there is only refreshment, light and peace in the sight of our Lord. It is that vision that provides us the only consolation that we seek. Our daughters were some of God's great blessings to us and we will never forget them. We will be always thankful for their lives and for what they contributed to us and our lives. No day passes that they do not enter our minds and prayers. To lose a child is one of life's greatest sorrows: to lose two of your children, so rapidly in time, breaks your heart in ways that cannot be clearly described to someone else. It is an experience that you can never get over.

My father had a saying that he often repeated ad nauseum: *"It's a great life, if you don't weaken."*

I often wondered what that meant, but I now know exactly what he meant. As I now know what Sen. Moynihan meant also. Life can be happy and certainly worthwhile, with many joyful and happy experiences that are so delightful to the senses and provide beautiful memories. But we also bear in mind as Jesus told us to follow Him, we see that His life contained much that was sad and painful and contained loss and sorrow as well. So, we must accept those moments in life as He did and keep the faith. It also brings to mind our Blessed Mother Mary, who also lost her child Jesus and the great sorrow that she experienced. She is our role model. And we will do our utmost to follow her example. Thy will be done!

December 19, 2016:
A love story ends for now! I lose My Sweetheart

September 17, 2016, marked the 61st anniversary of our wedding. Alice and I celebrated with a wondrous trip to Glacier National Park, Montana. It was a great journey that took us though this magnificent part of our country. The weather was sunny throughout our five days and the park was at its most glorious. We added this sojourn to so many others that we were able to take during our life together. We decided that this trip might have been near the best one of them all. Little did we realize that our happy days together were coming to an end.

We often spoke about which of us would die first. Each stated that we wanted to go first because the thought of living on without the other was just too imposing and difficult to accept. But we knew that that decision was not ours to make—it was God's. Thy will be done.

Shortly after the Thanksgiving holiday, when Alice again came through with a successful fantastic turkey dinner, feeding seventeen mouths, she began to complain of feeling weak. When this loss of energy failed to go away, she visited Dr. Byrd, her physician of many years. Early tests indicted possible anemia and he prescribed vitamins and an IV. A week later with Alice still feeling weak, Dr. Byrd ordered more blood tests. These showed a large increase in her calcium count. This required another IV which lowered the calcium. But after another week, her calcium count again rose to a high level. Dr. Byrd ordered another IV and a Cat Scan at Holy Family Hospital. This was on December 9. He ordered that the radiologist call him at home with the Cat Scan results. While Alice was receiving her IV, Dr. Byrd called and gave me the results of the Cat Scan. He told me that my wife has a cancerous tumor in the stomach area that was too far advanced to be treated. When I asked him how long she had to live, he told me it was a matter of weeks, maybe two or three, I was stunned and I broke out in tears. I decided not to tell Alice this until I was able to compose myself. Dr. Byrd then told me to have Alice admitted to the hospital for the night, and he would visit us at seven in the morning. After some problems, I was able to get Alice admitted and I spent the night with her at Holy Family. When she asked about the results of the Cat Scan, I was vague in my answer, but I did hint that it was not what we hoped for. But I could tell that my wife knew that she was dying.

At seven in the morning, Dr. Byrd arrived and I spoke to him privately, telling him that I had not told Alice everything, but that I believed she was aware that she was dying. Dr. Byrd went into the room and sat on the bed next to Alice. I also sat on the bed. Dr. Byrd then took Alice by the hand and told her that she had a cancerous tumor that neither he nor anyone could cure. He told her that she was going to meet the Lord very soon. I sat and watched and listened to this in a state of disarray. I was totally teared up. But Alice took this news with complete composure. She displayed courage and faith that astounded me. Then Dr. Byrd asked that we take each other by the hand and recite the Lord's prayer. This we did. When Dr. Byrd left, he told me he would stay in touch and he would attend Alice's funeral Mass. I stayed with Alice until it was time to take her home from the hospital. Her composure, in light of what she had just been through was incredible. But I could not withhold sobbing as I had fallen into a state that approached despair.

Alice died 10 days later. I took her home where with the aid of Hospice of Spokane, I set up a bed in the living room and tried to keep her as comfortable and pain free as I could. Thanks to Casey McClary and our son Jim, who drove over from Bellevue, we did the very best we could for Alice. But as the days went on, she became weaker and needed more help to walk. It became more difficult to keep her pain free and we had to move her from one position to another. I then had to use some of the pain killer medicines provided by Hospice. Nursing my dying wife was hard and cruel, but I felt that it was my duty as her husband. I was able to tell her how much I loved her and I know that she understood my feelings. These were the most difficult days of my life.

Finally, Alice asked me to move her to Hospice. I believe she realized that that was the right place for her and that was where she was going to die as did our daughter Kathleen. On Sunday, December 18, I had Alice Marie moved to the Hospice house. As they took her out, I knew that I might not see her again alive. I visited that day with her until 10 o'clock that night. She was asleep and pain free. I knew that her life was ending. I prayed my Rosary. Monday morning at six o'clock, I called Hospice, she was still sleeping. I got into my car and drove to Hospice. When I arrived, Bill Curry met me at the door. Alice had just passed away. I went into the room and kissed my wife for the last time. Again, I prayed the Rosary for her. When the man from Hennessey's arrived, I helped place Alice into the wrapping and into the hearse. I would not see her again. My sweetheart had passed on to the next world.

Adding to the sadness surrounding Alice's death, Dr. LeRoy Byrd who provided medical care of the first class to both Alice and me for over three decades and played such a personal role in Alice's final days, passed away himself on December 22, just three days after Alice. Imagine the shock I received when I saw his death notice and then, when I arrived at Hennessey's Funeral Home to see Dr. Byrd laid out in a room closely adjacent to Alice. I entered to pay my respects and spoke to Irene Byrd, his wife and his sons and daughters. I narrated to them Dr. Byrd's gentle care for my dying wife on the bed at seven in the morning at Holy Family Hospital. They were quite interested in this anecdote because it was so typical of LeRoy Byrd and what a unique human being he was. I mentioned that he had told me he would attend Alice's funeral Mass, but it was I who attended his Mass a few days after we had laid my wife to rest.

I first met Alice Curry at Rockaway Beach, Long Island, the summer of 1947. I had just graduated from Chaminade High School and would begin my freshman year at Fordham University in the Fall. I was never what you would call a "ladies man." I can't recall ever having a girlfriend during my high school years. I didn't go to the Senior prom because I didn't know any girl to ask. Besides, I couldn't afford it. But during my years at Fordham, I did date Alice, off and on. Problems were, one, she lived in the Bronx and I lived out on Long Island. Two, I rarely had enough money to take a girl out on a date. But I did work for the US Post Office at Christmas season and during the summer vacation. Most of this money went to pay my tuition at Fordham. But, I did have some that I saved for social reasons. Also, occasionally, Alice would buy some tickets to a Broadway show and invite me.

Many times, I would take Alice to the Yankee games and then have a beer later at the Concourse Plaza Hotel. Alice lived close to the Yankee Stadium, so these were not too costly dates. Alice attended my graduation from Fordham and on that day, I was commissioned as a Second Lieutenant in the USAF Reserve. I also had my orders for active duty, which meant we would be separated for an unknown period of time. As it turned out, this separation would be about four years. But in August 1955, I called her from Tyndall AFB, Florida, and proposed marriage. She eventually agreed and we entered into matrimony on September 17, 1955: a marriage which lasted over 61 years. Over my lifetime, God has bestowed many, many blessings on me, none of which I earned. But the greatest blessing I was given was my spouse, Alice Marie. When we were married, I was so happy and I knew that I had made a great choice. Little did I realize what a wondrous choice that was. It did not take a long period of time for me to realize that my wife was such a beautiful person, a human being so much better than I in so many ways.

And over the years, my love for her continued to grow. She was not only such a terrific wife, but she was also a superb mother and then later a generous loving grandma. As a military wife, a position calling for different challenging situations and attending to hard sacrifices, Alice took them in stride. She consistently amazed me and others as well.

And now, I must go on without her. I don't know how I will be able to do so. Life without Alice Marie does not seem possible. It will not be easy. I will ask the Lord to help me. He has always done so in the past. The Holy Spirit must join in to help. **Alice Marie, I love you, always!**

A Mass of Christian Burial was celebrated at Mary Queen Parish for Alice Marie Cunneely (December 28, 2016)

The following is the Eulogy delivered by Eugene P. Cunneely, Jr., her oldest son:

Good morning.

Thank you for being here today. My name is Gene Jr. and I am Alice's oldest son. Even if you did not know Alice a long time, she was still a good friend and you probably knew her well. She was easy to talk to and would become a friend almost immediately. She never gossiped, nor spoke ill of anyone. You would all agree that Alice was extremely pleasant, very generous, open, refreshingly genuine, had a great sense of humor, and was always a caring lady.

But Mom had other wonderful qualities as well. And, today I'd like to talk of two of these qualities, as I believe that they set her apart from all of us.

The first is her dedication to family. Alice was always dedicated to her family, whether it be her parents, children, daughters-in-law, or the grandkids.

Her dedication to family began in her early days in the Bronx of New York City. Both her parents were born in Ireland and processed through Ellis Island in the late 1880's.

Alice was the oldest of five kids. Her family was poor, but they didn't know it. Her Dad, Thomas Patrick Curry, tried his best, but was away a lot. Sadly, he was saddled by the bottle.

Katherine Keating, mom's mother, was a saint in her own rite.

Alice was born in 1927, two years before the Great Depression hit hard. Mom shared a bathroom with parents, three brothers, and her younger sister. Mom once told me a story that as a young girl, she had to share a bed with her Grandmother. And during the night, Grandma passed away. After that, she then shared the same bed with her sister McGee. Despite struggling to make the rent every month, Alice was able to attend St. Luke's Catholic Elementary School, many thanks to the Sisters of Charity. Mom was fortunate enough to attend an all-girls high school in the heart of Manhattan. Cathedral High School is still there today. Mom took the subway to downtown Manhattan every day. Kids didn't own SUV's in those days.

With an absent father, Alice was always willing to help out in any way possible. This included cooking, cleaning, watching the younger ones and anything else needed around the apartment. Many times, it was in a financial way. After high school in 1945, Alice went into the work force. One of her first jobs was at Woolworths 5 and 10 Cent store. This was followed by work at Harpers Publishing House. There, Mom was able to provide more money for the family.

Alice met Gene at Rockaway Beach in Queens. They dated for several years and in 1955 they were married at Tyndall AFB in Florida. As a new Air Force wife, Alice was again a rock of dedication, both for her husband as well for her five kids. She patriotically supported Dad in all his assignments, from tours in Korea to the volatile Middle East. Alice never complained once. Much volunteer work was expected and accomplished by Mom. Consoling young wives and helping start an orphanage in Korea were part of the job for Mom. Alice did more than her fair share of packing and unpacking, moving to various state-side air bases, such as McChord in Washington State to Delaware, and from Colorado to Canada. Not to mention Alaska, perhaps her favorite stop. I know that Dad was always relieved and had peace of mind with Mom back home "holding down the fort." I bet he even forgot the time in Colorado, when Mom had to dispatch a rattlesnake, who got too close to her kids.

While Dad was stationed in Saudi Arabia, Mom and five kids moved to Spokane. This is where they decided to put down roots after a 26- year military career. Alice loved beautiful Spokane. She made great friends here and loved her summers at the "granny shack" on Lake Pend Oreille. She was always a regular at the Kennel cheering for another Zag win.

Lastly, and most importantly, Mom's dedication to family was only surpassed by her great Catholic faith. I know no one who had a stronger love for Jesus than Mom. As Jesus instructed us to serve, rather than to be served, Alice's mission in life had the same purpose. It didn't matter which family, her New York family, her Air Force family, or her husband and five kids. She loved going to Mass at Mary Queen. And I know she enjoyed Father Vevik's messages in all of his homilies.

Alice did not graduate from any college, but she had New York street smarts, along with excellent judgment and common sense. In everything Mom accomplished in life, she did so with great dedication to family and through her faith. Mom had a wonderful life. She married a great man, traveled much of the world, including Rome and Ireland, and had a few good kids. And perfect grandchildren.

It's hard to believe she's not physically with us now. I thought she would be around forever. It has been said, "That with great faith, comes great bravery." Mom met the end with great courage and confidence in the afterlife. Most of all, Alice was an exemplary role model for all of us. We can all be happy and rejoice in knowing that Mom has now joined her parents, her three younger brothers, Jack, Thomas, and Martin and her close friends Nancy, Kate, DeeDee and Margarite and others, but especially her daughters, Mary Elizabeth, and Kathleen Anne. Heaven is filling up with a lot of great people. We will never forget her devotion to her families and to her faith. We will miss Alice greatly.

She was one of the best! We love you Mom.

A Surprise: I Receive Information About My Mother

Since I was about the age of 10 or 11, I did not see or hear from my mother. She had left our home when I was four and she had stopped her visits. In 1943, my mother wrote a letter to my brother Joseph that gave him permission to enlist in the Navy as he was under the age of 18. The letter came from Staten Island, a borough of New York City. To my knowledge, that was the last communication we received from her.

In November 2018, I received a phone call from a woman who identified herself as my cousin. Her name was Arlene Bernhardt Currie. Bernhardt was my mother's maiden name. She called from New Jersey and asked me if I was the son of Julia Bernhardt. I told her that I was. She asked me if I knew that I had a half-brother. I did not. She informed me that my mother had divorced my father and remarried in 1938 to a man named George Raab, Sr. They had a son named George Jr. born January 22, 1942.

The main reason Arlene called was that George Jr. had died and Arlene was the executor of his will. The attorney needed to obtain my signature on a document pertaining to the will as I was a relative of the deceased. I agreed to sign the document.

Arlene was very gracious and talkative and I took the opportunity to ask some questions about my mother and my half-brother. Arlene was the daughter of my mother's brother Louis, whom I barely recalled meeting though I was quite young at the time and he was single. My mother's husband died when George Jr. was only 10, and my mother went back to work. They were living in the Bronx. George Jr. did spend some time in the Army as an MP in Alabama. When he came out of service, he worked some time on Wall Street and then moved on to work at Macys. He passed away September 5, 2018. My mother June had died of a heart attack, November 11, 1987.

Long after my mother died, George Jr. told Arlene that he had two brothers. Arlene was very surprised. Arlene wrote that Aunt June was a very private person and never said anything in front of her. She asked George Jr. would he liked to have known you both. (Meaning myself and my brother Joseph). He said he would, but it was too late. She wrote that George would never have demanded to know anything from his mother about us as he thought it would upset her.

Arlene talked about our parent's split. Arlene could not imagine whatever possessed her aunt to give you boys up. I mentioned that there was large age difference between my father and mother. My mother liked music and dancing, and raising two young rascals, housework and cooking was not what she wanted in life. My dad was a hard worker and a good provider. He would not give up his sons and he took us back to his mother to raise us. And she did, although that poor woman was already taking care of two daughters and an alcoholic son, and she being in her seventies. My dad was very bitter about June leaving him like she did. But he carried on, never remarried and was a great grandfather to my five children. My brother went into the Navy and came through luckily surviving the largest naval battle in history, the Battle of Leyte Gulf. After the war, he married, became a New York policeman and then went to work on Wall Street where he was very successful: part of the Greatest Generation. Why my mother did not keep in contact with us, I will never understand. I am sure she you would have enjoyed her five grandchildren.

Arlene's last words were: "I would believe that she (June) had many, many regrets, but maybe little choices. Maybe she knew that your Dad would be the better choice. I am glad that you feel he was a great father, grandfather, and a good man."

I told Arlene over the phone, I liked her very much, and I thanked Arlene for filling in so much information about my mother that I would have never known. I wanted Arlene to know that I have never held a grudge against my mother for what she did. I can never be her judge. And I do pray for her every day, as she was the one who gave me my life. And for that I will always be grateful.

Seek Igloo

In Chapter 37, I write about a project that my Office undertook. At that time, I held the position of Deputy Chief of Staff for Plans, Programs and Manpower for Alaska Air Command. The project was given the title of Seek Igloo and its goal was to completely change the Command's Air Defense system. In our planning, this total revision would completely modernize the way AAC would identify aircraft penetrating the Alaskan ADIZ and intercept potential enemy aircraft. Additionally, Seek Igloo would be cost effective, saving the USAF over $40 million annually. Many of the

AAC Commanders and staff did not support this project as they felt it would be unsuccessful and might cause the end of Alaska Air Command as a full Command. While I was working with high-ranking Pentagon planners on this project, I knew that should Seek Igloo be approved, it would result in the closing of AAC. I kept this to myself.

On my retirement in 1977, the AAC Command Historian asked to interview me regarding Seek Igloo and what the future of AAC might be. I granted him the interview, which at the time was classified. It was declassified after three years. I asked for and received a copy of the interview from Eleventh Air Force in April 2010. Reading it again brought back many memories of a time when there was much going on in USAF and I was in the middle of a battle of the Generals, some who were trying to keep things from changing. And that was not possible.

Alaskan Air Command was disestablished August 9, 1990.

DEPARTMENT OF THE AIR FORCE
PACIFIC AIR FORCES

6 April 2010

Eleventh Air Force
Office of Historian
10427 Kuter Avenue
Elmendorf AFB AK

Col Eugene P. Cunneely (Ret)
N. 4618 Ely Road
Spokane WA

Dear Colonel Cunneely;

I found several documents related to the SEEK IGLOO Program, so I narrowed them down to the best ones to send you. I've included two documents that I think you will find useful.

The first is a copy of your interview with Grant Hales. The document automatically declassified in 2007, but I cleared it through the information security folks and the Eleventh Air Force Public Affairs Office, and it is cleared for public release. The second document is an overview briefing provided by ITT. The archive contains final briefs from all three contractors who bid on the project, but since all three looked very similar, I chose the simplest, easiest to read.

I found a copy of the SEEK IGLOO Final Study Report that includes a Staff Summary Sheet signed by you, from your trip to brief General Creech. The document is marked For Official Use Only and it is actually easier to release a declassified document than an old FOUO document. If I succeed in getting it released I will send it along as well

If you need further information, feel free to contact this office. My commercial phone number is 907-552-4763, or you may also contact me via e-mail at joe.orr@elmendorf.af.mil. Thanks for your service, and thanks for letting me help you.

JOE B. ORR, Civ, USAF
Supervisory Historian

[redacted]

27 July 1977

Colonel Eugene P. Cunneely
92nd Combat Support Group/CC
Fairchild AFB, Washington 99011

Dear Colonel Cunneely

The enclosed draft of our 17 June 1977 interview is provided for your review. Please feel free to make any comments you think are necessary on the draft.

Upon completion of your review please complete the enclosed release form, sign and date it.

Once again, thank you very much for your time and interest. My only regret is that there are not more senior managers within the Air Force of your caliber who make my position the interesting pleasure it can be rather than just a job.

Sincerely

SIGNED

GRANT M. HALES
Command Historian
Office of History

2 Atch
1. Interview Draft (1 cy) (S)
2. Release Form (U)

If ATTACHMENT #1 (is) ~~(are)~~ withdrawn or not attached, the classification of this correspondence will be downgraded to UNCLASSIFIED

[redacted]

[redacted]

DRAFT COPY OF INTERVIEW BETWEEN MR GRANT HALES,

AAC/HO AND COLONEL EUGENE P. CUNNEELY, AAC DCS/XP

ON 17 JUNE 1977.

Automatically Declassified
After 30 years IAW E.O. 11652
1 JAN 2007.
[illegible]
3WG/HO

Classified by AAC DCS/XP
Exempt from GDS, E.O. 11652
Exemption Category 3
Declassify on 2007

[redacted]

UNCLASSIFIED

17 June 1977

MR HALES: (U) You have been assigned as AAC DCS PLANS [redacted] almost four years now. You arrived in July of '73.

COL CUNNEELY: (U) That's right.

MR HALES: (U) During this 4-year period, there has been a major transition in the, at least, planning for the air defense radar structure in Alaska, and you have been very deeply involved in this throughout this period--as DCS Plans. What I would like to know, to begin with is, what the status of plans for modernization/improvement of the radar systems in Alaska were upon your arrival or were there any at all?

COL CUNNEELY: (U) As far as I could tell, there weren't any at all in 1973, and that was a surprise to me. Apparently, the previous planners and commanders had been thinking of going a different direction than just modernization of the current system. There were some proposals--one of which was given some circulation--the Arctic Crescent Program (1) I think you are probably familiar with that.

MR HALES: (U) Yes, I am.

COL CUNNEELY: (U) But, there really wasn't anything which was going to take the present system and change it to update it; that came as quite a surprise to me.

MR HALES: (U) Basically, then all that happened is the winding down of the original system built in the 1950s.

UNCLASSIFIED

COL CUNNEELY: (U) Closing down, and it was escalating in cost, and it was beginning to get more of a look from the Air Staff and DOD as far as the retention of it.

MR HALES: (U) This brings us to a somewhat coincidental occurrence in the same month, July 1973, when the then Chief of Staff of the Air Force, General [John D.] Ryan, visited Alaska and requested a review of the air defense structure.

COL CUNNEELY: (U) That was a key incident. General Ryan did visit Alaska and made a few stopovers and went back to the Pentagon and allegedly made the remarks (I thing this is pretty accurate) "There should be a re-look up there--'either we are going to fix up those installations or we ought to think about closing them.'" (That has been quoted quite frequently, I think it is contained in the document somewhere). He came through, I think, in either late June or July. Is that not correct.

MR HALES: (U) It was late June or early July from what I remember.

COL CUNNEELY: (U) Yes, As I remember, I got here the end of July. Some of those places don't look good early in the summer. We have the breakup, and they still look kind of crummy. I remember Galena, especially; it was very bad. One of the first tasks I got into as DCS Plans was to send a team out from DCS Plans, headed by the Director of Plans, to oversee the cleanup of Galena--which was a result of General Ryan's visit; a lot of things looked bad. General [Charles W.] Carson [then Commander of AAC] was also shocked. But, that was a key thing. General Ryan really spotlighted that we ought to do something about the radar sites in Alaska or seriously think about maybe shutting them down.

4

MR HALES: (U) Now this was at the Air Staff level, and other than to clean up and the sad condition of some of the installations, were you given any mandates from General Carson, Commander of AAC, at the time. Or, was there anything he envisioned for the future?

COL CUNNEELY: (u) ([redacted]) No, well let me take that back. One of the things General Carson did do was he tasked me to look at the possible closure of some radar sites in Alaska. He also asked me to do a study on whether or not we should retain Galena as a FOB (forward operating base) or if we could do without it. We made those two studies for him. Apparently, he asked for those studies just for himself and to be used in-house--possibly just as a trading item. Our study indicated: If we were to close out any radars, the most likely one to close out would be Cold Bay. He asked us to specifically look at Fort Yukon. We thought maybe Cold Bay was a better candidate to close. We also had received correspondence during the same time period, where Air Force was asking us to look at Kotzebue, the location of Kotzebue, between [Cape] Lisburne and Tin City always was spotlighted because they felt we could do without that one because of overlapping.

MR HALES: (U) The same rationale was used for closing down Northeast Cape.

COL CUNNEELY: (u) ([redacted]) That is correct. We have used that type rationale, and we have always had redress for retaining Kotzebue. It was our position that we would rather keep Kotzebue and give up something else. We did do those studies for General Carson. Galena has always been a problem--due to its cost, the yearly concern we have with the runway and the flooding and the shortness of the runway. Of course, everytime we study it, it comes out pretty much the same. We would like to do without it, but we don't know how.

MR HALES: (U) We do need a northern runway--

COL CUNNEELY: (U) We could use a runway up in central Alaska or maybe even possibly further out, but we don't have anything right now which appears to be a feasible place to go into so we will have Galena

MR HALES: (U) You need the runway to handle the high-performance aircraft.

COL CUNNEELY: (U) Yes.

MR HALES: (U) As a result of the comments of the Air Staff, a study was begun which became known as Saber Yukon.

COL CUNNEELY: (U) Yes.

MR HALES: (U) In the fall of 73, including people visiting from the Air Staff in late September and early October....

COL CUNNEELY: (U) That's right. The Director of Studies and Analysis was chartered to do the Saber Yukon study which, in effect, was a look at the future of the air defense posture in Alaska. A ~~some~~ team did come out, and I accompanied the team on a visit to several of the sites. We had numerous conversations and some correspondence answering questions back and forth prior to publishing of the report.

MR HALES: (U) The report itself didn't come to any conclusion

Z

7

[redacted]

COL CUNEELY: (U) That's right.

MR HALES: (U) During this time period--late 73 early 74--NORAD was involved with developing a new system for CONUS and the Canadian NORAD structure which is the Joint Surveillance System and Regional Operational Control Center concepts which Alaska NORAD Region and AAC were not included.

COL CUNNEELY: (U) That's true.

MR HALES: (U) At the same time an intrical part of the Regional Operations Control Center concept was used in the E-3 AWACs [Aircraft Warning and Control System] aircraft. Once again, when initial plans for its implementation came out in March of 74, AAC was missing again.

COL CUNNEELY: (U) That's right.

MR HALES: (U) As an historical note, the closure of Northeast Cape, Fire Island and a couple of other sites in the late 60s and early 70s was based on budgetary consideration and helped finance AWACS for Alaska. Here four or five years later, we are finally going to do it, and AAC is missing.

COL CUNNEELY: (U) That's right.

MR HALES: (U) This, of course, created tremendous stir within the command. Can you run through the reasons for this?

(u)

COL CUNNEELY: ([redacted]) Yes, that has always been quite a shock to me. I think we have finally gotten that under control; but, that was exactly what was happening. ADCOM/ADC at the time

[redacted]

was feeling pressures of the budget problems and was desperately seeking a less costly air defense system in the lower 48, CONUS. They were battling back and forth over several options. They settled on and finally solved the joint surveillance system option which would have created the four CONUS Regional Operational Control Centers which would have been paid for using BUIC [Backup Intercept Control] computer assets. It would have been a very austere peacetime system which would be supplemented by AWACs and by over-the-horizon back scatter radars which were in the planning at that time.

MR HALES: (u) ([redacted]) The over-the-horizon were replacements for the old DEW Line systems.

COL CUNNEELY: (u) ([redacted]) For the DEW Line to give them distance early warning of raids based on that, they would put their AWACs right on orbit, their peripherial radars would then be back up. Of course, the follow-on, long-range interceptor is also part of that planning. But, nowhere in that planning did they include the Alaskan ADCOM Region nor did they incidently include the Canadians. It was purely a renovation of the CONUS system. I thought that was rather short sighted on the part of ADCOM to think they could tear up the NORAD system like that and just renovate CONUS. As you pointed out, they made no provisions for Alaska in their planning. They hardly even alluded to us.

MR HALES: (U) Following that development, a number of letters, messages, phone calls were made to the Air Staff to try and incorporate AAC in these programs From what I have seen and the organizational standpoint, you had a major part in it.

COL CUNNEELY: (U) Yes.

9

MR HALES: (U) What were the major reasons for not including Alaska, or were there any given?

COL CUNNEELY: (U) The reasons which were given--and these were all given to me verbally, were: They feel Alaskan Air Command should do its own planning, and it is a dilimma--in a way. We are a major command [MAJCOM] and we do have our own planning and programming deputate. But, at the same time, Alaskan Air Command is also primarily in the air defense business we do have the Alaskan ADCOM Region and the Alaskan NORAD Region which, at that time, which is subservient to NORAD. So you had the ADC people who felt we shouldn't be doing any planning for the Alaskan Air Command--that's another MAJCOM. Now, since then, last year when ADCOM became a specified command, they really are truly the planning agency for air defense for Alaskan ADCOM Region. It is very muddy--in a way. But at that time, ADC planners did not feel it was their preogative to plan for Alaskan Air Command.

MR HALES: (U) It is the driving force of money always, then.

COL CUNNEELY: (U) Yes, exactly. There is so much money to go around to commands, and so they spend it on their own command. Now, I thought that was kind of short sighted, and I thought the NORAD people down there should have put a halter on that and said Hay, wait a minute, we have to go together as a NORAD-wise on any renovation of the air defense system in North America. We just can't take CONUS and change it so drastically and forget about Alaska, and that is eventually what happened when it got to the high-level Air Staff.

10

[redacted]

MR HALES: (U) Especially, in light of the fact it has just recently been verified that there is a need for the Alaskan system evolving from Saber Yukon.

CON CUNNEELY: (U) Yes.

MR HALES: (U) Was there a turning point or was it simply a matter of when it got to a high enough level in a larger view this changed?

COL CUNNEELY: (U) Well, what happened was ADC finally did sell the JSS program to the Air Force Council. In, I believe, about April of 74. The program was set up as the JSS program--only for CONUS. At that time, we got the Air Staff XOX [Director of Plans] to support us, Alaskan Air Command, to be added to the JSS program. This eventually came to be when I went up through the Air Force board structure in June, July and finally, I believe, September of 74, and I briefed the Air Force Council; it was approved to be added on to the JSS program as the fifth region. But, that was something we did ourselves. Now ADCOM/ADC did support us coming into the program.

MR HALES: (U) With their portions approved, they supported us--

COL CUNNEELY: (U) Yes, on that basis, they did support us.

MR HALES: (U) It included then the ROCC and the use of AWACs.

COL CUNNEELY: (u) ([redacted] This, at the time, only included the ROCC: however, the concept of operations for ROCC is built on the use of the AWACS. The ROCC is a peacetime as an ADC would be in CONUS. A peacetime system for soverignity. Remember, at this time we had gotten new direction in the PMD [Program Management Directive] on what air defense is really going to

[redacted]

be--primarily, it is going to be soverignty, airspace control and early warning. The words, "we are to be prepared to shoot down enemy bombers" were stricken; so, this has all happened in that period.

MR HALES: (U) This was in the Secretary of Defense Schlessinger's statement on air defense.

COL CUNNEELY: [redacted]) That's right. This all happened so what we are doing now is we are converting to a ground base peacetime system of the ROCC; in wartime the system would swap over and be done by the AWACS so that was part of the overall concept of operation. So basically by Alaskan Air Command getting into the ROCC program, we ultimately had to be included in the concept of operations for AWACS for Alaska, and we were. We had an excellent rationale for AWACS in Alaska in that we have the dual mission, and the AWACS is a dual capable aircraft--tactical and air defense. We have a dual mission here in Alaska so it was very favorably received; as a matter of fact, the present concept of operation as you know, Grant, specifies two AWACS for Alaska. That is because we told them we have a lot of airspace and a lot of territory to cover up here and I think its been preceived.

MR HALES: (U) In telling the Air Staff we have a large amount of airspace and terriroty, there is a basic problem which is often forgotten and that is there seems to be a consistent problem in justifying the situation of Alaska. There is a lack of understanding of logistical problems, tremendous amount of airspace and coastline which is larger than lower the 48 coastline. How much of a problem is this in dealing with staff people and boards such as what you have to deal with to get a program like this approved?

12

[redacted]

COL CUNNEELY: (u) [redacted]) Yes, it is a problem. It is hard for people to conceive, unless they have had a direct experience in Alaska. It is very difficult for them to perceive how much of a problem the defense of Alaska is. So almost every time we brief or write a paper, we try to build on that rationale and take a good look at it and see what we have got to face up there. Our proximity to Soviet launching bases, the amount of coastline which offers an opportunity for an enemy to strike us from that area; it will be very short early warning that we are going to get up here and the possibility that they would supplement the air attack with some small commando-type operations in Alaska. To someone sitting down in the lower 48, when we are talking about a state of the union, that sort of thing is difficult to perceive. You must always start with that.

MR HALES: (U) During the same time you and the staff were involved in getting AAC incorporated into the JSS/ROCC and AWACS program, there was a problem in technology--and generation of technology--between what exists in Alaska--being basically late 40s early 50s technology--and what exists in the lower 48, CONUS, in its being at least late 50s and early 60s in some cases. So then there was a problem, I assume, in tying in this modern organization with what existed here. There was definitely for change in the radar itself.

COL CUNNEELY: (u) [redacted] The thing which has always struck me as bing mysterious is that Alaska is still in basically a manual air defense system. We have the FYQ-9 which does a little computerization for display purposes only, but we are basically a manual air defense system in 1977, and we are going to stay a manual air defense system until 1981. We are basically going to be a manual air defense system for over 25-30 years.

[redacted]

Truly, they have just gone right by us. In the lower 48, they have been in SAGE since the late 50s-early 60s. They also had BUIC as backups which were a computerized system since the 60s. But, Alaska has stood in the manual system all that time and is going to stay there for 30 years. Yet, we are sticking out into a very strategic portion of the world. We have been very poor at telling our case to the Air Staff and making our requirements known. But, that's the way it is going to be, and believe it or not before they change this program, they are going to--the first grouping of new ROCCs are going to be built in the lower 48. We are going to be last--in Alaska. Even though Alaska is the one air defense region that is, on almost regular basis, seeing and intercepting Soviet aircraft. I have always thought thats--

MR HALES: (U) Alaskan Air Command/Alaska NORAD Region is always last.

COL CUNNEELY: (U) Yes, that's right. You know you want your most proficient system in that area. But, we are in the program; we are going to see automation, computerization and it is just going to be a long time in coming.

MR HALES: (U) Then there is the situation of the radars themselves. In early 1974, there was a required operational capability (ROC) designed.

COL CUNNEELY: (U) Yes.

MR HALES: (U) Initially, what this required operational capability was a point of discussion, to put it mildly, within the staff. The DCS Logistics radar maintenance people came up with a design and plan to basically modernize the FPS-93 radar system which won out intially.

COL CUNNEELY: (U) Yes.

MR HALES: (U) And that was turned down by the Air Staff.

COL CUNNEELY: (U) Well, let me tell you about that because it is an interesting period. Right after we were approved in the JSS Program, I approached the boss, [Major] General [Jack K.] Gamble [then Commander of AAC] with a proposal that I wanted to work on which would turn all of our radars into either unattended or minimally attended radars. The rationale I gave to him was: The system we are operating today is much too costly in terms of manpower and it is much too costly logistics support to keep the people out there fed, clothed, entertained. It is a tremendous logistics operation back and forth. I told him we could not continue such a costly operation without eventual budget restraints catching up with us, and saying we can't afford that any longer. We are going to have to close them down. General Gamble was not too receptive, at first, of that proposal. But, in the same time period, he was receptive to a proposal from logistics to see if we could not upgrade the radars to a solid state. I believe the General Dynamics people had a program they had sold to do such a thing in the Far East. Same

15

family of radars and they would solid state portions of it, and it would have reduced the radar maintenance by several positions. So, we did go in with a request. The initial request was for a grant of money for modernization of the radars under this proposal. We got into it much too late in the fiscal year. There was insufficient time for Air Staff to approve it. They came back and said, "we haven't said no yet, but give us a ROC," so that is the ROC you are referring to. We sent back the ROC, in fact it went in by twix [message] we were still, hopefully, trying to get it in before the 1 July cutoff so it was probably the only ROC in history that went in by twix. As you know--as history tells us, it was never approved because it ran into some trouble in the Air Staff. There were some people, particularly in Logistics who were not convinced this particular modification of the radar would do what it was hopefully going to do. There were some thoughts about costing, in any event, it was eventually never approved. It was never funded, and I remember the time--this was several months later when word came down to us that the funding for that ROC was not approved. I guess that would have been the FY 75 budget. When it came back that it was not approved, I ran down to General Gamble's office and I said, "General, they have not approved funding for that. How about taking another look at my proposal for minimally attended radars." It was then that General Gamble saw that we were not going to get the money for that modification that he agreed, for us [AAC DCS Plans] to go forward with a formal proposal to what is now the Seek Igloo program. Because we already had a ROC in for modification, the Seek Igloo Program has been built on that ROC and they don't say the same thing at all, as you pointed out in your question. If you read the ROC,

16

you don't see anything about Seek Igloo in there at all. How it got tied in was we went back said, "the preferred solution for that ROC is now minimally attended radar."

MR HALES: (U) The modern state of the art radars.

COL CUNNEELY: (U) Modern state of the art radars. That ROC was only a vehicle by which we formally made the proposal to the Air Staff.

MR HALES: (U) So you were already on the table it was simply a matter of changing the wording to fit what you wanted.

COL CUNNEELY: (U) Yes. Whenever you go in with a ROC, you say this is a preferred solution; then you give alternate solutions. Well, the original preferred solution of the General Dynamics modification went by the wayside so we said now our preferred solution is minimally attended radar, and we costed it out. Of course, as you know, it gave a very favorable cost savings picture.

MR HALES: (U) And this was approved approximately a year later, June of 75.

COL CUNNEELY: (U) Yes, just about a year later. We saw we weren't going to get the money and General Gamble said, "if we aren't going to get the money for that, go ahead with the minimally attended concept."

MR HALES: (U) This takes us roughly up through June of 75. We have been included in the air defense-wide program, the

17

NORAD-wide program, now have means of improving the radars themselves in Alaska.

COL CUNNEEL(?): (U) Yes.

MR HALES: (U) It is now 1977, two years after that and no new radars; as you were just saying, we are going to be the last one to get the JSS/ROCC. Presumably, there are budget limitations or there are problems along the way as far as getting these programs actually going. Being approved and having them are two entirely different things.

COL CUNNEELY: (U) Yes. Well, you are exactly right and it is frustrating and the programming business is a very frustrating business these days. JSS is the key; without JSS/ROCC you can't get the minimally attended radars. Thats number one and we have seen the JSS program slip quite a bit.

MR HALES: (U) Why?

COL CUNNEELY: (U) Well, after we got in to it, in October of 1974, the Canadians came on board. When the Canadians came on board, it became a whole new ball game. The original idea of using those BUIC assets that were laying around had to go out the window because the BUIC assets were not capable of meeting the Canadian's needs. So then we had to re-do the JSS Program from the ground up to design and develop the requirements which would meet the Canadian's needs. Remember, the Canadians want--not just a peacetime system, but a wartime system. The Canadians are not sold on AWACS yet. So we had to meet the Canadians very, very rigid requirements and Alask's slightly less than Canada's

18

[redacted]

rigid requirements and the CONUS, and that threw the whole thing into a completely series of meetings and conferences--

MR HALES: (U) Certainly, throwing in a foreign government and a whole new layer of bureaucracy didn't help matters either.

COL CUNNEELY: (U) Yes, that's right. That thing has been going on and as you know, we are just ready to go to RFP [Requirement Funding Panel] this fall on JSS so that thing did slip 18 months or so. The minimally attended radar program [Seek Igloo] I would have liked to have seen it come in a little bit sooner than it is presently programmed. Electronic Systems Division [of Air Force Systems Command] has taken a very conservative approach to it. They are not as aggressive on that program as I would have liked to have seen. I am convinced having talked to all the contractors who are the leading builders of modern radars that they could provide Alaska with a good minimally attended radar at an earlier date than what we have programmed--and possibly we will once we get the contract we may find the minimally attended radars will be built a little bit sooner than are presently programmed.

MR HALES: (U) You have a conservative time table then.

COL CUNNEELY: (U) Yes, at this time.

MR HALES: (U) Not only for implementation, but for the contractor to perform his work.

COL CUNNEELY: (u) ([redacted]) Yes, I think so--probably ESD [Electronics System Division wouldn't agree with that but all things taken into consideration unless we run into some major slippages, we are talking about FY 82 before we've got the ROCC all of our minimally attended radars out there. The AWACS sitting on the ramp out here and hopefully

[redacted]

19

[redacted]

F-15s somewhere around here. (1984)

MR HALES: (U) Of course, the aircraft situation is entirely different.

COL CUNNEELY: (U) Yes, that's right. We haven't got there yet.

MR HALES: (U) One other thing in this respect, the MAR or minimally attended Seek Igloo radar system with the NORAD-wide system is timing of them coming together. I have seen discussions on trying to schedule them so they come together at basically the same time in the interest of saving money. Is this working out.

COL CUNNEELY: (S-GDS-84) Not completely as it stands right now--the program--the scheduled program--when we have our ROCC here at Elmendorf, we probably will only have about six of our radar sites converted the minimally attended configuration.

MR HALES: (U) That is out of 13.

COL CUNNEELY: (u) ([redacted]) Out of 13. Now, as I mentioned, I am hopeful that once we drop a contract to industry on building and installing those radars maybe we will see a better picture. Because as you know you can't get the minimally attended concept into being until the ROCC is operational. You still need your radar operators and controllers at the site. So, as soon as the ROCC is built, we can bring the controllers back here.

MR HALES: (U) Which will remote the signal back to the ROCC.

[redacted]

20

[redacted]

COL CUNNEELY: (U) Yes, then you truly get to the minimally attended configuration.

MR HALES: ([redacted]) As a sidelight to remoting of the signal back to the ROCC at Elmendorf, communications is a big factor so building the satellite system by RCA Alascom now--that has been going on for a couple of years now is the key to the system being able to work.

COL CUNNEELY: (u) ([redacted]) It is and by everything told to me by communicators we are going to get very excellent communications from them--a lot better than we have--more reliable. So we should be in pretty good condition then.

MR HALES: (U) And as a kind of corrolary to this, the problems of manning these remote sites have been a constant attempt on the part of the Alaskan Air Command and I know for many years into the mid-60s if not early 60s. Contracting of the base support--type functions really don't require military personnel, civil engineering, cooks, other mundane type duties. This was once again being studied by Alaskan Air Command as part of the Fresh Ideas program in early 1976 when the Base Management Advisory Group (BMAG) under the direction of the Air Force Chief of Staff came to the same conclusion that AAC had been coming to for a number of years.

COL CUNNEELY: (U) Yes.

MR HALES: (U) And then, because of this being approved at the top level instead of working its way up, there was movement in this direction.

[redacted]

COL CUNNEELY: (U) Yes.

MR HALES: (U) Rather rapidly.

COL CUNNEELY: (U) That is well put.

MR HALES: (U) What exactly has been our part in relationship to BMAG study and the aftermath.

COL CUNNEELY: (U) Well, as you mentioned, BMAG came along at the time we started the Fresh Ideas study, and we said lets forget Fresh Ideas for a while, and we did participate in the BMAG study. The BMAG study came up with the same things we had come up with in our studies that contracting would be less costly as compared to what it takes us to operate the system under the present military/civilian rates. A bonus would be the elimination of a great number of remote unaccompanied tours. The BMAG study said the same thing we had. It verified those figures right down--just kind of updated our figures. The difference, of course, was that the BMAG was directed by the Chief of Staff, and he was prepared to implement the results of that study without any further staffing at any level. The Alaskan Air Command position opposed contracting as you know. [Lieutenant] General [James E.] Hill [then Commander of AAC] opposed it based on what he felt was--he had two ideas in mind. He felt you were getting into contracting military-essential positions and repercussions might be felt worldwide. However, the Chief of Staff, I am sure, had his mind made up. The advantages of elimination of those remote tours were, I think, the driving force behind his decision implement at this time. Of course, we have seen this now in what we are getting out of Air Staff out of the Chief of Staff now to continue the contract of the forward operating bases.

MR HALES: (U) Galena and King Salmon as well as th Air Force base on Shemya island.

COL CUNNEELY: (U) There is no question in my mind that they are next.

MR HALES (U) Were these proposed after the 13 AC&W sites?

COL CUNNEELY: (U) They were originally included in the BMAG and were dropped ou of the BMag study at an early point. I think [Major] General Thompson [Head of the BMAG] dropped them out at our request because we indicated it was too much to bite off the first time around. We wanted to do more study on it so we did drop them out. But, he did indicate to us in a visit here that he felt that those particular bases should be added on as soon as possible. The Chief of Staff supported him on that.

MR HALES: (U) There has been some problems in that with the change of command last October 1976 [Lieutenant] General M.L. Boswell's [Commander of AAC as of 15 October 1976] previous relationship with General [David C.] Jones [AF Chief of Staff] he has had an inside line, in effect, so presenting you with certain problems in the Air Staff in not being their marching orders.

COL CUNNEELY: (U) Yes. General Boswell worked directly for General Jones and speaks to him directly--has never varied his position that we are going to contract those three bases next year to the maximum extent. Now that word has not filtered to all levels of the Air Staff, and we have been having some coordination problems; again, there are certain elements--not only in the Air Staff-but probably here at headquarters who still may feel that is an unwise way to go. But, I feel the decision is made, and we ought to get about implementing it. Again, I think he is using the same rationale he used on the AC&W sites. Its too clostly to do it the way we are doing it, and the benefits of reducing those unaccompanied tours to the morale of the personnel force, particularly in certain career fields is a bonus. It is something we should strive for.

MR HALES: (U) The turmoil and cost of simply moving people let alone--

COL CUNNEELY: (U) The cost of moving, the turmoil, the tearing up of families, I thing the Chief of Staff decided there is a better way to do it.

MR HALES: (U) At the same time, the contracting has not affected the hard-core operations.

COL CUNNEELY: (U) No.

MR HALES: (U) --they are remaining military so--

COL CUNNEELY: (U) That's right. The actual alert force, the maintenance of the aircraft, the loading of the alert aircraft, the operations section which handles the scrambling will all stay military. Those are truly military-essential positions.

MR HALES: (U) They would be key positions if the aircraft from Alaska were deployed elsewhere during wartime.

COL CUNNEELY: (U) That's right also. Almost everything else support wise and it looks like security police force is going to be civilianized.

24

[redacted]

MR HALES: (U) I have seen that done in SAC.

COL CUNNEELY: (U) It has been done. It has been done in many parts of the world. I do think it is a good idea. You know, I go back to what I said earlier in our interview that I was flabbergasted by how costly this operation is in Alaska and how much it eats up in terms of manpower and support of manpower. We have a good mission--an important mission--but we should be looking for the slickest, cheapest, most efficient way of doing it. We should break down traditional barriers to get there because we have got budget problems, and we're going to have big budget problems in the future. We just can't keep doing business the way we've been doing it for the last twenty some odd years and survive. We have got to find better ways of doing it and still meet mission, and we will. The contractors will do their job if we get good contractors, and we monitor those contractors properly; we'll still be doing the mission out there.

MR HALES: (U) There are a couple things we touched on briefly as far as the radar systems are concerned that we really haven't discussed. How much of an involvement do we have with the over the horizon back scatter [OTHB] as far as placement on the DEW Line, primarily in Alaska. We have some involvement, though minor.

COL CUNNEELY: (u) ([redacted]) The original plan called which called for an over

[redacted]

[redacted]

the horizon--looking over the pole, of course, fell out rather early. Technically, it wasn't feasible due to the effects of the aurora --I don't know technically--(sun spots, etc.) However, then we went to a plan of one on the northwest coast of the United States and one on the northeast coast.

MR HALES: (U) Pave Paws--near Beale AFB, California.

COL CUNNEELY: ([redacted]) Yes, that would look out and would presumably cover a great portion of the Gulf of Alaska and the Aleutian Chain--that could have been of great use to us for early warning of anything which might be coming through those areas out along the chain and anything that may be coming around us and penetrating--lets say, the Alaskan pan handle. So we were very much interested in OTHB and there was--there could have been, if it work as was forecasted quite a bit of capability for us there. Now, OTHB has dropped out of the funding and what its future is going to be I don't know. But, it doesn't look too good right now, and that is very significant-not so much to us as to ADCOM.

MR HALES: (U) You have touched on the other part of my question as the discussion of radar for the Gulf of Alaska, located in Prince William Sound area near Valdez and Cordova.

COL CUNNEELY: ([redacted]) Yes. [Lieutenant] General [James C.] Sherrill, when he was the region commander, he was also the CINCALCOM, was very interested in getting some radar coverage in Valdez area. We made some overatures about that to the Air Staff, and he did also. We didn't get much response. What we did, on that subject, was when the

[redacted]

redefinition of the air defense mission--stressing soverignty and airspace control, we pointed out that we do have a lot of area of the United States airspace uncovered in the Alaskan panhandle area. We also had a lot of discussions with FAA [Federal Aviation Administration] Alaska, who are quite interested in getting some radar coverage which would provide them close control of air traffic between Anchorage down through Juneau and into the Seattle area. So our proposition also included FAA. We also approached the Canadians and said "you have in your projections into the future you will want soverignty radars on your west coast of Canada, the northwest territories and British Columbia"; and they did. What we have suggested in our ROC is that we would try to satisfy the requirements of the Canadians, the FAA and the Alaskan Region with a series of radars which would cover the panhandle of Alaska--including Cordova, Yakatat, Sitka and down in that area there. The ROC is in, the Canadians have studied it and have tentatively agreed, it has been on the agenda of the permanent board of joint defense, we have submitted it as a follow-on program to Seek Igloo in that we want to be sure we don't put the cart before the horse while priorities are still the Alaskan radar--the ones which do provide us defense and early warning in Alaska. But, we think if we do get a buy of new radars--minimally attended types we could add those three on and the Canadians will add one down in the Queen Charlotte area, and we will eventually have a series of radars that will cover all the way down to the west coast of North America. Were talking 1982 to 1985--somewhere in there.

27

[redacted]

MR HALES: (U) Assuming it is approved.

COL CUNNEELY: (U) Yes, I think if Seek Igloo goes it will go. It will be a good follow-on.

MR HALES: (U) One other thing I would like to touch on briefly although in many ways is equally important, is the aircraft. We have F-4s which really aren't interceptor. We have a limited number, and we are planned to get more in the near future. We have 24 with two NOA [nonoperationally authorized] and we will get 12 more.

COL CUNNEELY: (U) Right, we will get 12 more beginning this fall.

MR HALES: (U) The additional aircraft are helpful because of the logistical problems and the vast areas we have talked about. But, they still aren't the answer because of the growing capability of the other side [Soviet Union] and the limitations of the F-4 because of its not being designed as an interceptor in the first place. I have seen repeatedly year after year when ALCOM put their submission to the Joint Chiefs of Staff many and AAC on its own to the Air Staff for the need for better interceptor-type of aircraft. The F-15 you alluded to earlier is the most logical candidate of what is available at the current time. Is there any discussion in this area at the time.

COL CUNNEELY: (U) ([redacted]) Yes, there is a lot of discussion but, unfortunately, Split

[redacted]

28

[redacted]

this year were were again not included in the JSOP [Joint Strategic Operation Plan]--Alaska was not included in the FOI [follow-on interceptor]. It is very disappointing to us. You are exactly correct. The F-4, as versitile as the aircraft is, is not sufficiently long-legged enough to cover the early warning areas in Alaska on the bases we have. We have asked and we have no answer for, and we have ask Air Staff to include us in the FOI buy. The FOI has been tentatively, at least, identified by Air Staff as the F-15.

MR HALES: (U) Is it going to be like a B model which will be a light varient on the F-15 coming into the inventory now.

COL CUNNEELY: (u) ([redacted] Well, the decision has not been made. You know ADCOM originally said we want to enhence the F-15, put some air defense type hardware on it. I think probably they have backed off on that now because they were running into budget problems. I think they would be happy to get the F-15 in any shape or form. I think we would also be delighted to get the F-15 in any shape or form. As the FOI. We are still identifying the FOI, F-15 in our objective force planning. We want the FOI here as early as we can get it. Our eventual position that we would like to have is a squadron of FOIs and a squadron of air ground aircraft, hopefully, maybe F-16s in 1984. So a squadron of F-15s for air defense, a squadron of F-16s for air to ground.

MR HALES: U ([redacted]) Will a follow-on target type aricraft to replace the T-33 be necessary once the modern radar is into place?

[redacted]

COL CUNNEELY: (U) Well, we have agreed with the ADCOM ROC for a follow-on target aircraft to the T-33. We feel, yes, we need a good realistic faker aircraft for our aircrews and our ground environment to train against. I have to be honest; it doesn't look very hopeful right now. Because the T-33s days are numbered. I know people have been saying that for years, but surely, they can't go on forever.

MR HALES: (U) They have been in Alaska since 1948.

COL CUNNEELY: (U) Good heavens, They must have been the first ones off the line, I guess.

MR HALES: (U) Of course, the T-33 is nothing more than a refinement of the trainer version of the P-80, [redacted].

COL CUNNEELY: (U) That's right.

MR HALES: (U) The were here in 48, seeing if they could fly in Alaska.

COL CUNNEELY: (U) They are still flying out there, and they are doing real well, believe it nor not. But the Air Force programmers would like to see it phased out, and right now, there is nothing on the books to replace it. We have been looking at T-38s or F-5s, I am skeptical whether those aircraft would fill the bill up here or not.

MR HALES (U) Is there anything that you have really tried to do or would have liked to have done which you haven't done?

COL CUNNEELY: (U) Well, sure--

MR HALES: (U) There are a lot of things which need to be done that we have talked about but is there anything I haven't mentioned, in particular?

COL CUNNEELY: (U) Well, I think, in four years here, I have seen a lot of changes come. I think most of them were for the good. I think we are getting more for our dollar in Alaska today than we were in 1973. A lot of people disagree with me about ALCOM, but I think we are still able to do the job up here in the configuration we are in without ALCOM. I think we have to take into consideration the threat. I have always tried in plans and programs and manpower to realistically look at the threat to Alaska before I took a position on anything. I feel that if you don't do it that way, then you're really cheating the system if you try to see or build a threat which doesn't exist or try to justify programming action to fight shadow targets. I think Alaska is a good place to put forces--flying forces, ground forces; it is a great place to train with the airspace; there is a lot of ground space up here.

MR HALES: (U) Reductions continued to a certain extent and conversion into more combat forces, do you think there is a viability for the headquarters itself, as a major command.

COL CUNNEELY: (U) Well, I do but I don't see it over night. I don't see it happening momentarily. I think what we will see in the Air Force in the next few years are some pretty big changes which may override this. I think we are going split to see more in terms of single managers of systems--single managers for fighter aircraft, well we already have a single manager for airlift, right? We have MAC for military airlift, we have SAC which runs all the bombers, tankers , and I think we are going to see single manager for fighters, I think TAC will take over.

MR HALES: (U) Not talking just about the Air Force but generally throughout DOD. Like MAC for a specified command.

COL CUNNEELY: (U) We may even see some of that. I think we have just got to keep going in these directions. I think it is very hard for guys who have been in the Air Force as long as I have, or even shorter periods than I have to realize it is a new ball game now. We are running into a situation where we are just not going to ever again get the fat money we used to get. Now even with the cycles we have, we always got money to do this and do that and to subsidize a lot of things which are nice to have. And we are not going to continue to have that, and a lot of big thing are going to change. Contracting, which we talked about, broke a lot of hearts. General Gamble, he damn near cried about that. It just tore his heart out to see that sort of thing happening. There is a lot more crying going to be done before it is over.

MR HALES: (U) Yes. With personnel costs being what they are it is going to be a fat area.

COL CUNNEELY: (U) War is getting expensive. Getting ready

32

UNCLASSIFIED

for war is expensive. There is just so much Congress is going to give us. So the poor guy who is setting in the Chief of Staff position is going to say, "I'm sorry guys but we are . . . " If you look at the things that were chopped out of the budget this year; significant things-things that, I know guys swore we have got to have-biggies. They really believed it and I am sure they were good but-- and we are sitting here with minimally attended radars and were in there were in the budget. Why., because we're saving money. They say "Hay, they are streamlining. This one is eventually going to save us money--we'll fund it."

MR HALES: (U) We may be able to buy another F-16 or whatever--equipment--

COL CUNNEELY: (U) That's right, and I think the management of the Air Force is going to be forced in that direction, Grant. I think eventually, AAC is going to be swallowed up in that sort of thing. The only thing which keeps it going is the political influence on the Alaskan contingent up here. There isn't anybody, including General Boswell who doesn't agree that we've got over kill up here in terms of the Alaskan Air Command, a three-star general and they know that. So, when the time is right and other things fall into place and the right secretary [of Defense] can come up here and say "Hay, look what I've done for you." I've put squadron of F-4s and I've done this and I've done that.

-END-

UNCLASSIFIED

Article from *Sourdough Sentinel*

The *Sourdough Sentinel* was an Elmendorf Air Force Base news publication. The following article was published in the mid 1970's.

The Air Force is always concerned with getting the most defense for the best price. The emphasis on defense spending nationally increases that concern. Alaskan Air Command, working with the 11th Tactical Control Group and other key staff agencies found a way to please even the most hostile critic for defense spending.

Thanks to the genius of a Colonel at Alaskan Air Command and some sharp folks on his staff. These guys came up with a program to replace all the old radar sites with brand new up to date space age equipment. Col. Cunneely took his Project to Air Defense Command, NORAD, Canadian Air Force, and finally the Pentagon where all the senior General make the decisions.

Col. Cunneely was the Deputy Chief of Staff for Manpower. In that position, Cunneely understood clearly that the Alaskan Air Command was heavily over populated. Also, he saw that Alaskan Command needed to be transformed and updated. Cunneely's plan would reduce personnel, bring in faster fighter aircraft, and completely get rid of the ancient radars, now the most important aspect of the Minimally Attended Radars. Computers track targets with new planar array radar. The sites now can be maintained by two site technicians, where prior, some 350 military people had been before. The ROCC almost immediately can give vital seconds to the aircrews, who must go out to meet incoming planes.

Col. Cunneely, who led the battle in selling the MAR program, undertook many hours of travel to allied Commands for meetings, plus at least 12 visits to the Pentagon. Many who were skeptical, were concerned that altering the radar sites would put our northern defenses in a vulnerable position. But Cunneely was relentless. Gradually, he convinced more folks, that replacing the old equipment with brand new hardware, plus reducing the Alaska Air Command numbers would be immensely beneficial for Top Cover Defense. In his many briefings, Cunneely emphasized that the MAR program will not only be much more effective in several ways. And also, Cunneely made it most important, the MAR program will bring forth huge savings, close to $40 million a year. Such savings to the USAF budget was not easy to neglect. Cunneely met with a great number of engineers and Commanders. These people were experienced and wise. In the final, Cunneely gave a Command briefing on the MAR program in the Pentagon to senior Four-Star Generals. The Generals gave the MAR program huge support.

In retrospect, I certainly was pleased and I knew for sure that what we accomplished was a great move forward. However, although I had done everything I could to pass the MAR program. And, I was quite sure that it would be successful. At the same time, something I never spoke about, I knew for sure that what I had accomplished included some things that were surprising and would make many people unhappy. I knew was that the MAR Program would eventually shut down the Alaskan Air Command, for good.

After finishing my work on MAR, I certainly received much support and backing due to my successes. However, I would no longer be flying and I felt badly about that. I was a Command Pilot and completed over 5,000 hours. I now began to realize that my flying career would soon be ending. And, of course my military career would also be ending soon.

One of the Greatest Generation

This is about my brother Joe. His full name is Joseph Laurence Cunneely, Jr. He was named after my father, and remarkably, his birthday was identical to our father's: October 26. October 26 happened also to be the date on which my father and mother were wed. My brother was a difficult child. For a reason I could never discern, he seemed to engage in fist fights with other kids almost on a daily basis. When we were living in Kew Gardens, Joe and I were enrolled in P.S. 99. I was in kindergarten, and he was in the fourth grade. We walked to school together and my brother always was protective of me. But it was not unusual that during the lunch hour, my brother and some other classmate would engage in a brawl that would draw a crowd of shouting students. These would end with a teacher breaking up the fight and my father coming to the school and taking my brother home. These occurrences took place so often that my father transferred Joseph to the parochial school in Forest Hills. In this school, he hoped that the Nuns would impose some needed disciple on my brother. To get to Our Lady Queen of Martyrs, my brother had to take the LIRR train one stop each morning and walk one block to the school. He did not enjoy this transfer. Later when our family broke up and we moved to Elmhurst, we attended P.S. 89. We walked close to a mile to this school and my brother resumed his bad behavior. But as he grew older and attended Newtown High School, Joe competed in track and performed very well. He matured into a good student and a true gentleman. My brother and I became close friends and my admiration for him grew. And that continued throughout our lives until he died in 1986.

And then came the war.

My brother graduated from high school in June 1942, at the age of 17. The United States had entered the war the previous December. Had he waited until he turned 18, he would be drafted. But my brother decided he wanted to enlist in the Navy right away. It was a sad and frightening day when my father and I accompanied my brother to Manhattan and saw him off to the Great Lakes Training Station in Illinois. We could only pray that he would survive what would clearly be a long and terrible struggle that would take many lives before it ended. My brother saw action in the Pacific on the USS Kitkun Bay (CVE 71), a Casablanca Class escort aircraft carrier. Following is a description of the history of this carrier's action in which my brother Joseph, a young man still in his teens, like many others, participated, serving his country. Following is a history of the USS Kitkun Bay and the action it saw in the Pacific War.

When my brother Joe returned from the War, it was a great relief and a happy time for our family. In WWII, America lost millions of its young men in battle. A whole generation, including fathers, husbands, and young men just out of high school like my brother. He rarely spoke about his experiences, rather, it was his goal to get a job, marry his girlfriend Patricia Halloran and settle down to a normal life. He accomplished that and more. After a few jobs, he studied and took the policemen's competitive exam and came in with a very high score. After a couple of years as a patrolman, he was assigned to take charge of the Police Athletic League.

Joe managed that organization and expanded it to cover many different sports. He was very effective with kids. However, when he was promoted to Sergeant, he returned to Precinct duty. He had a distinguished career with several decorations but decided to retire at the 20-year mark. He then went to work on Wall Street, where he quickly learned the business and was most successful. By then, he and wife Pat, with the financial help of our father, purchased a home on the waterfront in East Patchogue, Long Island, where he lived until he died. By that time, my brother had been a most successful investor in the market and was doing very well enjoying his life on the water and his beautiful home. But then he was stricken with cancer that eventually took his life. The saddest episode in his life was a breakup with our dad that was most hurtful to both of them. (I cover this in one of my book's chapters.)

Losing my brother occurred about one year after the death of our father. So, within a short period, both of my remaining close relatives passed away. One was 93 years of age, and the other was only 61. My brother epitomized the men and women of what is now referred to as the "Greatest Generation." I believe that he and they were indeed, rightly named. At a huge sacrifice, both to those who survived as well as to those who did not, what they accomplished was truly unbelievable in scope. They defeated the Fascist tyranny of Hitler and the treachery and dastard cruelty of the Japanese warlords. Then returning home, they rebuilt their lives, used the GI Bill to educate themselves, married, raised families and participated in returning to the normalcy they savored. They also aided in rebuilding the very same nations that they had defeated in war.

My brother was only one of millions of Americans who were a part of this marvelous generation. He was my 'big brother,' my protector as a child, a wise advisor in my later years and certainly my closest friend. I am so proud of him, and I miss him. His life was too short. He loved his family, including his father and his nephews and nieces. I will always remember him, and I still pray for his soul, daily. May he rest in peace!

The following attachment is a history of the USS Kitkun Bay, the ship my brother Joseph served on in the Battle of Leyte Gulf. This battle, which is described as the greatest naval battle of World War II, took place in October 1944. My brother, who had just turned 18 years of age, was at the helm the Kitkun Bay when it was struck by a Japanese suicide bomber killing one man and wounding 16. In a most interesting coincidence, this episode was witnessed by our first cousin Louis Reed, an officer aboard the USS Almaack, a class attack cargo ship that was eight miles off the carrier's stern. My cousin was well aware that his cousin Joe was aboard the USS Kitkun Bay and feared greatly that the carrier might have been gravely damaged and his cousin killed. Miraculously, the Kitkun Bay survived as did my brother.

My cousin Louis Reed served honorably in the Navy having seen action in the landings in Sicily, and at Saipan, Guam, Luzon and Iwo Jima. Continuing his service in the Naval Reserve, he retired in the grade of Captain. He then became a New York City fireman. His brother James served honorably in the U.S. Army. All of these men, my brother and two cousins, are included in the category of the Greatest Generation having served in the big war, and then returning home, finished their education, married their sweethearts, raised children and went on to successful careers in civilian life. This nation has to be proud to have men such as these when they were needed the most.

My brother Joe and I, as kids in New York.

Joseph Cunneely, Jr.

Joseph L. Cunneely, Jr., of East Patchogue, passed away on August 10.

A retired vice president of Fidata Trust of New York City, he was also a retired sergeant of the New York City Police Department, a naval veteran of World War II, and a graduate of City College.

Mr. Cunneely was also involved for several years with the Police Athletic League, as well as its director of activities for Queens County.

He was currently chairman of the board of governors of the Swan Yacht Club and had the distinction of being elected commodore of the Yacht Club for six terms.

Survivors include his wife, Patricia, brothers, sisters and spouses: Eugene and Alice of Washington, Dorothy and William Anderson of Lake Grove; Ellen and Daniel Halloran of Flushing; Kathleen Roth of Connecticut, and Barbara and Paul Ruebenacker of New Jersey; and 22 nieces and nephews.

Religious and Swan Yacht Club memorial services were held at Ruland Funeral Home. A Mass of Christian Burial was celebrated by the Reverend Hugh Cannon at Mary Immacualte Church, Bellport, with interment following at Calverton National Cemetery.

USS Kitkun Bay (CVE 71)

- formerly ACV 71, formerly AVG 71 -
- later CVHE 71 -
- decommissioned -

Sorry, no coat of arms available.

USS KITKUN BAY was the 17th CASABLANCA - class escort aircraft carrier. Decommissioned on April 19, 1946, and stricken from the Navy list on May 8, 1946, the KITKUN BAY was sold on November 18, 1946, to Zidell Machinery & Supply Co., Portland, Oreg. She was scrapped early in 1947.

Embarked Squadrons:

Period	Squadron (Aircraft)
May - August 1944	VC-5 (12 FM and 8 TBM)
August - November 1944	VC-5 (14 FM and 12 TBM)
November 1944 - February 1945	VC-91 (17 FM and 11 TBM)
May - September 1945	VC-63 (FM and TBM)

back to top go to the end of the page

History of USS KITKUN BAY:

KITKUN BAY, originally designated as an AVG, was classified as ACV 71 on 20 August 1942 and reclassified as CVE 71 on 16 July 1943. Laid down 3 May 1943 she was launched 8 November 1943 by Kaiser Shipbuilding Co., Inc., Vancouver, Wash., under a Maritime Commission contract ; sponsored by Mrs. Edward A. Cruise, and commissioned 15 December 1943, Captain J. P. Whitney in command.

After a shakedown along the Pacific coast, KITKUN BAY departed San Diego 28 January 1944 on a replenishment voyage to the New Hebrides bases. After loading passengers, planes, and other cargo, she sailed for home 18 February via Pearl Harbor and arrived San Diego 6 March. Upon her return, the planes of VC-6 were brought aboard for training and assignment. With Rear Admiral Harold B. Salada, Commander, Carrier Division 26 embarked, she sailed 1 May for Pearl Harbor and the completion of her training exercises.

On 31 May her task unit sortied forth to escort the bombardment and transports units of Task Group 52.17 to Saipan. On 13 June her planes shot down their first enemy aircraft and the next day began the bombing and strafing of enemy positions in the Marianas. KITKUN BAY's planes alternated flying support missions for the

Saipan landings and air cover for ships east of this island. Eight enemy planes were splashed in attacks on her formation on the 17th and her own guns downed three more the next day. Early July brought a brief respite at Eniwetok but 14 July she resumed support sorties at Saipan, Tinian, and from 2 to 4 August at Guam.

Withdrawn, she steamed to Espiritu Santo, New Hebrides, for upkeep before sailing to the Solomons for additional practice in support of amphibious operations. Heading westward on 8 September her task unit escorted an assault force to Peleliu and Angaur Islands in the Palaus group and provided cover from the 15th to 21st. Withdrawn to Manus, Admiralty Islands, she made preparations for the invasion of Leyte, P.I., and her finest hour.

Departing 12 October she soon joined Rear Admiral Clifton Sprague's task unit "Taffy 3" composed of 6 CVE's and their screen of escorts. On 20 October KITKUN BAY began launching strikes against Leyte. These operations conducted from a position east of Samar Island continued until early in the morning of the 25th when Japanese warships were sighted on the northwest horizon. Admiral Kurita's powerful Center Force had passed through the San Bernadino Straits unnoticed, hoping to destroy the supply ships off Leyte. Not designed to exchange gunfire with surface warships "Taffy 3" launched what planes were ready and turned southward behind a smoke screen. In the ensuing 2-hour running battle, the courageous maneuvers and skillful action of its screen, the diverting attacks by its planes, the astute orders of its command officers averted annihilation. In the forefront of the circular formation KITKUN BAY escaped any direct hits as the shells splashed ever closer until 0925 when the enemy suddenly broke off the engagement and retired. The less fortunate GAMBIER BAY and three escort ships went down fighting valiantly; while suffering some gunfire damage, the Center Farce lost three cruisers as a result of the attacks of the aircraft of "Taffies 2 and 3."

The final phase of the epic Battle of Leyte Gulf was the retaliatory air strikes by both sides. Before the "Fore noon" watch had expired, KITKUN BAY had splashed a suicidal Betty but had also been crashed by a Zeke which struck the port catwalk killing 1 man and wounding 16. The losses for the day also included two planes and their crews. The next day she sailed for Manus in the Admiralty Islands for replenishment and repairs.

Arrived 1 November, she departed Manus the 7th for Pearl Harbor where VC-5 was replaced by VC-91. Despite a submarine attack en route, KITKUN BAY returned safely to Manus 17 December. New Year's Day 1945 dawned with CVE 71 steaming as part of Task Unit 77.4.3 (Lingayen Transport Cover Group) bound for the invasion of western Luzon. After passing through Surigao Straits, the convoy underwent a series of air attacks. Air cover destroyed seven enemy planes but at 1857 an Oscar got through and crashed KITKUN BAY's portside amidships at the waterline. Almost simultaneously a 5-inch shell struck her starboard side. The resultant fires and flooding were brought under control but 16 were dead and 37 wounded. The following day with a list and only one engine operating she withdrew and proceeded by stages first to Leyte, Manus, Pearl Harbor and arrived San Pedro, Calif., 28 February.

Two months later she sailed again for the Western Pacific. After a training period in the Hawaiian Islands, she departed 15 June for Ulithi and duty with the 3rd Fleet. On 3 July KITKUN BAY sortied forth with other escorts and ships of the "train" for support of the fast carriers operating off the coast of Japan. Mid-August she was reassigned to Task Force 44 gathering at Adak, Alaska, to escort Admiral F. J. Fletcher, COMNORPAC, who had been designated to receive the formal surrender of the Japanese in northern Honshu and Hokkaido. Arrived off Honshu 7 September, she remained in the area until the 27th, seeing to the feeding and transportation of American prisoners of war. Detached to participate in Operation "Magic Carpet," she first debarked 554 troops at San Francisco 19 October. Additional voyages to Pearl Harbor and Okinawa concluded 12 January 1946 at San Pedro, Calif.

KITKUN BAY entered Puget Sound Naval Shipyard, Bremerton, 18 February and decommissioned 19 April. Sold 18 November 1946 to Zidell Machinery & Supply Co., Portland, Oreg., she was scrapped early in 1947.

In addition to the Presidential Unit Citation KITKUN BAY earned six battle stars during World War II.

USS Kitkun Bay launches FM-2 fighters during Battle off Samar

USS Kitkun Bay is attacked by Japanese suicide bomber in a fatal dive.

Made in United States
Orlando, FL
04 November 2026

90203567R00325